AF389102

Novel Analytical Methods in Food Analysis

Novel Analytical Methods in Food Analysis

Editors

Philippe Delahaut
Riccardo Marega

MDPI • Basel • Beijing • Wuhan • Barcelona • Belgrade • Manchester • Tokyo • Cluj • Tianjin

Editors
Philippe Delahaut
Analytical Laboratory
CER Groupe
Marloie
Belgium

Riccardo Marega
Analytical Laboratory
CER Groupe
Marloie
Belgium

Editorial Office
MDPI
St. Alban-Anlage 66
4052 Basel, Switzerland

This is a reprint of articles from the Special Issue published online in the open access journal *Foods* (ISSN 2304-8158) (available at: www.mdpi.com/journal/foods/special_issues/methods_food_analysis).

For citation purposes, cite each article independently as indicated on the article page online and as indicated below:

LastName, A.A.; LastName, B.B.; LastName, C.C. Article Title. *Journal Name* **Year**, *Volume Number*, Page Range.

ISBN 978-3-0365-5816-5 (Hbk)
ISBN 978-3-0365-5815-8 (PDF)

Contents

About the Editors

Philippe Delahaut

Philippe Delahaut graduated in veterinary medicine in 1974 from the University of Liège. In collaboration with Department of Obstetrics of the Faculty of Veterinary Medicine, he created the Applied Hormonology Laboratory in Marloie (Belgium), which aimed at developing laboratory techniques for early diagnosis of pregnancy and of radioimmunological assay, under the supervision of Prof Ectors and Beckers. Subsequently, he was largely involved in the development of immunological assays for the detection of residues of anabolic substances in foodstuffs of animal origin. These techniques have been widely used by official control laboratories. With experience in the development of reagents and in particular antibodies, he participated to several European projects in calls FP5, 6, and 7 frameworks (Biocop, Moniqa, Conffidence), as well as Eurostar (Go With The Flow). He is the author of more than 100 publications, as well as several presentations at international congresses/regulatory bodies'symposia.

Riccardo Marega

Riccardo Marega obtained his master degree in medicinal chemistry and pharmaceutical technologies at the University of Trieste (Italy) in 2005, by working on the implementation of functional carbon nanotube derivatives for biomedical applications under the supervision of Prof. Maurizio Prato. He then moved to a pharmaceutical company (Eurand, now Actavis) in the Research Area of Trieste, to pursue a Ph.D. program in Molecular Sciences under the supervision of Prof. Maurizio Prato and Dr. Erminio Murano. There, he was involved in the synthesis of carbohydrate-drug-CNT nanohybrids as anticancer platforms, along with the development of diffusion-based NMR spectroscopy for the study of functionalized CNT derivatives. In 2010 he joined Prof. Bonifazi's group at the department of Chemistry as a post-doctoral fellow, being mainly involved in the synthesis, characterization, and biological evaluation of magnetic CNT derivatives as multi-functional anti-cancer nanomaterials, and in the surface derivatization of biocompatible substrates promoting directed cellular migration. Later, he was awarded with a FRS-FNRS post-doctoral fellowship in the same institution. For around one year he was affiliated with the Physics Department at UNamur and with the Private research Centre CER Groupe, where he later become an associate researcher holding a permanent position as a R&D Project Manager. Since March 2022 he is R&D Team Leader for Integrated Bioproduction in the same Institution. He has been involved in the redaction of about 30 scientific manuscripts, and his results have been presented at internationally-relevant scientific congress, such as Hybrid Materials 2015 (Sitges, Spain) and NT12 (Brisbane, Australia), among others. He is now active in extramural fundraising at both national and international level, and manages a small team of laboratory assistants, Master and Ph.D. candidates, to pursue tasks at different technological readiness level (3-9).

foods

MDPI

Editorial

Novel Analytical Methods in Food Analysis

Philippe Delahaut * and Riccardo Marega

Analytical Laboratory, CER Groupe, Rue du Point du Jour 8, 6900 Marloie, Belgium; r.marega@cergroupe.be
* Correspondence: expevet@gmail.com; Tel.: +32-495-30-44-06

Citation: Delahaut, P.; Marega, R. Novel Analytical Methods in Food Analysis. *Foods* **2022**, *11*, 1512. https://doi.org/10.3390/foods11101512

Received: 20 April 2022
Accepted: 10 May 2022
Published: 23 May 2022

Publisher's Note: MDPI stays neutral with regard to jurisdictional claims in published maps and institutional affiliations.

Food analysis is a discipline with a huge impact on both economical and medical aspects of modern societies, meaning that it is at the cornerstone between industrial, medical, and regulatory needs.

The development of analytical methods in food matrices has always been difficult due to the large variety of their physicochemical properties (e.g., physical state, lipid content, pH, among others), which can change analyte structure and extraction efficiencies (e.g., Mallard reactions) due to different processing throughout preparation and distribution (e.g., fermentation, heating, mechanical stress).

On the one hand, such complexity can be tackled by a combination of sample preparation protocols and use of analytical instrumentation that is typically available in specialized laboratories (in terms of personnel and equipment, e.g., mass spectrometry analysis of food allergens). On the other hand, there is a great demand for the "decentralization" of analytical food methods by means of protocol simplification and on-site analysis (e.g., portable immunoassays). Furthermore, the simultaneous detection of multiple analytes at the same time (multiplexing) is an ongoing trend in the development of methods and instruments that increase throughput while lowering costs and operator intervention.

Integration of biological reagents (antibodies, aptamers), materials (nanoparticles, nanotubes), technologies (microspotting, microfluidics), and physical principles (spectroscopy and spectrometry) is today consolidating at both the academic and industrial levels, aiming at the exploration and control of the vast chemical space intersecting with food analysis.

Marchand et al. propose a strategy combining non-targeted and targeted lipidomics MS-based approaches to identify disrupted patterns in serum lipidome upon growth promoter treatment in pigs [1]. Evaluating the relative contributions of the platforms involved, the study aims at investigating the potential of innovative analytical approaches to highlight potential chemical food safety threats. The strategy enabled highlighting specific lipid profile patterns involving various lipid classes, mainly in relation to cholesterol esters, sphingomyelins, lactosylceramide, phosphatidylcholines, and triglycerides. Thanks to the combination of non-targeted and targeted MS approaches, various compartments of the pig serum lipidome could be explored, including commonly characterised lipids (by Lipidyzer™ platform kits), triglyceride isomers (by triglyceride platform methods) and unique lipid features (by non-targeted LC-HRMS). Thanks to their respective characteristics, the complementarity of the three tools could be demonstrated for public health purposes, with enhanced coverage, level of characterization, and applicability.

Pietschmann et al. discuss how the misuse of antibiotics as well as incorrect dosage or insufficient time for detoxification can result in the presence of pharmacologically active molecules in fresh milk [2]. Hence, in many countries, commercially available milk has to be tested with immunological, chromatographic, or microbiological analytical methods to avoid consumption of antibiotic residues. They thus report on a novel, sensitive and portable assay setup for the detection and quantification of penicillin and kanamycin in whole fat milk (WFM) based on competitive magnetic immunodetection (cMID). Their results demonstrate the suitability of cMID-based competition assay for reliable and easy on-site testing of milk.

Damiani et al. report on the origin discrimination of Argentinian honeys as a case study to compare the capabilities of three spectroscopic techniques as fast screening platforms for honey authentication purposes [3]. Each sample was fingerprinted by FT-MIR, NIR and FT-Raman spectroscopy. The results obtained in their work suggests the major potential of FT-MIR for fingerprinting-based honey authentication and demonstrate that accuracy levels that may be commercially useful can be reached.

Kuragano et al. developed a microliter-scale high-throughput screening (MSHTS) system for Aβ42 aggregation inhibitors using quantum-dot nanoprobes [4]. This study aimed at elucidating whether the MSHTS system could be applied to the evaluation of processed foods. Therefore, they examined Aβ42 aggregation inhibitory activity of salad dressings, including soy sauces. They demonstrated that non-heat-treated raw soy sauce exhibited higher Aβ42 aggregation inhibitory activity than heat-treated soy sauce, and concluded that MSHTS system can be applied to processed foods.

Schelm et al. report on the development of methods for detecting possible adulterations on truffles [5]. A real-time PCR (polymerase chain reaction) assay allowing the detection and quantitation of Asian black truffles in Tuber melanosporum up to 0.5% was developed. In addition, a capillary gel electrophoresis assay was designed, which allows for the identification and quantitation of different species. The methods can be used to ensure the integrity of truffle products.

Jafari et al. discuss in their review the increasing demand for portable and handheld devices to provide rapid, efficient, and on-site screening of food contaminants [6]. Recent technological advancements in the field include smartphone-based, microfluidic chip-based, and paper-based devices integrated with electrochemical and optical biosensing platforms. Furthermore, the potential application of portable mass spectrometers in food testing might bring (in the future) the confirmatory analysis from the laboratory to the field. To this end, the analytical performance of these devices and the extent they match the World Health Organization benchmark for diagnostic tests (i.e., the Affordable, Sensitive, Specific, User-friendly, Rapid and Robust, Equipment-free, and Deliverable to end-users (ASSURED) criteria) was evaluated critically. A five-star scoring system was used to assess their potential to be implemented as food safety testing systems. The main findings highlight the need for concentrated efforts towards combining the best features of different technologies, to bridge technological gaps and meet commercialization requirements.

Bergwerff et al. discuss in their review how food microbiology is deluged by a vastly growing plethora of analytical methods [7]. The context is that the highest risk of food contamination comes through the animal and human fecal route, with a majority of food-borne infections originating from sources in mass and domestic kitchens at the end of the food-chain. Whatever the scientific and technological excellence and incentives, the decision-maker determines this implementation after weighing mainly costs and business risks.

Gavage et al. report on the recent accessibility and technological improvements of high resolution mass spectrometry (HRMS) for the analysis of different classes of contaminants and residues [8]. This kind of instrument is often considered as a research tool, but the wide range of potential contaminants and residues that must be monitored, is increasing. Their review aims, through a series of relevant selected studies and developed methods dedicated to the different classes of contaminants and residues, to demonstrate that HRMS can reach detection levels in compliance with current legislation and is a versatile and appropriate tool for routine testing.

Tsagkaris et al. critically review the available screening methods for pesticide residues based on optical detection during the period 2016–2020 [9]. Optical biosensors are commonly miniaturized analytical platforms introducing the point-of-care (POC) era in the field. Various optical detection principles have been utilized, namely colorimetry, fluorescence (FL), surface plasmon resonance (SPR), and surface enhanced Raman spectroscopy (SERS). Overall, despite being in an early stage facing several challenges (i.e., long sample preparation protocols or interphone variation results), such POC diagnostics pave a new road into the food safety field in which analysis costs will be reduced and a more intensive testing will be achieved.

Walpurgis et al. propose a narrative review with an overall aim of indicating the current state of knowledge and the relevance concerning food and supplement contamination and/or adulteration with doping agents and the respective implications for sportspeople drug testing [10]. The identification of a doping agent (or its metabolite) in sports drug testing samples constitutes a violation of the anti-doping rules defined by the World Anti-Doping Agency. Reasons for such adverse analytical findings (AAFs) include the intentional misuse of performance-enhancing/banned drugs. While the sensitivity of assays employed to test pharmaceuticals for impurities is in accordance with good manufacturing practice guidelines allowing to exclude any physiological effects, minute trace amounts of contaminating compounds can still result in positive doping tests. In addition, food was found to be a potential source of unintentional doping, the most prominent example being meat tainted with the anabolic agent clenbuterol.

The research manuscripts and reviews reported in this special issue are thus representative examples of the complexity, variety and demands of the food analysis domain. Indeed, spectrometry-based methods (mass, vibrational) remain the reference ones for laboratory-scale analytical demands, while immunoassays are still the most common base for portable assay development for on-site applications. While laboratory scale methods are gaining sensibility and operability thanks to hardware and software optimizations (data-treatment and analysis), portable ones are exploring the use of more alternative signal transducers (e.g., nanoparticles) and on the integration of the analytical result with the data intrinsically originating from hand-held devices (automated time, localization and cloud based analytical data transmission). At last, it has to be noted how methods that were initially designed to reply a specific analytical request are then used to provide answers to fields that are different from the original ones. In this respect, the examples reported herein concerning how to check the lipidome profile in order to detect β-agonist use in animals and on how to relate dietary consumption with athlete's scores after anti-doping test are explanatory of the current trend of scope broadening and wide applicability of some methods. Taken together, these manuscripts highlight how the creativity of the authors in particular, and of the scientific community in general, is yielding novel analytical methods and responses to longstanding problems in food analysis.

Author Contributions: Writing—original draft preparation, R.M.; writing—review and editing, P.D. All authors have read and agreed to the published version of the manuscript.

Funding: This research received no external funding.

Data Availability Statement: Not applicable.

Conflicts of Interest: Philippe Delahaut is an author in one of the review articles published in this special issue.

References

1. Marchand, J.; Guitton, Y.; Martineau, E.; Royer, A.-L.; Balgoma, D.; Le Bizec, B.; Giraudeau, P.; Dervilly, G. Extending the Lipidome Coverage by Combining Different Mass Spectrometric Platforms: An Innovative Strategy to Answer Chemical Food Safety Issues. *Foods* **2021**, *10*, 1218. [CrossRef]
2. Pietschmann, J.; Dittmann, D.; Spiegel, H.; Krause, H.-J.; Schröper, F. A Novel Method for Antibiotic Detection in Milk Based on Competitive Magnetic Immunodetection. *Foods* **2020**, *9*, 1773. [CrossRef] [PubMed]
3. Damiani, T.; Alonso-Salces, R.M.; Aubone, I.; Baeten, V.; Arnould, Q.; Dall'Asta, C.; Fuselli, S.R.; Pierna, J.A.F. Vibrational Spectroscopy Coupled to a Multivariate Analysis Tiered Approach for Argentinean Honey Provenance Confirmation. *Foods* **2020**, *9*, 1450. [CrossRef] [PubMed]
4. Kuragano, M.; Yoshinari, W.; Lin, X.; Shimamori, K.; Uwai, K.; Tokuraku, K. Evaluation of Amyloid B42 Aggregation Inhibitory Activity of Commercial Dressings by A Microliter-Scale High-Throughput Screening System Using Quantum-Dot Nanoprobes. *Foods* **2020**, *9*, 825. [CrossRef] [PubMed]
5. Schelm, S.; Siemt, M.; Pfeiffer, J.; Lang, C.; Tichy, H.-V.; Fischer, M.; Kuragano, M.; Yoshinari, W.; Lin, X.; Shimamori, K.; et al. Food Authentication: Identification and Quantitation of Different Tuber Species via Capillary Gel Electrophoresis and Real-Time PCR. *Foods* **2020**, *9*, 501. [CrossRef] [PubMed]

6. Jafari, S.; Guercetti, J.; Geballa-Koukoula, A.; Tsagkaris, A.S.; Nelis, J.L.D.; Marco, M.-P.; Salvador, J.-P.; Gerssen, A.; Hajslova, J.; Elliott, C.; et al. ASSURED Point-of-Need Food Safety Screening: A Critical Assessment of Portable Food Analyzers. *Foods* **2021**, *10*, 1399. [CrossRef] [PubMed]
7. Bergwerff, A.A.; Debast, S.B. Modernization of Control of Pathogenic Micro-Organisms in the Food-Chain Requires a Durable Role for Immunoaffinity-Based Detection Methodology—A Review. *Foods* **2021**, *11*, 832. [CrossRef] [PubMed]
8. Gavage, M.; Delahaut, P.; Gillard, N. Suitability of High-Resolution Mass Spectrometry for Routine Analysis of Small Molecules in Food, Feed and Water for Safety and Authenticity Purposes: A Review. *Food* **2021**, *10*, 601. [CrossRef]
9. Tsagkaris, A.S.; Pulkrabova, J.; Hajslova, J. Optical Screening Methods for Pesticide Residue Detection in Food Matrices: Advances and Emerging Analytical Trends. *Foods* **2021**, *10*, 88. [CrossRef] [PubMed]
10. Walpurgis, K.; Thomas, A.; Geyer, H.; Mareck, U.; Thevis, M. Dietary Supplement and Food Contaminations and Their Implications for Doping Controls. *Foods* **2020**, *9*, 1012. [CrossRef] [PubMed]

 *foods*

Review

ASSURED Point-of-Need Food Safety Screening: A Critical Assessment of Portable Food Analyzers

Safiye Jafari [1,2], Julian Guercetti [3,4], Ariadni Geballa-Koukoula [5], Aristeidis S. Tsagkaris [6], Joost L. D. Nelis [7], M.-Pilar Marco [3,4], J.-Pablo Salvador [3,4], Arjen Gerssen [5], Jana Hajslova [6], Chris Elliott [7], Katrina Campbell [7], Davide Migliorelli [2], Loïc Burr [2], Silvia Generelli [2,*], Michel W. F. Nielen [5,8] and Shana J. Sturla [1,*]

[1] Department of Health Sciences and Technology, ETH Zürich, Schmelzbergstrasse 9, 8092 Zürich, Switzerland; jafaris@ethz.ch
[2] CSEM SA, Center Landquart, Bahnhofstrasse 1, 7302 Landquart, Switzerland; davide.migliorelli@csem.ch (D.M.); loic.burr@csem.ch (L.B.)
[3] Nanobiotechnology for Diagnostics (Nb4D), Institute for Advanced Chemistry of Catalonia (IQAC) of the Spanish Council for Scientific Research (CSIC), Jordi Girona 18-26, 08034 Barcelona, Spain; Julian.guercetti@iqac.csic.es (J.G.); pilar.marco@cid.csic.es (M.-P.M.); jpablo.salvador@iqac.csic.es (J.-P.S.)
[4] CIBER de Bioingeniería, Biomateriales y Nanomedicina (CIBER-BBN), Jordi Girona 18-26, 08034 Barcelona, Spain
[5] Wageningen Food Safety Research, Wageningen University and Research, P.O. Box 230, 6700 AE Wageningen, The Netherlands; ariadni.geballakoukoula@wur.nl (A.G.-K.); arjen.gerssen@wur.nl (A.G.); michel.nielen@wur.nl (M.W.N.F.)
[6] Department of Food Analysis and Nutrition, Faculty of Food and Biochemical Technology, University of Chemistry and Technology Prague, Technická 5, Dejvice, 166 28 Prague 6, Czech Republic; Aristeidis.Tsagkaris@vscht.cz (A.S.T.); jana.hajslova@vscht.cz (J.H.)
[7] Institute for Global Food Security, School of Biological Sciences, Queen's University, 19 Chlorine Gardens, Belfast BT9 5DL, UK; J.Nelis@qub.ac.uk (J.L.D.N.); chris.elliott@qub.ac.uk (C.E.); katrina.campbell@qub.ac.uk (K.C.)
[8] Laboratory of Organic Chemistry, Wageningen University, Stippeneng 4, 6708 WE Wageningen, The Netherlands
*   Correspondence: silvia.generelli@csem.ch (S.G.); sturlas@ethz.ch (S.J.S.)

Citation: Jafari, S.; Guercetti, J.; Geballa-Koukoula, A.; Tsagkaris, A.S.; Nelis, J.L.D.; Marco, M.-P.; Salvador, J.-P.; Gerssen, A.; Hajslova, J.; Elliott, C.; et al. ASSURED Point-of-Need Food Safety Screening: A Critical Assessment of Portable Food Analyzers. *Foods* **2021**, *10*, 1399. https://doi.org/10.3390/foods10061399

Academic Editors: Philippe Delahaut and Riccardo Marega

Received: 19 April 2021
Accepted: 12 June 2021
Published: 17 June 2021

Publisher's Note: MDPI stays neutral with regard to jurisdictional claims in published maps and institutional affiliations.

Abstract: Standard methods for chemical food safety testing in official laboratories rely largely on liquid or gas chromatography coupled with mass spectrometry. Although these methods are considered the gold standard for quantitative confirmatory analysis, they require sampling, transferring the samples to a central laboratory to be tested by highly trained personnel, and the use of expensive equipment. Therefore, there is an increasing demand for portable and handheld devices to provide rapid, efficient, and on-site screening of food contaminants. Recent technological advancements in the field include smartphone-based, microfluidic chip-based, and paper-based devices integrated with electrochemical and optical biosensing platforms. Furthermore, the potential application of portable mass spectrometers in food testing might bring the confirmatory analysis from the laboratory to the field in the future. Although such systems open new promising possibilities for portable food testing, few of these devices are commercially available. To understand why barriers remain, portable food analyzers reported in the literature over the last ten years were reviewed. To this end, the analytical performance of these devices and the extent they match the World Health Organization benchmark for diagnostic tests, i.e., the Affordable, Sensitive, Specific, User-friendly, Rapid and Robust, Equipment-free, and Deliverable to end-users (ASSURED) criteria, was evaluated critically. A five-star scoring system was used to assess their potential to be implemented as food safety testing systems. The main findings highlight the need for concentrated efforts towards combining the best features of different technologies, to bridge technological gaps and meet commercialization requirements.

Keywords: food safety; portable food analyzer; point-of-need; ASSURED criteria; portable mass spectrometer; optical biosensor; electrochemical biosensor; microfluidic device; lab-on-a-chip; smartphone-based biosensor

1. Introduction

Food safety issues pose serious public health risks worldwide, accounting for 420,000 deaths each year, according to the World Health Organization (WHO) [1]. For instance, a recent food poisoning incident in Uganda in 2019 resulted in 311 illness cases and five fatalities. To identify the cause, the World Food Program halted its Super Cereals aid to many countries suffering from famine. After international investigations, the cause was identified to be Tropane alkaloids, namely atropine and scopolamine contamination by mass spectrometry (MS) and infrared spectroscopy coupled to chemometrics [2]. Considering the complexity of the food chain today, efficient, and reliable food safety systems are crucial to ensure consumer safety and minimize health risks and economic losses. The standard methods for chemical food safety testing rely largely on high-performance liquid chromatography or gas chromatography coupled with MS (LC-MS or GC-MS) [3]. The standard methods for microbiological/pathogen food safety are ISO reference methods based on the culture plate analysis. Although these laboratory-based techniques are highly reliable, they are time-consuming, expensive, and require trained personnel. These requirements limit their wide application across the various stages of complex food supply chains, thus limit the testing frequency and increase contamination risks. Therefore, there is an urgent need to develop portable food analyzers for on-site point-of-need food safety screening. Such tools are expected to ensure consumer safety, particularly in resource-limited settings by allowing for increases in sampling frequency and reducing costs.

Recently, biosensors have gained increasing attention as screening methods for on-site food safety testing [4,5]. Biosensors are often integrated into small bioanalytical devices, which provide rapid, selective, and sensitive detection of analytes in samples. They consist of three main components:

- The biorecognition element, in general, an antibody, aptamer, or enzyme that binds specifically to the target analyte;
- The transducer converts resulting signals, which can be optical, electrochemical, magnetic, calorimetric, etc.;
- The readout system is used to visualizes the result.

Different types of biosensors have been combined with either paper-based or chip-based microfluidics to form lab-on-a-chip devices, which provide a powerful tool for point-of-need food testing. Although these devices provide rapid and user-friendly analysis, based on performance criteria set out in EU regulations (EC) No 882/2004 and (EU) No 519/2014, they are considered screening tests. As screening methods, they are meant to differentiate between large numbers of compliant (negative) samples and a few suspects (positive) samples. According to EU regulation 2002/657/EC, suspect results require follow-up by a confirmatory instrumental method to declare those samples either compliant or non-compliant [6]. In this regard, recent advancements in portable mass spectrometers could eventually enable confirmatory MS analysis to be performed on-site [7]. As the nearest technology to the gold standard, we decided to investigate the status of portable MS technology in the food safety field to have a comprehensive picture of the portable food analyzers. While the performance of portable MS could already meet screening analysis requirements, instruments cost between 100,000–300,000 euros which is more than 1000 times higher than that of biosensing devices. Therefore, a clear goal should be to achieve performance specifications sufficient to qualify the portable mass spectrometer for on-site confirmatory analysis.

In this review, we aim to provide key insights into the technological advancements of miniaturized and integrated food analyzers based on state-of-the-art electrochemical and optical biosensing platforms. The ASSURED criteria correspond to Affordable, Sensitive, Specific, User-friendly, Rapid and Robust, Equipment-free, and Deliverable to end-users. These criteria were set by the WHO for point-of-care diagnostic test performance evaluation in resource-limited settings and are complementary to the performance criteria for screening and confirmatory methods in the EU regulation 2002/657/EC [6,8]. Therefore, we considered them as the basis of comparing the performance of portable food safety analyzers in this review article. Affordability is considered as the cost of the test and the additional equipment necessary for realizing the test. In the ASSURED criteria, Sensitivity is defined as true positive rate and Specificity as true negative rate. However, data on true positive and negative rates, for the examples evaluated for this study, were often not provided with validation studies based on the regulatory guidelines for such an assessment (EU regulation (EU) No 519/2014). Therefore, we chose to consider the limit of detection (LOD) compared to the maximum residue limit (MRL) indicated in EU regulations as a proxy for Sensitivity, and the selectivity (the analyte signal relative to interferences of similar molecules or other contaminants) as a proxy for Specificity [6]. User-friendliness is evaluated by simplicity, automation, and minimum training required for performing the test. To complete the evaluation, we addressed Rapidness as total analysis time, Robustness as test vulnerability toward minor changes in assay conditions, Equipment-free as the need for extra equipment, and Deliverable to end-users as the accessibility of the technology to the public.

For each of these terms, a five-star grading was used to compare different classes of portable devices as shown in Table 1. The grading was in part subjective to our assessment of the reviewed papers in this work to reflect the status of portable food safety analyzers. The assessed overall score was calculated as the average of stars for all the terms in the ASSURED criteria. Based on our grading, paper-based colorimetric and smartphone-based devices are the most promising technologies for on-site food analysis. This ranking is following the trend observed in devices already on the market, namely paper-based test strips. An in-depth discussion of the scoring of each of the technologies in Table 1 is reported in the following sections. Finally, the potential role of portable MS analyzers for food analysis has been evaluated and the challenges of bringing the confirmatory MS technique from the laboratory to the field are discussed. While biosensing devices as screening tests are complementary to confirmatory methods (LC- or GC-MS), portable MS devices could potentially replace both methods.

Table 1. The Affordable, Sensitive, Specific, User-friendly, Rapid and Robust, Equipment-free, and Deliverable to end-users (ASSURED) criteria grading assigned in this comparative evaluation of different types of portable food analyzers. Five stars were given to the device with the highest performance in each criterion compared with other devices. As an example, the paper-based colorimetric device is the cheapest of the portable analyzers, so it was assigned five stars in the Affordability criteria. (star ★).

Criteria	Affordable	Sensitive	Specific	User-Friendly	Rapid and Robust	Equipment-Free	Deliverable to End-user	Overall Score
Paper-based colorimetric	5	1	4	5	5	5	5	4.3
Smartphone-based optical	4	2	4	4	2	4	4	3.4
Smartphone-based electrochemical	2	5	4	3	3	2	3	3.1
Microfluidic chip-based electrochemical	2	5	4	2	3	1	2	2.7
Microfluidic chip-based optical	2	4	4	2	3	1	2	2.6
Paper-based electrochemical	3	2	4	3	3	2	2	2.6
Raman/IR -based	2	1	5	3	4	2	2	2.6
Portable MS as screening	1	5	5	1	3	1	1	2.4

2. Portable Optical Food Analyzers

Optical biosensing platforms are based on measuring changes in properties of light, such as intensity, wavelength, polarization, or propagation direction as the biosensor response. These changes can be monitored using different methods, such as colorimetric, fluorescence, surface plasmon resonance, infrared, and Raman spectroscopy [9]. The integration of these optical techniques with paper-based, microfluidic chip-based, and smartphone-based platforms has been considered for on-site food analysis. Optical transducers are versatile, spanning read-outs ranging from those distinguished by the naked eye in colorimetric detection to involving spectrometers and refractometers. Colorimetric detection with the naked eye is the cheapest biosensing platform, however, it can only provide qualitative analysis. Using an imaging system such as a reader or a smartphone improves the sensitivity to a semi-quantitative level but introduces more cost and decreases user-friendliness. On the other hand, optical detection based on spectrometry and refractometry, which provides quantitative analysis, is usually performed using lab-based benchtop UV-Vis and fluorescence spectrometers [10]. Recently, there has been a trend toward the development of miniaturized spectrometers with a similar performance to benchtop devices at a lower cost. While this goal is particularly challenging, considering the fabrication of miniaturized gratings and reflective optics, and decreasing the path length, some of these devices are already available on the market at a few thousand dollars [11].

2.1. Paper-Based Optical Food Analyzers

Paper-based optical food analyzers with simple colorimetric detection are the most common type of portable food analyzers (Table S1). Simple paper-based assays as test strips are commercially available for various food contaminants such as aflatoxins, marine biotoxins, and pesticides. The two main categories of these devices are lateral flow assays (LFA) and microfluidic paper-based analytical devices (µPADs), both providing rapid, low-cost, and user-friendly on-site analysis [12,13]. While the sample flow in LFA only moves in one direction, microfluidic channels in µPADs can direct the flow in complex patterns enabling multi-step procedures [14].

From the ASSURED criteria, these devices meet ideally the affordability (five stars), specificity (four stars), and rapidness (five stars) criteria. Also, they are equipment-free (five stars) and commercially available, thus immediately and easily deliverable to end-users (five stars). However, they only receive one star in sensitivity since the results are qualitative or semi-quantitative due to the intrinsic limitations of the technology. Their overall score is 4.3 (Table 1), top-ranked amongst the portable food analyzers when all criteria are weighted equally.

One of the main reasons for the popularity of paper-based devices is the paper itself. The cellulose membrane not only provides an immobilization platform for biorecognition elements, such as antibodies or aptamers but also acts as a transportation and reaction platform for the sample and other reagents. The LFA mainly uses nitrocellulose as the support material, owing to its excellent protein adsorption properties for the immobilization of biomolecules. Their rapidness was demonstrated in a study combining LFA with nucleic acid extraction, amplification, and colorimetric detection of *Escherichia coli* and *Streptococcus pneumonia* in milk and spinach [15]. Compared to time-consuming conventional culture plate assays, which require more than five hours, this hybrid LFA provided the results within one hour. Another rapid LFA example was developed for the detection of hazelnut allergens in cookies with a carbon nanoparticle-labeled antibody. The qualitative result in this example was displayed in 30 s [16].

Multiplexing can further increase the performance of paper-based assays and bring it closer to the traditional microplate assays with high-throughput analysis capability [17]. In this regard, rapid multiplex detection of 10 different foodborne pathogens in different food matrices (dairy products, marine products, beverages, snacks, and meats) was achieved within 20 min using a disc with multiple paper-based LFA devices [18]. The geometry of this hybrid assay permits simultaneous sample injection into all the LFAs, making it simple and user-friendly (Figure 1a). Another example is a hybrid paper-lab-on-a-chip (paper-LOC) injector for carbofuran screening in apple extracts [19]. The paper-LOC device is a low-cost and multiplexed device, with a price of 0.30 euro for analyzing two samples. However, the device has low sensitivity with an LOD of 0.050 mg·kg^{-1}, which is 50 times higher than the MRL for carbofuran (0.001 mg·kg^{-1}) set by the EU commission regulation (part A of Annex I to Reg. 396/2005). This device also features integrated sample handling, with sample and reagent injection using silicone tubing (Figure 1b).

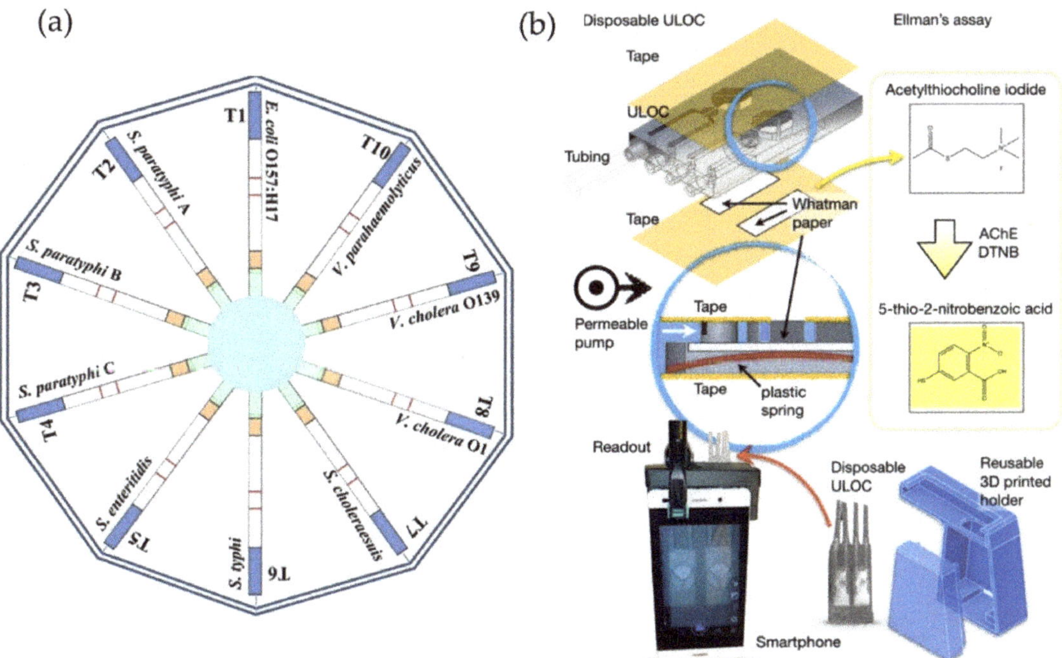

Figure 1. Examples of hybrid paper-based optical food analyzers: (**a**) a disc containing 10 different LFA assays for foodborne pathogen detection. Reproduced with permission from [18] Scientific reports, Copyright (2016), under CC BY 4.0; (**b**) a hybrid paper-LOC device provides semiquantitative carbofuran screening using a smartphone as a detector. Reproduced with permission from [19] Sensors, Copyright (2019), under CC BY 4.0.

Although colorimetric detection with the naked eye is a simple approach with a high level of user-friendliness, enabling non-experts to perform on-site food analysis, measurements are not always accurate and/or sensitive enough. Furthermore, the colorimetric signal can easily be affected by (ambient) light conditions, color metamerism, and colored food matrices set another challenge for creating a successful color-based detection system [20]. To circumvent such limitations, paper-based assays have been combined with various spectroscopic detection systems. Chemiluminescent detection was utilized to detect dichlorvos, an organophosphate insecticide in vegetables, in a wide linear range from 0.006–2 mg·kg^{-1} and with an LOD equal to 0.0016 mg·kg^{-1}, which is lower than the regulatory limit of 0.01 mg·kg^{-1} [21]. Localized surface plasmon resonance was also used for biogenic amine detection for monitoring salmon freshness. In this case, nanoparticle-embedded papers served as gas sensors, providing high particle transfer efficiency and a strong resonance reflectance dip [22]. While these examples showcase the potential for integration of spectroscopic and paper-based systems, they increase the cost and reduce the user-friendliness of paper-based devices.

2.2. Microfluidic, Chip-Based Optical Food Analyzers

Microfluidics systems provide precise control and manipulation of small amounts of fluids using microchannels [23]. The microchannels are fabricated mainly using transparent polymers such as polydimethylsiloxane (PDMS), and poly(2,5-dimethoxyaniline) (PDMA), adhered to a glass substrate [24]. This relatively new field became more popular with the development of 3D printing technologies providing easy and cost-effective manufacturing of the prototypes. The microfluidic devices can mimic reactors to carry out sample preparation, filtration, dilutions, and detection, which results in a reduction of handling errors and a subsequent increase in analytical robustness [25]. These advantages are of great importance in the development of an integrated portable food analyzer (Table S2). Considering the ASSURED criteria, microfluidic chip-based optical food analyzers provide sensitive and specific analysis with four stars for each of these parameters. However, they require additional equipment, such as a built-in camera, optical filters, illumination sources, and pumps, which reduces affordability (two stars) and limits accessibility (two stars). Compared to other devices, they obtain three stars in rapid and robust and two stars in user-friendly terms, resulting in an overall score of 2.6. The liquid handling automation, the potential for high-throughput, and integration of sample preparation in these devices might increase their appeal for on-site analysis.

In general, optical biosensing is a multi-step procedure, involving extraction of contaminants from the food sample, transport to the biorecognition element, and optical transduction. Integrating these steps into a single microfluidic chip-based device can effectively improve the applicability of optical food analyzers in real-life settings. Toward this goal, an integrated microfluidic device was fabricated for simultaneous extraction, preconcentration, and detection of ochratoxin A in wine. The analysis time was less than 30 min with an LOD of 0.26 $\mu g \cdot kg^{-1}$, which is 8 times lower than the EU regulatory limit of 2 $\mu g \cdot kg^{-1}$ (Figure 2) [26]. The PDMS-based device consisted of two consecutive modules performing a two-phase extraction and immunoassay detection, but a microscope was needed for quantitative analysis. The authors suggest replacing the microscope by integrating the device with an on-site fluorescence photodetector. Moreover, the measurement was performed under continuous flow using syringe pumps, which provides automation but reduces user-friendliness for non-expert users.

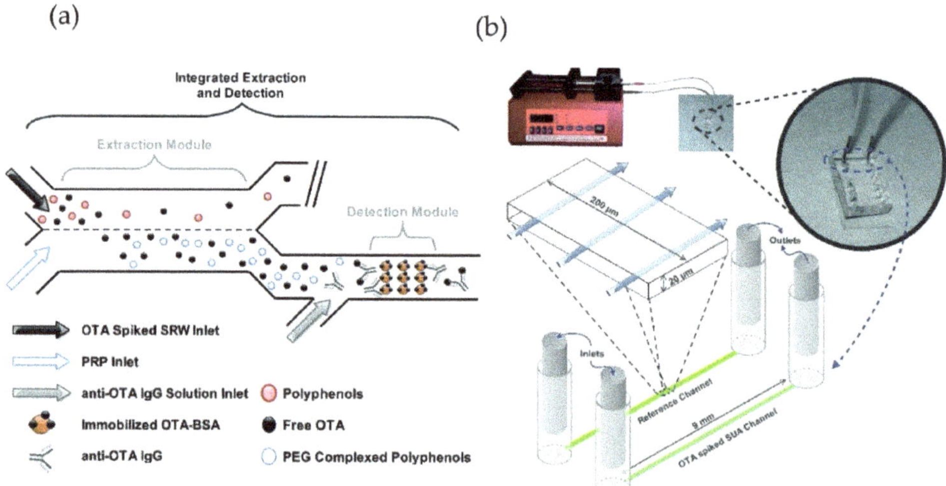

Figure 2. An example of a microfluidic chip-based optical food analyzer used for ochratoxin A (OTA) detection in red wine. (**a**) The integrated microfluidic strategy for performing an aqueous two-phase extraction with PEG-rich phase (PRP); (**b**) After performing an indirect competitive fluorescence-linked immunosorbent assay, the measurements are performed with a fluorescence microscope under continuous flow. Reproduced with permission from [26] Lab on a Chip, Copyright (2014), under Royal Society of Chemistry.

Many lab-on-a-chip platforms in the food safety field are designed for the detection of pathogens, which requires integrating the sample preparation, isothermal amplification, and detection processes in a single device. The reported studies combine nucleic acid recognition, polymerase amplification, and fluorescence imaging to detect bacteria like *Salmonella typhimurium* and *Escherichia coli* with limits of detection of 4 to 10 cells·μL^{-1} in milk and cultured media [27,28]. These devices mostly use centrifugal platforms in the shape of a compact disc, consisting of different chambers for extraction, amplification, and detection steps. The fluid flow between chambers, the reaction, and the mixing procedures are controlled by centrifugal force, eliminating the need for pumps, valves, and tubes. Although these devices provide sensitive analysis below the MRL set by EU regulations, they require a centrifuge and a miniaturized spectrometer. To improve the sensitivity even further, one outstanding study reported integrating a supercritical angle fluorescence microlens array in a microchip. The microlens has a high fluorescence collection efficiency, which resulted in higher sensitivity and LOD of 1.6 copies·μL^{-1} of pathogenic DNA. The result was comparable to the conventional spectrometers' performance but with a higher noise level and a lower signal to noise ratio [27].

Other optical detection methods, such as surface plasmon resonance (SPR) were also employed for optical microfluidic chip-based food analysis. While SPR provides label-free optical detection, it has a lower sensitivity compared to fluorescence-based detection, particularly for the detection of small molecules, such as mycotoxins due to the low refractive index change. In some studies, the SPR sensitivity was improved by signal amplification using nanomaterials. An integrated gold chip with gold nanoparticles was reported for the detection of aflatoxin B1 in wheat with an LOD of 0.094 μg·kg^{-1}, which is 20 times lower than the EU regulatory limit of 2 μg·kg^{-1} [29]. This device used gold nanoparticles to amplify the SPR signal, thus improving the sensitivity. Another study implemented a 3D dextran layer on a nanostructured imaging surface plasmon resonance chip for signal amplification. This device was used for the detection of deoxynivalenol and ochratoxin A in beer with LODs of 17 ng·mL^{-1} and 7 ng·mL^{-1}, respectively. Interestingly, the study reported a preliminary in-house validation with 20 beer samples. While this device was able to detect deoxynivalenol below regulatory limits for beer, pre-concentration of the sample was required before detection of ochratoxin A [30]. The chip could be reused 450 times, which reduces the cost of each test. Another analogous strategy for reducing cost was reported involving a regenerable glass biochip for detection of ochratoxin A in coffee beans, reusable up to 20 times [31]. With this chemiluminescence-based device, researchers were able to detect ochratoxin A with an LOD of 7 μg·kg^{-1} and a total analysis time of 12 min. However, the use of an immunoaffinity column for sample enrichment introduced more cost and decreased user-friendliness. In terms of the total analysis time, most of the integrated devices provide the results in less than one hour, including sample preparation [32–35].

2.3. Smartphone-Based Optical Food Analyzers

Integrating optical detection into smartphone-based biosensors takes advantage of the smartphone optical components, such as the light source and the camera as a photodetector. The crucial aspect here is that a smartphone not only detects but also provides the location (GPS), time, and wireless data transfer to stakeholders, thus enabling the geo-temporal mapping of (food) contamination issues. From the ASSURED criteria, these devices improve user-friendliness and accessibility of on-site food testing (4 stars). When compared to other portable food analyzers, they were rated with four stars in terms of affordability, user-friendliness, equipment-free, and deliverable to end-users. They have a slightly better sensitivity (two stars), compared to equipment-free paper-based optical devices. Their overall score, therefore, is 3.4, which places them amongst the top three most promising portable food analyzers.

Smartphone-based optical food analyzers are mainly used with colorimetric detection, which results in a picture to be analyzed using a smartphone app based on RGB, CieLAB,

HSV, or greyscale. The functioning of these color spaces for optimal sensitivity and error reduction has been investigated extensively. It was found that individual RGB channels often produce optimal colorimetric performance, although other color spaces such as CieLAB or even novel artificial channel combinations of various color spaces combined may equally perform very well in certain assays [36]. The reviewed smartphone-based colorimetric detection strategies are based on the immunoassay test strips [37,38], nanoparticle aggregation [39,40], or enzyme inhibition assays [41,42]. Although the limits of detection of the colorimetric smartphone-based food analyzers reported in the literature are fit for screening methods based on the food safety regulations, they only provide qualitative or semi-quantitative results.

Alternative approaches such as fluorescence and chemiluminescence have been explored to provide quantitative results for the detection of mycotoxins in corn and tetracyclines and quinolones in milk, respectively [37,43]. With these methods, pesticide detection in spinach with LOD 5 to 10 times lower than the regulatory limit was achieved, affirming them as fit for purpose [44]. Furthermore, the fluorescence method is preferred over chemiluminescence for pathogen detection since fluorescence imaging directly detects the fluorescent-labeled molecules and does not require an enzymatic reaction to produce the detection signal. Smartphone-based fluorescence assays were reported with detection limits of 58 colony-forming unit (CFU)$\cdot$mL^{-1} in fruit juice and 1 CFU$\cdot$mL^{-1} in yogurt for *Salmonella typhimurium*, and 10 CFU$\cdot$mL^{-1} for *Escherichia coli* in egg samples, which were all lower than the EU regulation limits [45,46].

All the assessed optical smartphone-based devices required external equipment and accessories. Mainly 3D printed modules are produced to place the smartphone in a fixed position and acquire the data in a reproducible manner. This could reduce the effect of external variations, like illumination conditions and user handling errors, while improving the sensitivity. An optosensing 3D printed platform was developed for quantitative colorimetric detection on the smartphone (Figure 3a). This device was validated with a UV-Vis spectrometer for the detection of streptomycin in honey and milk with an LOD of 0.009 mg$\cdot$kg^{-1} [47]. The anti-streptomycin aptamer-conjugated gold nanoparticles were used as the colorimetric indicator. The ratio of the absorbance at 625 nm to that at 520 nm was measured as the optosensor signal. One of the main advantages of 3D printing is the possibility of realizing rapid prototyping of disposable microfluidic chips at a low cost. A few papers reported estimations of the final cost of the device. For example, the above-mentioned article reported a price of 13 US dollars per device and 5 US dollars per test [47]. Another study reported a total cost of 5.20 US dollars for colorimetric detection of aflatoxin B1 in corn samples (Figure 3b) [48]. The detection was based on an indirect competitive immunoassay in the PDMS microfluidic channel. Then, the chip was aligned in the 3D-printed optical accessory attached to a smartphone. The image captured by the smartphone camera was directly processed using a custom-developed Android app. The reported price is comparable to the 96 microwell plate Enzyme-linked Immunosorbent Assay (ELISA) or 3–10 US dollars per sample. These numbers suggest a high commercial potential for smartphone-based optical food analyzers. Since the inter-phone variability has proven to be a major hurdle for the implementation of these devices, the evaluation of different smartphone brands has been reported in several studies. The device-independent color space and randomized combined channel approaches were used for smartphone-based image analysis to reduce the interphone variations [36,49].

The food matrix also has a defining role in the development of a smartphone-based optical food analyzer. Liquid matrices such as milk, different types of beverages like juices and cider are the preferred matrices, as they do not require complex extraction procedures (see Table S3). This is mainly due to the complexity of integrating sample preparation with a detection platform in a single device, which limits the applicability of the developed food analyzers. In general terms, it has been observed that further effort in development is needed to simplify the complex extraction protocols for non-expert users.

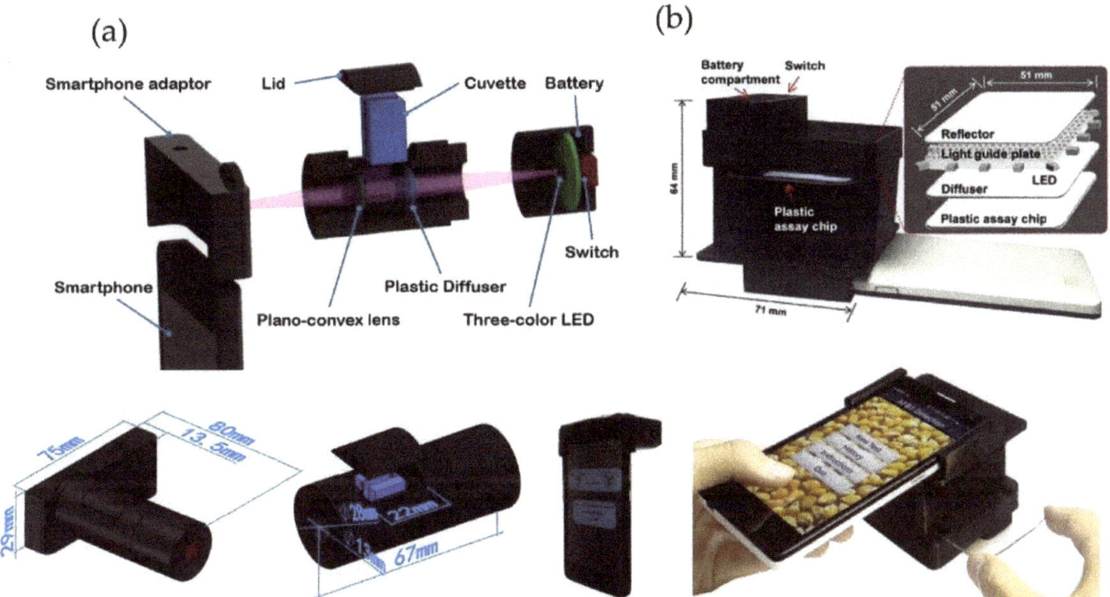

Figure 3. Examples of smartphone-based optical food analyzers: (**a**) The 3D printed smartphone optosensing platform for detection of streptomycin in honey and milk. The anti-streptomycin aptamer-conjugated gold nanoparticles are used as the colorimetric indicator. The ratio of the absorbance at 625 nm to that at 520 nm was measured as the optosensor signal. Schematic overview of the internal structure and dimensions. Reproduced from [47] Analytica Chimica Acta Copyright (2017), with permission from Elsevier; (**b**) The integrated Smartphone-App-Chip system for aflatoxin B1 detection in corn. The detection is based on an indirect competitive immunoassay in the microfluidic channel. Then the assay chip is aligned in the 3D-printed optical accessory attached to a smartphone. The image captured by the smartphone camera is directly processed using a custom-developed Android app. Reproduced with permission from [48] Analytical Chemistry, Copyright (2017), American Chemical Society.

Even though the time of the procedure varies in most of the cases analyzed, the measurement time ranges between five minutes in the best cases and up to 45 min [37,44]. Extra time may be required for sample preparation depending on the composition of the matrix and can take from 10–45 min.

2.4. Raman and IR-Based Portable Food Analyzers

Although vibrational spectroscopy, infrared (near-IR and mid-IR), and Raman spectroscopy, has been mostly applied to reveal food adulteration or verify the origin and authenticity of food, there is a trend towards its implementation in the food safety field (Table S4) [50,51]. Considering the ASSURED criteria, these devices are rapid and robust (four stars) and provide high analytical specificity (five stars). They are relatively user-friendly (three stars) as the analyses are non-invasive and require minimal sample preparation. This feature is a big advantage when compared to multi-step sample preparation required for other devices, from grinding the solid sample and using organic solvent for extraction, to filtration and dilution. Moreover, portable IR and Raman spectrometers have recently become widely available, further increasing the applicability of such platforms in bringing food analysis to the field [52]. While a handheld NIR device such as SCiO is commercially available for only 600 euros, portable Raman spectrometers are generally available at a cost of few thousand euros [53]. Therefore, they are assigned only two stars in affordability and equipment-free terms. Also, they suffer in terms of sensitivity (one star), as the quantitative analysis is still a challenge. Their overall score is 2.6.

Surface-Enhanced Raman Spectroscopy (SERS) can either directly screen for an analyte or can be complementary to other screening methods such as bioanalytical assays to enhance specificity and sensitivity. Signal enhancement of the Raman scattering is necessary for a range of applications and can be achieved using different colloidal or solid nanocomposite substrates [54]. The solid substrates can be immobilized on various surfaces, for example, paper or hydrogels [55,56]. The paper-based SERS substrates can be used at the same time as a sample collection tool to swab the surface of a sample. Although combining a bioanalytical method with SERS can drastically reduce the attained LOD, it increases the method's complexity and cost. For example, coupling a lateral flow immunoassay to SERS resulted in 3 orders of magnitude lower LOD (reaching the pg· mL^{-1} level) compared to the colorimetric naked-eye detection for the antibiotics neomycin and quinolone in milk [57].

It is important to underline that acquiring quantitative results using SERS remains a challenge, even though more and more studies report their results as quantitative (see Table S4). The use of anisotropic nanoparticles (nano-cubes, nano-rods, and nano-stars) and internal standards has positively affected SERS quantification capabilities [58]. In the study reported by X. Li et al., water molecules were used as the internal standard, since their Raman scattering signal was quite stable [59]. In another study, 4-methylthiobenzoic acid (4-MBA), a Raman active probe molecule, was embedded in gold-silver core-shell nano-cubes and exploited as the internal standard to attain quantitative results [60].

Another challenge is that SERS can mostly detect analytes on the surface of food, which does not correspond to the whole amount of a contaminant in a food matrix. One example is pesticide residues, which can be found on the surface or in the inner parts of fruit, depending on their polarity. In such cases, the LOD is in general expressed using the unit "ng·cm^{-2}", which is not in line with the MRL units (mg·kg^{-1}) (Figure 4) [61,62]. In some other studies, an extraction method was used before SERS screening, an approach that eliminates one of the most desirable characteristics of SERS: being a non-invasive method with no sample preparation [63].

Figure 4. SERS application for rapid in-situ pesticide detections in fruits and vegetables. The nanocellulose decorated with Ag nanoparticles (Ag/NC) substrate was used to detect pesticides (thiram and thiabendazole) on apple and cabbages samples. Reproduced from [62] Carbohydrate Polymers, Copyright (2019), with permission from Elsevier.

Near-IR (NIR) spectroscopy is another method that has been applied for food contaminant screening with minimum sample preparation. The typical NIR spectrum consists of complex, broad, and superimposed absorption bands of low intensity. Consequently, NIR provides mostly qualitative results and the application of chemometric tools is necessary to achieve an effective screening. Among various chemometric models, partial least square regression and linear discriminant analysis have been applied for the screening of both pesticides and mycotoxins [64,65]. These models are based on a linear relationship between the NIR spectrum intensity and concentration. They use multi-variable data analysis (multiple wavelengths) with as many as 20–1000 variables to achieve quantification.

The mid-IR spectrum consists of strong vibrational bands in the molecular fingerprint region. The mid-IR spectral intensity is a thousand-fold higher than in the NIR, thus achieving higher sensitivity of the measurements. However, the mid-IR is limited to solid food samples with low moisture content, due to the strong water absorption in this wavelength range. Fourier transform infrared (FT-IR) spectroscopy in mid-IR was reported for the identification of different pathogenic bacteria [66]. The data was obtained using hyperspectral imaging and the data analysis was performed by image segmentation with machine learning and artificial neural networks. The sample preparation included growth and cultivation in an agar plate and subsequent dilution and drying of the sample on an imaging substrate. Another study reported *Botrytis cinerea* fungus detection on tomato leaves in both visually symptomatic and pre-symptomatic plants [67]. The detection accuracy was 100% using chemometrics data analysis at different stages of the disease with a semi-portable spectrometer. A total number of 240 spectra was acquired from 16 leaf samples for accuracy studies. The high detection accuracy for both true-positive and true-negative rates of 100% was also reported for the detection of aflatoxin B1 and total aflatoxins in 90 peanut oil samples. Samples with the concentration higher than $20~\mu g \cdot kg^{-1}$ were identified as positive for both aflatoxin B1 and total aflatoxins detection. This threshold is higher than the EU MRLs for both aflatoxin B1 and total aflatoxins as $2~\mu g \cdot kg^{-1}$ and $4~\mu g \cdot kg^{-1}$, respectively [68]. Quantitative detection was achieved using FT-IR for the detection of clenbuterol in beef meat with an LOD of $2~\mu g \cdot kg^{-1}$, which is 20 times higher than the regulatory limit of $0.1~\mu g \cdot kg^{-1}$ (Commission Regulation (EC) No 1312/96) [69]. Despite a 100% accuracy, the number of samples was limited, and the sensitivity of these studies was not fit for the purpose concerning regulatory requirements.

3. Portable Electrochemical Food Analyzers

Electrochemical biosensors have a great potential in portable analysis owing to the ease of miniaturization and integration of the transducer while providing quantitative results. The electrochemical transducer consists of an electrochemical cell connected to a potentiostat/galvanostat, which performs the measurements. The traditional electrochemical setup is a benchtop device with the standard electrodes in millimeter dimensions. However, the electrochemical cell has been miniaturized at low cost with screen printing or inject printing of the 3 electrodes, reducing the required sample volumes to few microliters [70,71]. Apart from the commercially available miniaturized potentiostats, there has been an increasing trend towards the development of open source and cost-effective portable potentiostats for resource-limited applications [72,73]. The open-source potentiostat equipped with wireless connectivity to the smartphone developed by Whiteside's group is a recent example of such an instrument [74]. The total cost of a single device was 61.5 US dollars, which can be reduced to 15 US dollars if produced in large numbers. This is considerably cheaper compared to the cost of commercial laboratory potentiostats, which typically range in a few thousand dollars.

No matter which technique is used, the key feature of electrochemical detection is providing intrinsic quantitative results [75]. Moreover, the electrochemical signal is not affected by ambient illumination conditions and the color or turbidity of the sample. This is an extra advantage in the food safety field, where extracts often remain colored or exhibit a certain level of turbidity, which can interfere with optical measurements [20,76]. However, electrochemical detection can be affected by electrode surface fouling and/or poisoning caused by food matrix components, affecting the robustness of measurements. The fouling issue can be circumvented using magnetic beads to extract the analyte from the matrix. This technique was successfully applied to detect domoic acid in shellfish below MRL. The cost-effective nanomaterial carbon black (approximately one US dollar per kg) was used to reduce background noise and improve sensitivity [77,78].

Alongside the miniaturization of the sensing platform, other procedures such as extraction, separation, and washing steps should be miniaturized and automated to bring the laboratory-based food safety tests to the field. The electrochemical biosensor's ease of miniaturization and integration makes them a suitable choice to be incorporated in either paper-based or chip- microfluidic devices, designed for automation and miniaturization of the sample extraction and handling processes [79]. Moreover, the potentiostat connectivity to the smartphone is an excellent way to make the electrochemical food safety analysis more user-friendly, minimizing the need for user training [80]. Here, examples of portable electrochemical biosensors in the food safety field are categorized in terms of their integration formats into paper-based, chip-based, and smartphone-based microfluidic devices.

3.1. Paper-Based Electrochemical Food Analyzers

Integrating an electrochemical biosensor into a paper-based device can provide more sensitive and quantitative results compared to colorimetric equivalents (lateral flow assay and μPAD) [81]. The microfluidic paper-based electrochemical device, or μPED, was introduced quickly after the μPAD both by Henry's and Whiteside's groups independently in 2009 and 2010 [82,83]. Since then, the μPEDs have been developed mainly as diagnostic tests, while their application in the food safety field remains quite limited, with only seven studies addressing the detection capabilities of the developed μPED devices in food matrices (Table S5) [84]. Despite being quite new, this integration strategy is particularly promising for the food safety field, with its potential to combine affordability of paper-based assays and sensitivity of the electrochemical analysis. Although the integrated device provides improved sensitivity, ranked two stars over the optical paper-based devices with only one star, this comes at the expense of affordability, being equipment-free and deliverable to end-user terms, due to the need for a potentiostat and electrodes. The electrochemical devices are also less user-friendly compared to optical paper-based devices with only three stars. Thus, their overall score is 2.6.

As affordability and user-friendliness are the key features of colorimetric paper-based devices, the competitiveness of the μPED depends directly on the development of affordable, user-friendly portable potentiostats and low-cost electrodes. Although there are some good examples of low-cost portable potentiostats in the literature, they still lack true user-friendly interfaces and are not comparable to colorimetric assays. However, the electrodes could be integrated into the paper substrate at a very low cost using different techniques, such as screen-printing, inkjet printing, and microwires [85]. The study performed by Araujo et al. is a good example of minimizing the cost of carbon electrode integration in the paper through laser scribing pyrolysis with a CO_2 laser [86]. This method is not only low cost at 0.025 US dollars per device, but also environmentally sustainable, not requiring a multi-step printing process using conductive inks and other chemicals, as the screen-printing technique does. Moreover, the pyrolysis produces a porous non-graphitizing carbon material composed of graphene sheets and aluminosilicate nanoparticles, increasing the active surface area by around six times, leading to an outstanding electrochemical performance.

A common challenge in portable food analyzer development, namely the necessity for additional validation and benchmarking in complex food matrices, mostly is neglected in μPED development as well. Only two studies benchmarked their devices against a well-established method [87,88]. Marín-Barroso et al. developed a μPED for the detection of gliadins in flour [88]. The device coupled screen-printed electrodes modified with carbon nanofibers to a paper-based immunoassay and used chronoamperometry for the detection, achieving an LOD of 3 mg·kg^{-1}, which is more than three times below the EU regulatory limit. The device was benchmarked against the official method for gliadin quantification in flour (ELISA assay) and the authors reported a correlation between the developed method and the ELISA with an R^2 close to one. The selectivity of the sensor was determined by challenging it with albumin, casein, glutenin from wheat, β lactoglobulin, and folic acid. Only casein led to a 40% increase in signal, which was attributed to the cross-reactivity of the antibody against this target [89]. The complexity of the sample preparation, particularly for solid food samples, is another challenge common to several detection devices and it undermines the simplicity of the electrochemical detection systems. While three μPEDs were developed to detect contaminants in solid food, other studies focused on juices, milk, or alcoholic beverages [88,90,91]. Some of these studies mention LODs that are slightly above EU regulations. This includes a study on the detection of *Escherichia Coli* in cucumber and beef using a label-free impedimetric technique [90]. The reported LOD value was 1500 CFU·g^{-1} while EU regulation in ground beef stipulates at least two out of five samples must have a maximum of 50 CFU·g^{-1} (Commission Regulation (EC) No 2073/2005).

The total assay time in the assessed μPED devices, like any other biosensing platform, depends on the assay type. The assays using conventional antibodies as recognition elements [88,90,92,93] never featured an analysis time below 30 min (not counting the extraction procedures), while the enzymatic assays can often be completed in under 15 min [94]. Finally, an interesting case was the development of origami-based μPEDs with pre-loaded reagents, allowing a simplified detection protocol (Figure 5) [94]. This device was used for multiplex pesticide detection (paraoxon, 2,4-dichlorophenoxyacetic acid, and atrazine) at ppb levels in river water samples. The detection was based on an enzyme-inhibition assay. The multiplex analysis was achieved by folding and unfolding different parts of the paper-based origami structure. The measurement was performed with no need for adding reagents or sample preparation. Another example of such devices has been developed for the detection of ethanol in beer [95]. A comprehensive protocol detailing the manufacturing of such a device has recently become available [96].

3.2. Microfluidic Chip-Based Electrochemical Food Analyzers

The integration of electrochemical biosensors within the microchannels of a chip-based microfluidic device is appealing due to the miniaturization capability of the electrochemical platform. Thus, it has been widely applied for the development of point-of-care and lab-on-a-chip devices for medical diagnostics. Moreover, an increasing trend toward the development of microfluidic solutions for food safety and environmental analysis has been observed [97,98]. Typically, electrochemical detection is incorporated within a microfluidic system by patterning electrodes on a flat glass or silicon substrate. Then a polymer-based layer of microchannels adheres to this substrate. Analogous to the microfluidic chips used for optical detection, the microchannels are fabricated mainly using polymers such as polydimethylsiloxane (PDMS), and Poly(2,5-dimethoxyaniline) (PDMA) due to the ease of fabrication and low cost [99].

The integration of microchannels and the use of flowing liquids can reduce the sample volume significantly, down to a few microliters, and increase the diffusion rate of the analyte molecules to the biorecognition element leading to a faster binding and biosensor response. Alternatively, the electroactive molecules can reach the electrode faster, increasing the electrochemical signal and improving the analytical performance of the device [100]. However, these advantages depend on the precise optimization of the flow rate and microchannel design. Apart from portability, other advantages of a chip-based electrochemical food analyzer include ease of integration and automation of multi-step procedures, high-throughput, and faster analysis. Therefore, these devices are attributed five stars for sensitivity, four stars for specificity, and three stars for rapid and robust consideration. However, they are not equipment-free and mostly require not only a potentiostat but also a pump, which provides them with only one star, exactly like the microfluidic chip-based optical devices. The need for additional equipment adds to the cost when compared to paper-based colorimetric and even electrochemical test strips, affecting the affordability of these devices (two stars) only slightly better than chip-based optical devices due to the lower price of a potentiostat compared to a miniaturized spectrometer. Thus, their overall score is 2.7. To circumvent the drawbacks, different high-throughput, automation, and multiplexing strategies have been explored when developing this class of devices (Table S6).

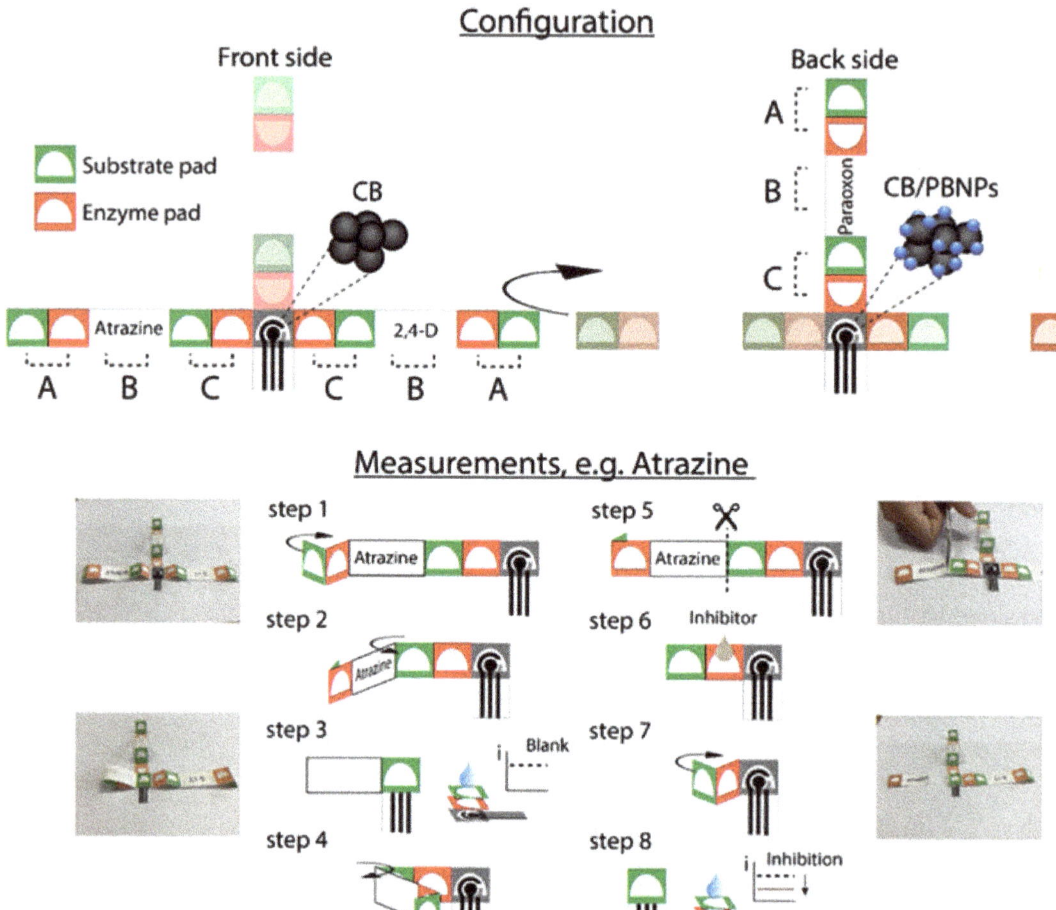

Figure 5. An example of a paper-based electrochemical food analyzer. The three-dimensional origami paper-based device was used for the detection of several classes of pesticides in river water based on the enzyme-inhibition assay. The multiplex analysis was achieved by folding and unfolding different parts of the paper-based origami structure. The measurement was performed with no need for adding reagents or sample preparation. Reproduced from [94] Biosensors and Bioelectronics, Copyright (2019), with permission from Elsevier.

Like the optical biosensing platforms previously discussed, the electrochemical biosensors also involve multi-step procedures, ranging from extracting the analyte from food samples to incubation with the eventual biorecognition elements and transport to the electrode. The microfluidic chip-based systems provide an easy way to integrate different modules and automate the overall procedure. Integration and automation of the multi-step biosensing process can effectively increase reproducibility and throughput while decreasing analysis time and cost. A very good example, showcasing the integration of different modules is a film-based integrated device developed by Park et al. for the detection of foodborne pathogens *Staphylococcus aureus* and *Escherichia coli* [101]. The device can perform multiple functions such as gene amplification, solution mixing, and electrochemical detection. The bacterial lysis was performed off-chip, then the extracted cell lysate was introduced to the device. After the amplification step, the target pathogen gene was detected by square wave voltammetry within 25 s, while gel electrophoresis detection requires about 30 min.

The fully automated electrochemical devices AutoDip and MiSense are the most advanced prototype examples, showing the applicability of microfluidic chip-based devices in the food safety field. The AutoDip is a fully integrated microfluidic platform with a user-friendly automated sampler, based on the ball-point pen mechanism (Figure 6a) [102]. This device consists of a disposable reagent module and an external actuator, controlling the incubation and washing steps by consecutive dipping of the electrode into the reagents. Therefore, the off-chip sample treatment is minimized with no need for microchannels, valves, or external pumps. As a showcase, the authors used the AutoDip device with a commercially available acetylcholinesterase biosensor (AC1.AChE, BVT Technologies, Czech Republic), to detect the pesticide chlorpyrifos with an LOD of 0.033 mg·kg^{-1} in apple samples. Although this was below the MRL for this pesticide at the time of publication (0.5 mg·kg^{-1}), the MRL has recently been lowered to 0.01 mg·kg^{-1} [103].

(a) (b)

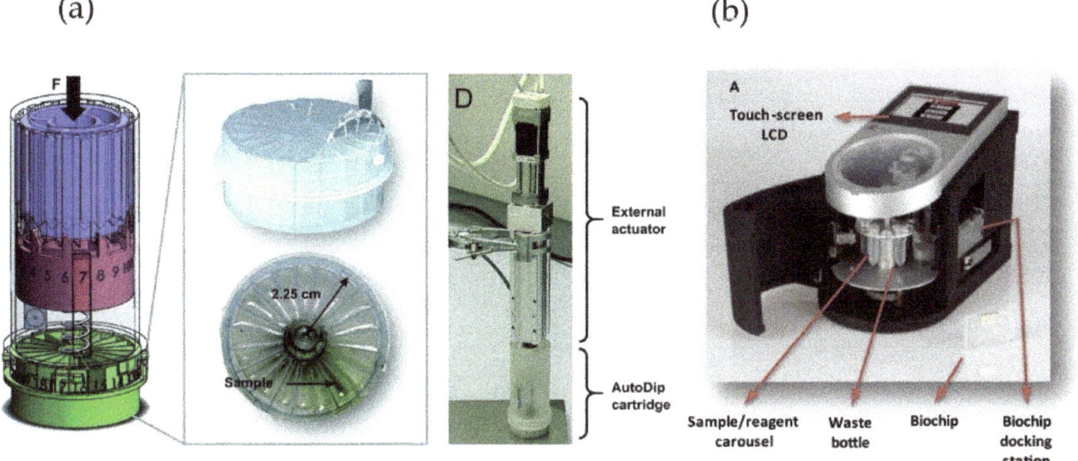

Figure 6. Examples of microfluidic chip-based electrochemical food analyzers. (**a**) AutoDip design and evaluation model. The reagent module is pre-filled with reagents, while an external vertical actuator is controlling the incubation and washing steps by consecutive dipping of the electrode into the reagents. Reproduced with permission from [102] lab on a chip, Copyright (2015), under Royal Society of Chemistry; (**b**) MiSense device designed for rapid and reliable detection of aflatoxin B1. The biochip attachment to the docking station creates a microfluidic channel on the top of the electrodes. Reproduced from [104] Talanta, Copyright (2016), with permission from Elsevier.

MiSense is an automated platform consisting of a biochip in an integrated microfluidic system (Figure 6b) [104]. The biochip is comprised of an array of 6 gold working electrodes with shared reference and counter electrodes on a silicon dioxide substrate. The device includes a biochip docking station, a pump, microfluidic tubing, reagent containers, and a waste bottle. It was used for aflatoxin B1 detection in wheat and fig samples with a limit of detection of 2 µg·kg^{-1}, which is the same as the MRL for aflatoxin B1 for cereals and below the MRL in fig samples [105]. The accuracy of the analysis was confirmed with a comparative study to the analogous ELISA test; correlation of the results resulted in a relationship with an R^2 value of 0.96. The analysis time using MiSense was 25 min, shorter compared to 2–3 h for the ELISA test. Although MiSense is a portable platform providing fast and fully automated detection, the extraction procedure is not automated and involves the use of a solid-phase extraction cartridge, which adds to the time and cost of each test.

Several additional studies achieved a higher degree of automation by the creative design of the microfluidic platforms. One study developed a microfluidic chip-based device not only to detect the atrazine in orange juice but upon detection, to remove the contaminant from the juice, an interesting feature for industrial application [106]. The automated microfluidic platform is a hybrid polydimethylsiloxane–polyester chip, including a micromixer channel for efficient reagent mixing. After detection, atrazine was removed from the sample via anodic oxidation in the degradation chip. Other studies incorporated magnetic control functionality to the electrochemical microfluidic platform [107–109]. In these devices, the immunomagnetic separation minimized the non-specific response and reduced the time for immuno-reaction with active magnetic mixing [110].

Another key feature of electrochemical chip-based devices is the possibility of multiplexing measurements on disposable chips, resulting in cheaper and faster analysis. The simplest way to increase throughput is using multiple working electrodes for parallel electrochemical detection of a single analyte in multiple samples [111]. For example, in two studies, eight working electrodes were used for simultaneous measurement of *citrus tristeza* virus in citrus and *Salmonella typhimurium* in milk [112,113]. The cost of the disposable chip for citrus testing for eight simultaneous measurements was 1.99 US dollars per device, lower than the ELISA cost of 8.30 US dollars per microwell. The multiplexing of electrochemical measurements on microfluidic devices was achieved using dual-channel [114], separate sensing areas [115], or multiplex arrays [116] in different food matrices. From these studies, the multiplex device developed by Crew et al. for amperometric detection of six organophosphate pesticides is particularly interesting, since a neural network program was used for modeling the response [116]. This facilitated the interpretation of the analytical results, reducing the level of user training. The device was transported in a standard vehicle for field testing and powered from a car battery via the lighter socket. While the on-site testing was performed only with 4 different water samples, all the samples were identified negative for pesticide contamination with a 7.5% coefficient of variation between measurements.

One of the drawbacks of the microfluidic chip-based devices potentially limiting their on-site food safety application is the need for pumps and additional equipment. Several strategies have been used to circumvent this drawback. Lu et al. developed a dual-channel indium tin oxid (ITO)-microfluidics with capillary-driven PDMS channels for simultaneous electrochemical detection of two mycotoxins, fumonisin B1 and deoxynivalenol, in corn extracts [114]. The difference in the size of the inlet and outlet ports provided a capillary pressure difference, which drives the sample through the channel towards the outlet without a pumping unit. The LODs of 135 μg·kg^{-1} and 175 μg·kg^{-1} were achieved respectively for fumonisin B1 and deoxynivalenol, which are lower than the MRLs for these toxins [105]. Another good example is a low-cost, pump-free, capillary flow-driven microfluidic chip developed for the detection of *Salmonella* [117]. The flexible device is made of two polyethylene terephthalate (PET) layers, one containing the microchannels and the other substrate for the inkjet-printed electrodes and electrowetting valves. The last example is a self-pumping lab-on-a-chip device for automated detection of botulinum toxin in 15 min [118]. This PDMS-based integrated device consists of a mixer, an electrochemical sensing zone, and a capillary pump. The capillary pump provides steady flow, eliminating the need for externally powered pumps, simplifying the overall operation.

Other notable studies of electrochemical chip-based microfluidic devices developed for food analysis include the use of novel biorecognition elements such as aptamers and molecularly imprinted polymers [119–121]. The microfluidic chip developed by Lin et al. was particularly innovative with a dual recognition platform using both molecularly imprinted polymers and aptamers, to detect carbofuran in vegetable and fruit samples [121].

Dual recognition improved the selectivity in complex food matrices. The PDMA microchannel included two functional areas. At the first functional area, the molecularly imprinted polymer adsorbs the carbofuran from the food extract, then the carbofuran is desorbed with the eluent buffer, flowing to the second functional area, to be captured by the aptamer on the electrode. Also, label-free impedimetric chips are of particular interest, since they provide a single-step measurement [122,123]. These devices are mainly reported for pathogen detection in food samples [124,125]. Considering that pathogens are much larger molecules than toxins, for the same number of bound molecules per surface area, they can increase the charge transfer resistance on the electrode much more. This means the system is more sensitive and less affected by impedimetric measurement drawbacks, such as nonspecific binding or environmental noise.

3.3. Smartphone-Based Electrochemical Food Analyzers

Smartphone-based biosensing devices have emerged as new bioanalytical tools in recent years, paving the road to citizen science owing to the smartphone's wide accessibility, mass production, and capabilities for integration into lab-on-a-chip systems [126]. The electrochemical detection is of particular interest to be integrated into a smartphone-based device, being independent of the smartphone's model variability. The smartphone in these examples was only used as the control unit of the potentiostat and as a user interface, displaying eventual commands and results. The improved accessibility (deliverable to end-users) and user-friendliness are strong features of this class of devices (three stars). However, their user-friendliness is graded lower than smartphone-based optical food analyzers, as the electrochemical training is more complicated than image analysis for non-experts. Their sensitivity and specificity are dictated by the electrochemical methods applied and were given five and four stars, respectively. Their affordability is lower (two stars) than their optical counterpart, the smartphone-based optical analyzers. This situation is due to the need for a potentiostat, which also reduces its performance as equipment-free, which is rated with two stars. Thus, their overall score is 3.1, still placing them in the top 3 most promising portable food analyzers, along with paper-based and smartphone-based optical food analyzers. While smartphone-based electrochemical devices have been developed for medical diagnostics and environmental analysis, there are very few examples of smartphone-based electrochemical biosensors for food safety testing [127–131].

One of the two best examples of user-friendly and simple smartphone-based electrochemical food analyzers involves integrated exogenous antigen testing (iEAT) (Figure 7a) [132]. This device was used to detect food protein antigens with a detection limit of 0.1 mg·kg^{-1} in less than 10 min, at a cost of only 3 US dollars per test. The target protein antigens were gliadin in wheat, Ara h1 in peanut, Cor a1 in hazelnut, casein in milk, and ovalbumin in egg white. The integrated iEAT system consists of a disposable extraction kit, an electrode chip, a pocket-size potentiostat, which connects to a smartphone through a Bluetooth connection. The battery can be charged wirelessly with the smartphone. For data analysis, a smartphone App communicates with the iEAT device through Bluetooth and the data is uploaded to a cloud server. It can take photos of the users and analyze foods, set the detection channels and allergen types, display the measurement results, store the measurement time and location, and track the food intake history. The eventual measured presence of hazardous levels of antigens is displayed with a simple warning label, minimizing the need for user training. Although the system includes multi-step procedures, it is simple enough for non-expert users.

Another interesting device is the lab-on-a-glove developed by Wang's group for the detection of organophosphorus compounds on fruit and vegetable surfaces (Figure 7b) [133]. The disposable glove biosensor consists of a sampling active area, printed on the thumb finger, while the enzyme-based biosensing detection area is printed on the index finger. The glove is coupled with a miniaturized potentiostat capable of real-time wireless data transmission to a smartphone. At the reported development stage, the results were qualitative and the user interface on the smartphone only showed the measured current plot. Recently, the same research group developed a quantitative glove biosensor for fentanyl (opioid) detection with an LOD of 10 μM [134]. Further development for a more user-friendly user interface would be needed before widespread application.

(a)   (b)

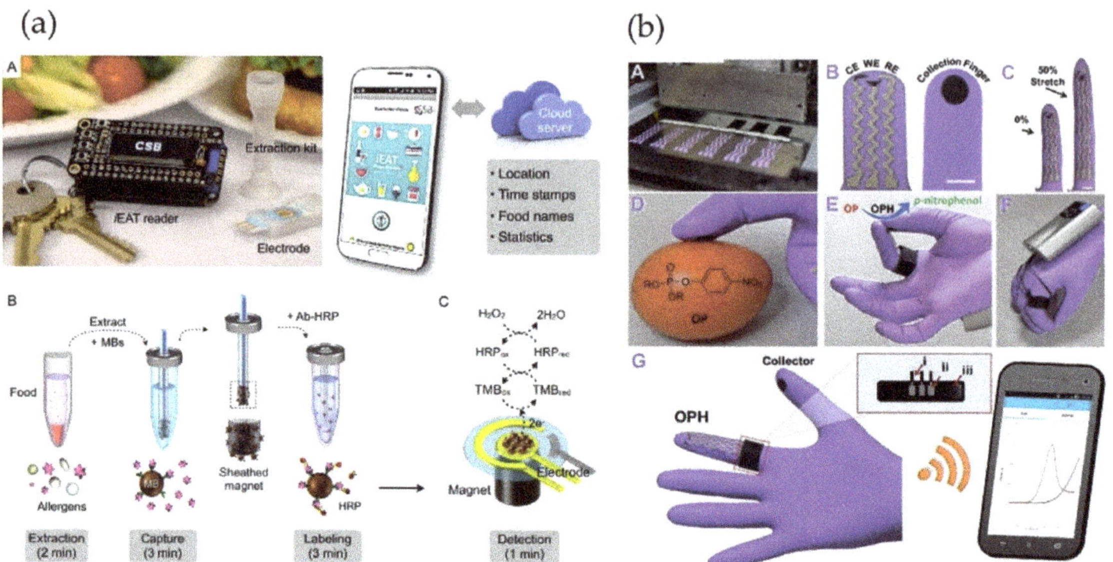

Figure 7. Examples of smartphone-based electrochemical food analyzers. (**a**) The iEAT platform. The keychain-sized detector is linked with a smartphone App. After extraction of the antigen from the sample, the detection is achieved by mixing HRP-labelled magnetic beads with the substrate (TMB, 3,3′,5,5′- tetramethylbenzidine) and moving it to the electrode. Reproduced with permission from [132] ACS Nano, Copyright (2017), American Chemical Society; (**b**) The lab-on-a-glove platform. The glove biosensor consists of a sampling finger and a sensing finger, containing the immobilized enzyme. Reproduced with permission from [133] ACS Sensors, Copyright (2019), American Chemical Society.

4. Portable Mass Spectrometry for Food Analysis

Conventional MS instrumentation is used primarily for confirmatory analysis in food safety applications. Nonetheless, recent studies report developments in the field of portable MS devices, which could be or are presented for applications in food analysis. We used the ASSURED criteria to assess the performance of the MS instrumentation in food safety screening applications. In terms of sensitivity and specificity, the MS instrumentation is an all-star (five stars), compared to all other screening assays described. The unequivocal identification is the main reason why MS is used as a confirmatory method according to the EU regulation 2002/657/EC [6]. The selection to monitor specific ions, characteristics for each substance, and the robust ion ratios, can guarantee the detection of specific food contaminants and differentiate between similar substances if they are not optical isomers. The rapidness is highly dependent on the ion source employed. In most portable MS applications for food safety monitoring, an ambient/direct ion source is employed, which significantly reduces the analysis time, however, the robustness of the ambient ion sources is not always guaranteed (three stars). Finally, the affordable, user-friendly, equipment-free, and deliverable to end-user characteristics are all weak (one star), contributing to an overall rating of 2.4. Mass spectrometers are significantly more expensive than the other screening assays described. They rely on relatively large and heavy instruments compared to screening assays and, with only a few exceptions, they require trained staff to operate the instrument and assess the result (Table 1).

Portable mass spectrometers are mainly applied in industry, environmental control, and forensics. Systems dedicated to food analysis are not commercially available yet.

However, portable mass spectrometry (MS) is expected to find its way to on-site testing for food quality and safety parameters in the future. Historically, the first portable mass spectrometer was developed by John Hipple in 1942 for routine gas analysis [135]. Since then, a portable MS for gas analysis in the industry has evolved to the point of having a palm-sized MS system, weighing less than 2 kg, having a volume of fewer than 2 L, and operating with a 5 W battery, which is a significant improvement considering the conventional bulky benchtop instrumentation of up to 200 kg [136]. To transit from benchtop to portable MS instrumentation, all the components, including the mass analyzer, the ion source, the vacuum system, and the energy supply need to be miniaturized. Furthermore, other aspects such as sample clean-up, preparation, and separation need to be adjusted [137]. At the same time, these systems should be comparable in performance and capabilities with conventional benchtop instrumentation to be able to reach the relevant food safety detection levels and confirmation [138]. In practice, this means that compromises between performance, throughput rates, and cost need to be made [137].

Focusing on the mass analyzer, in a review published in 2016 by Snyder et al., over 30 miniaturized MS systems were discussed, which included different types of mass analyzers [139]. Among them, the quadrupole and ion trap mass analyzers are those used mostly in the field of food analysis. In contrast, miniaturization of the magnetic sector and time of flight (TOF) mass analyzers was quite a challenge, as their resolution depends on the length of the path the ions travel [140–142], and no application in the food safety field has been cited. For analyte ionization, atmospheric pressure ionization (API) methods, such as electrospray ionization (ESI), AP chemical ionization (APCI), and AP-Matrix-assisted laser desorption ionization (AP-MALDI) have been used in portable MS systems [137]. Moreover, with ambient or direct ionization (AI) sources, samples can be directly ionized with minimum or no sample preparation under ambient conditions, without any chromatographic separation. AIMS have been already applied in various food-related analytical methods [143] and can be easily coupled with miniaturized MS systems [144].

The ion trap mass analyzers reported for food safety applications are based on different generations of the Mini, a rectilinear ion trap analyzer developed at Purdue University (West Lafayette, IN, USA) equipped either with Low-Temperature Plasma (LTP), or Discontinuous Atmospheric Pressure Inlet (DAPI) ambient ionization. The Mini 10.5 is the first generation rectilinear ion trap analyzer that weighs 10 kg and can be operated on a 70 W battery [139]. This device was used by Huang et al. for the detection of melamine in whole milk, milk powder, and fish [145]. The authors chose heated air as a plasma carrier instead of helium to reduce the cost and to make it more portable without loss of sensitivity. The method is high-throughput; up to two samples can be analyzed per minute, and it has an LOD of 0.25 ppm for melamine in the milk sample, less than the EU regulatory limit of 1 ppm [146]. The Mini 10.5 was also used for direct analysis of two pesticides, diphenylamine, and thiabendazole in apples and oranges. Although this method did not provide sufficient quantification, it could differentiate between organic and non-organic fruits based on the detection of targeted pesticides [147]. Wiley et al. and Janfelt et al. used the mini 10.5 mass analyzer to detect food-safety-related compounds with a focus on several pesticides, including atrazine. Even though these two studies achieved adequate sensitivity compared to the benchtop Thermo LTQ linear ion trap MS, they did not perform real sample analysis, and the reported analysis suffered from low reproducibility [148,149].

The Mini 11 is the next-generation analyzer. At 5 kg, it weighs half as much as the Mini 10.5 and can be operated for two hours on a 35 W battery [139]. The Mini 11 was used in combination with ESI-MS, DESI-MS, and LTP-MS, to detect the origins of milk, fish, and coffee beans, respectively. The accuracy of the classification was high, enabling the characterization of adulterated food groups [150].

The next portable ion trap analyzer is the Mini 12 which can be operated with a 50 W battery [139]. Although the Mini 12 is heavier than the previous generations at 15 kg, it provides a high level of user-friendliness enabling non-expert users to perform the analysis.

As demonstrated by Li et al., a peel of a conventionally grown orange was inserted in a paper spray cartridge, then the spray solvent was added and the ions were generated and detected [151]. In another study that involved the use of the Mini 12 by Pullian et al., direct leaf spray ionization was used to detect qualitatively the fungicide chlorothalonil in maple tree leaves even 5 days after its application [152]. Finally, in the last example of this ion trap analyzer, direct ionization slug-flow microextraction (SFME) nanoESI was applied to detect the plasticizer Bis(2-ethylhexyl)phthalate (DEHP) with an LOD of 5 ppm in spiked fruit punch, and of bisphenol A, a plastic monomer with hormone-like properties forbidden for use in infant formula bottles, with an LOD of 10 ppm in spiked milk sample [153]. A commercially available portable linear ion trap MS system has been developed by PurSpec Technologies, which weighs 20 kg and can be operated with a 100 W battery [154]. It was applied with SFME nanoESI for the detection of multiple fentanyl compounds, directly from beer, milk, or cola, with an LOD of 10 ppb, featuring low chemical noise [155].

Apart from ion trap analyzers, a single quadrupole MS has been implemented in portable food safety analysis as well. A good example is a technique developed by Zhang et. al. using a portable pyrolysis gas chromatography (GC) quadrupole MS with a total analysis time of only five minutes for detection of microplastics in seawater [156]. The high temperature selected for the pyrolysis (715 °C) ensured the full decomposition, ionization, and subsequent identification based on both generated ions and ion ratio. This technique was applied for in situ analysis of seawater, but it could be used, after further modification, for fish samples on fishing boats. A portable single quadrupole MS was also used in combination with a solid-phase microextraction (SPME) transmission mode followed by direct analysis in real-time (DART) ionization to identify pesticides in grape juice. In total, 3 pesticides were quantitatively analyzed with an LOD of 10 ppb, namely pyrimethanil, pyraclostrobin, and azoxystrobin, and 4 more were quantitatively analyzed with an LOD of 5 ppb, namely cyprodinil, metalaxyl, imazalil, and atrazine. The SPME-DART-portable single quadrupole MS approach was also applied to detect the origin of milk samples [157]. The complete analysis time was less than two minutes, which is significantly reduced compared to the standard benchtop SPME analysis with chromatographic separation [158,159]. Recently, Blokland et al. used a transportable single quadrupole MS system with different ionization methods, from which coated blade spray (CBS) showed the most promising results for the analysis of liquid food samples or extracts of solid foods [160].

Finally, a cutting-edge advancement in portable quadrupole MS development was the successful miniaturization of a triple quadrupole (QqQ) mass analyzer, being the gold standard in conventional food safety analysis by GC- or LC-MS. This development could pave the way for the on-site adaptation of standard methods used by routine food safety laboratories. The researchers demonstrated the application of the portable QqQ system using a standard LC column for thiabendazole detection in spiked apple pulp, achieving a total run time of six minutes and an LOD of 10 ppb, which is comparable to that of conventional benchtop instruments [161].

Other parts of the MS spectrometer that need to be adjusted for portable applications include the power supply and vacuum system. Nonetheless, significant improvements have been made in the field, as with the ionization part of the MS systems. For example, the power of the ion source, as Josha et al. demonstrated recently, could be supplied even by a USB interface plugged into a smartphone [162].

The ASSURED criteria are not directly/strictly applicable in MS analysis, since they are intended to assess screening assays. However, the use of MS, in terms of sensitivity and selectivity is superior compared to screening assays, thus on-field applications using portable MS could lead to improved food-safety monitoring. Most portable MS techniques described exploit simple and rapid direct or ambient ion sources without time-consuming chromatographic separation, which is already an improvement compared to confirmatory methods in routine laboratory analysis. From the techniques reviewed, most have focused initially on the detection of pesticides. Surprisingly, no techniques have been developed for the detection of natural toxins, even though they are a considerable risk for consumer's health and a huge financial burden for the food industry [163]. This omission might be due to the fact that for some toxins a higher level of sensitivity is needed to reach the EU regulatory limit at $\mu g \cdot kg^{-1}$ compared to some pesticides at $mg \cdot kg^{-1}$, which is also considered the biggest challenge for portable MS development [105,164]. Also, the chemical structure of some of the pesticides favors high ionization efficiency, leading to lower detection limits. Improvements in instrumentation, miniaturization of the ion analyzer, pumping system, energy source, user-friendly interface, and development of AIMS source with minimal to no sample preparation highlight a path for further development.

As an intermediate solution to bridge the gap between portable biorecognition-based rapid screening and potential portable GC- or LC-MS for confirmatory analysis, direct MS without any chromatographic separation might be considered following the development of a dedicated LFA. In this context, the selectivity of the chromatographic separation is replaced by immuno-trapping on an LFA, where the antibodies isolate the analyte of interest. By subsequent dissociation of the analyte from the LFA, the retrieved solution can be directly analyzed by ESI or DART-MS. This approach offers possibilities as a confirmatory method maintaining the advantages of screening, such as rapid development and ease of use, with those of confirmation, providing unequivocal identification of the substance [165,166].

5. Conclusions

The increasing demand for reliable, rapid, and on-site food safety analysis is a motivating factor toward the development of fully integrated and automated portable food analyzers. Based on the ASSURED criteria, key trends in technological advancements have been identified that increase analysis sensitivity, combine complementary technologies, and involve full automation and integration of portable devices. Paper-based optical food analyzers are the simplest and most common platforms for on-site qualitative and semi-quantitative analysis, and there is a trend toward the development of affordable and portable readers to enhance their sensitivity and provide quantitative data. The development of hybrid devices, particularly combining LFA with electrochemical detection is the next leap toward meeting the ASSURED criteria by combining the complementary features of different platforms. Microfluidic chip-based electrochemical and optical food analyzers are good examples of integrated and automated platforms in portable food analysis. The design of fully automated and integrated devices, which include all necessary steps from sample preparation to detection, will bring them closer to the point-of-need analysis. Finally, although the portable MS analyzer is gaining momentum, it still lags in terms of affordability, robustness, and user-friendliness needed for point-of-need food safety analysis.

Despite many technological advancements, the new devices described in this review are still at the proof-of-concept stage. Although these devices usually have strong performance metrics within some of the ASSURED criteria, none of them meet all the requirements to be the ideal fit for purpose. Main limitations include lack of quantitation, automation, integration of sample preparation, and user-friendliness. Researchers should focus on both the ASSURED criteria and regulatory guidelines when developing such devices. While there are some notable examples involving the measurement of relevant analytes below regulatory limits in real samples, in most studies the platforms have not been validated in a real-life setting based on regulatory guidelines. This represents a critical gap for their implementation as food safety screening tools. Possible reasons for the lack of validation and benchmarking of portable food analyzers may relate to the required high sensitivity and reproducibility at analyte concentrations below the MRLs in complex food matrices. While real sample analysis is satisfactory in many cases, achieving a false negative rate equal to or lower than 5% in different food matrices, as required for validation studies, is a difficult task, requiring substantial additional research effort. Considering that validation and benchmarking studies play a crucial role in ensuring the applicability of new devices, stakeholders, granting bodies, and the food safety research community could more explicitly communicate the importance of and recognize the value of validation studies, and support and encourage technology developers to perform these studies.

Supplementary Materials: The following are available online at https://www.mdpi.com/article/10.3390/foods10061399/s1, spreadsheet Table S1: paper-based optical food analyzers, spreadsheet Table S2: microfluidic chip-based optical food analyzers, spreadsheet Table S3: smartphone-based optical food analyzers, spreadsheet Table S4: Raman and IR-based portable food analyzers, spreadsheet Table S5: paper-based electrochemical food analyzers, spreadsheet Table S6: microfluidic chip-based electrochemical food analyzers.

Author Contributions: Conceptualization, writing-original draft preparation of abstract, introduction, optical and electrochemical section introduction, microfluidic chip-based electrochemical food analyzer, and smartphone-based electrochemical food analyzer, co-writing all other sections, review, and editing, S.J.; writing—review and editing, and supervision, S.J.S.; writing—review and editing, and supervision, S.G.; writing—review and editing, M.W.F.N.; writing-original draft preparation of portable optical food analyzer, J.G.; writing-original draft preparation of MS section, A.G.-K.; writing-original draft preparation of Raman and IR-based portable food analyzers, A.S.T.; writing-original draft preparation of paper-based electrochemical food analyzer, J.L.D.N.; review and editing, M.-P.M.; review and editing J.-P.S.; review and editing, A.G.; review and editing, J.H.; review and editing, C.E.; review and editing, K.C.; review and editing, D.M.; review and editing, L.B. All authors have read and agreed to the published version of the manuscript.

Funding: This project has received funding from the European Union's Horizon 2020 research and innovation program under the Marie Sklodowska-Curie grant agreement No 720325, FoodSmartphone.

Data Availability Statement: The data presented in this study are available in [the supplementary material, spreadsheet Tables S1–S6].

Conflicts of Interest: The authors declare no conflict of interest. The funders had no role in the design of the study; in the collection, analyses, or interpretation of data; in the writing of the manuscript, or in the decision to publish the results.

References

1. World Health Organization (WHO). *WHO Estimates of the Global Burden of Foodborne Diseases: Executive Summary*; WHO: Genenva, Switzerland, 2015.
2. Haughey, S.A.; Chevallier, O.P.; McVey, C.; Elliott, C.T. Laboratory Investigations into the Cause of Multiple Serious and Fatal Food Poisoning Incidents in Uganda during 2019. *Food Control* **2021**, *121*, 107648. [CrossRef]
3. Wang, X.; Wang, S.; Cai, Z. The Latest Developments and Applications of Mass Spectrometry in Food-Safety and Quality Analysis. *TrAC Trends Anal. Chem.* **2013**, *52*, 170–185. [CrossRef]
4. Griesche, C.; Baeumner, A.J. Biosensors to support sustainable agriculture and food safety. *TrAC Trends Anal. Chem.* **2020**, *128*, 115906. [CrossRef]
5. Lu, Y.; Yang, Q.; Wu, J. Recent Advances in Biosensor-integrated enrichment methods for preconcentrating and detecting the low-abundant analytes in agriculture and food samples. *TrAC Trends Anal. Chem.* **2020**, *128*, 115914. [CrossRef]
6. European Commisdion. *Commission Decision of 12 August 2002 Implementing Council Directive 96/23/EC Concerning the Performance of Analytical Methods and the Interpretation of Results*; European Commission: Brussels, Belgium, 2002.
7. Ma, X.; Ouyang, Z. Ambient ionization and miniature mass spectrometry system for chemical and biological analysis. *TrAC Trends Anal. Chem.* **2016**, *85*, 10–19. [CrossRef]
8. Kosack, C.S.; Page, A.L.; Klatser, P.R. A guide to aid the selection of diagnostic tests. *Bull. World Health Organ.* **2017**, *95*, 639–645. [CrossRef] [PubMed]
9. Chen, C.; Wang, J. Optical Biosensors: An exhaustive and comprehensive review. *Analyst* **2020**, *145*, 1605–1628. [CrossRef] [PubMed]
10. Lechuga, L.M. Optical biosensors. In *Comprehensive Analytical Chemistry*; Elsevier: Amsterdam, The Netherlands, 2005; Volume 44, pp. 209–250. ISBN 9780444507150.
11. Yang, Z.; Albrow-Owen, T.; Cai, W.; Hasan, T. Miniaturization of optical spectrometers. *Science* **2021**, *371*. [CrossRef]
12. Martinez, A.W.; Phillips, S.T.; Butte, M.J.; Whitesides, G.M. Patterned paper as a platform for inexpensive, low-volume, portable bioassays. *Angew. Chem. Int. Ed. Engl.* **2007**, *46*, 1318–1320. [CrossRef]
13. Fu, L.-M.; Wang, Y.-N. Detection methods and applications of microfluidic paper-based analytical devices. *TrAC Trends Anal. Chem.* **2018**, *107*, 196–211. [CrossRef]
14. Carrell, C.; Kava, A.; Nguyen, M.; Menger, R.; Munshi, Z.; Call, Z.; Nussbaum, M.; Henry, C. Beyond the lateral flow assay: A review of paper-based microfluidics. *Microelectron. Eng.* **2019**, *206*, 45–54. [CrossRef]

15. Choi, J.R.; Hu, J.; Tang, R.; Gong, Y.; Feng, S.; Ren, H.; Wen, T.; Li, X.; Wan Abas, W.A.B.; Pingguan-Murphy, B.; et al. An integrated paper-based sample-to-answer biosensor for nucleic acid testing at the point of care. *Lab Chip* **2016**, *16*, 611–621. [CrossRef] [PubMed]
16. Ross, G.M.S.; Bremer, M.G.E.G.; Wichers, J.H.; Van Amerongen, A.; Nielen, M.W.F. Rapid antibody selection using surface plasmon resonance for high-speed and sensitive hazelnut lateral flow prototypes. *Biosensors* **2018**, *8*, 130. [CrossRef]
17. Tsagkaris, A.S.; Uttl, L.; Pulkrabova, J.; Hajslova, J. Screening of carbamate and organophosphate pesticides in food matrices using an affordable and simple spectrophotometric acetylcholinesterase assay. *Appl. Sci.* **2020**, *10*, 565. [CrossRef]
18. Zhao, Y.; Wang, H.; Zhang, P.; Sun, C.; Wang, X.; Wang, X.; Yang, R.; Wang, C.; Zhou, L. Rapid multiplex detection of 10 foodborne pathogens with an up-converting phosphor technology-based 10-channel lateral flow assay. *Sci. Rep.* **2016**, *6*, 21342. [CrossRef] [PubMed]
19. Tsagkaris, A.S.; Pulkrabova, J.; Hajslova, J.; Filippini, D. A Hybrid lab-on-a-chip injector system for autonomous carbofuran screening. *Sensors* **2019**, *19*, 5579. [CrossRef] [PubMed]
20. Nelis, J.L.D.; Tsagkaris, A.S.; Dillon, M.J.; Hajslova, J.; Elliott, C.T. Smartphone-Based Optical Assays in the Food Safety Field. *TrAC Trends Anal. Chem.* **2020**, *129*, 115934. [CrossRef]
21. Liu, W.; Guo, Y.; Luo, J.; Kou, J.; Zheng, H.; Li, B.; Zhang, Z. A molecularly imprinted polymer based a lab-on-paper chemiluminescence device for the detection of dichlorvos. *Spectrochim. Acta Part A Mol. Biomol. Spectrosc.* **2015**, *141*, 51–57. [CrossRef]
22. Tseng, S.-Y.; Li, S.-Y.; Yi, S.-Y.; Sun, A.Y.; Gao, D.-Y.; Wan, D. Food quality monitor: Paper-based plasmonic sensors prepared through reversal nanoimprinting for rapid detection of biogenic amine odorants. *ACS Appl. Mater. Interfaces* **2017**, *9*, 17306–17316. [CrossRef]
23. Picó, Y. *Chemical Analysis of Food: Techniques and Applications*, 1st ed.; Elsevier: Amsterdam, The Netherlands, 2012.
24. Stroock, A.D. Microfluidics. In *Optical Biosensors*; Elsevier: Amsterdam, The Netherlands, 2008; pp. 659–681. ISBN 9780444531254.
25. Connelly, J.T.; Kondapalli, S.; Skoupi, M.; Parker, J.S.L.; Kirby, B.J.; Baeumner, A.J. Micro-total analysis system for virus detection: Microfluidic pre-concentration coupled to liposome-based detection. *Anal. Bioanal. Chem.* **2012**, *402*, 315–323. [CrossRef]
26. Soares, R.R.G.; Novo, P.; Azevedo, A.M.; Fernandes, P.; Aires-Barros, M.R.; Chu, V.; Conde, J.P. On-chip sample preparation and analyte quantification using a microfluidic aqueous two-phase extraction coupled with an immunoassay. *Lab Chip* **2014**, *14*, 4284–4294. [CrossRef]
27. Hung, T.Q.; Chin, W.H.; Sun, Y.; Wolff, A.; Bang, D.D. A novel lab-on-chip platform with integrated solid phase pcr and supercritical angle fluorescence (saf) microlens array for highly sensitive and multiplexed pathogen detection. *Biosens. Bioelectron.* **2017**, *90*, 217–223. [CrossRef]
28. Choi, G.; Jung, J.H.; Park, B.H.; Oh, S.J.; Seo, J.H.; Choi, J.S.; Kim, D.H.; Seo, T.S. A centrifugal direct recombinase polymerase amplification (direct-rpa) microdevice for multiplex and real-time identification of food poisoning bacteria. *Lab Chip* **2016**, *16*, 2309–2316. [CrossRef]
29. Bhardwaj, H.; Sumana, G.; Marquette, C.A. A label-free ultrasensitive microfluidic surface plasmon resonance biosensor for aflatoxin b1 detection using nanoparticles integrated gold chip. *Food Chem.* **2020**, *307*, 125530. [CrossRef]
30. Joshi, S.; Annida, R.M.; Zuilhof, H.; van Beek, T.A.; Nielen, M.W.F. Analysis of mycotoxins in beer using a portable nanostructured imaging surface plasmon resonance biosensor. *J. Agric. Food Chem.* **2016**, *64*, 8263–8271. [CrossRef]
31. Sauceda-Friebe, J.C.; Karsunke, X.Y.Z.; Vazac, S.; Biselli, S.; Niessner, R.; Knopp, D. Regenerable immuno-biochip for screening ochratoxin a in green coffee extract using an automated microarray chip reader with chemiluminescence detection. *Anal. Chim. Acta* **2011**, *689*, 234–242. [CrossRef]
32. Wang, Y.; Gan, N.; Zhou, Y.; Li, T.; Hu, F.; Cao, Y.; Chen, Y. Novel label-free and high-throughput microchip electrophoresis platform for multiplex antibiotic residues detection based on aptamer probes and target catalyzed hairpin assembly for signal amplification. *Biosens. Bioelectron.* **2017**, *97*, 100–106. [CrossRef] [PubMed]
33. Weng, X.; Neethirajan, S. Paper-based microfluidic aptasensor for food safety. *J. Food Saf.* **2018**, *38*. [CrossRef]
34. Fernández, F.; Hegnerová, K.; Piliarik, M.; Sanchez-Baeza, F.; Homola, J.; Marco, M.P. A label-free and portable multichannel surface plasmon resonance immunosensor for on site analysis of antibiotics in milk samples. *Biosens. Bioelectron.* **2010**, *26*, 1231–1238. [CrossRef] [PubMed]
35. Chalyan, T.; Potrich, C.; Schreuder, E.; Falke, F.; Pasquardini, L.; Pederzolli, C.; Heideman, R.; Pavesi, L. AFM1 detection in milk by fab′ functionalized si3n4 asymmetric mach–zehnder interferometric biosensors. *Toxins* **2019**, *11*, 409. [CrossRef]
36. Nelis, J.L.D.; Zhao, Y.; Bura, L.; Rafferty, K.; Elliott, C.T.; Campbell, K. A randomized combined channel approach for the quantification of color- and intensity-based assays with smartphones. *Anal. Chem.* **2020**, *92*, 7852–7860. [CrossRef]
37. Li, Z.; Li, Z.; Zhao, D.; Wen, F.; Jiang, J.; Xu, D. Smartphone-based visualized microarray detection for multiplexed harmful substances in milk. *Biosens. Bioelectron.* **2017**, *87*, 874–880. [CrossRef]
38. Ye, Y.; Wu, T.; Jiang, X.; Cao, J.; Ling, X.; Mei, Q.; Chen, H.; Han, D.; Xu, J.J.; Shen, Y. Portable smartphone-based QDs for the visual onsite monitoring of fluoroquinolone antibiotics in actual food and environmental samples. *ACS Appl. Mater. Interfaces* **2020**, *12*, 14552–14562. [CrossRef]
39. Zhong, L.; Sun, J.; Gan, Y.; Zhou, S.; Wan, Z.; Zou, Q.; Su, K.; Wang, P. Portable smartphone-based colorimetric analyzer with enhanced gold nanoparticles for on-site tests of seafood safety. *Anal. Sci.* **2019**, *35*, 133–140. [CrossRef] [PubMed]

40. Wu, Y.-Y.; Liu, B.-W.; Huang, P.; Wu, F.Y. A novel colorimetric aptasensor for detection of chloramphenicol based on lanthanum ion–assisted gold nanoparticle aggregation and smartphone imaging. *Anal. Bioanal. Chem.* **2019**, *411*, 7511–7518. [CrossRef] [PubMed]

41. Hosu, O.; Lettieri, M.; Papara, N.; Ravalli, A.; Sandulescu, R.; Cristea, C.; Marrazza, G. Colorimetric multienzymatic smart sensors for hydrogen peroxide, glucose and catechol screening analysis. *Talanta* **2019**, *204*, 525–532. [CrossRef] [PubMed]

42. Guo, J.; Wong, J.X.H.; Cui, C.; Li, X.; Yu, H.Z. A smartphone-readable barcode assay for the detection and quantitation of pesticide residues. *Analyst* **2015**, *140*, 5518–5525. [CrossRef]

43. Machado, J.M.D.; Soares, R.R.G.; Chu, V.; Conde, J.P. Multiplexed capillary microfluidic immunoassay with smartphone data acquisition for parallel mycotoxin detection. *Biosens. Bioelectron.* **2018**, *99*, 40–46. [CrossRef]

44. Cheng, N.; Song, Y.; Fu, Q.; Du, D.; Luo, Y.; Wang, Y.; Xu, W.; Lin, Y. Aptasensor based on fluorophore-quencher nano-pair and smartphone spectrum reader for on-site quantification of multi-pesticides. *Biosens. Bioelectron.* **2018**, *117*, 75–83. [CrossRef]

45. Wang, S.; Zheng, L.; Cai, G.; Liu, N.; Liao, M.; Li, Y.; Zhang, X.; Lin, J. A microfluidic biosensor for online and sensitive detection of salmonella typhimurium using fluorescence labeling and smartphone video processing. *Biosens. Bioelectron.* **2019**, *140*, 111333. [CrossRef]

46. Zeinhom, M.M.A.; Wang, Y.; Song, Y.; Zhu, M.-J.; Lin, Y.; Du, D. A portable smart-phone device for rapid and sensitive detection of E. Coli O157:H7 in yoghurt and egg. *Biosens. Bioelectron.* **2018**, *99*, 479–485. [CrossRef] [PubMed]

47. Liu, Z.; Zhang, Y.; Xu, S.; Zhang, H.; Tan, Y.; Ma, C.; Song, R.; Jiang, L.; Yi, C. A 3D Printed smartphone optosensing platform for point-of-need food safety inspection. *Anal. Chim. Acta* **2017**, *966*, 81–89. [CrossRef]

48. Li, X.; Yang, F.; Wong, J.X.H.; Yu, H.-Z. Integrated smartphone-app-chip system for on-site parts-per-billion-level colorimetric quantitation of aflatoxins. *Anal. Chem.* **2017**, *89*, 8908–8916. [CrossRef] [PubMed]

49. Ross, G.M.S.; Salentijn, G.I.; Nielen, M.W.F. A critical comparison between flow-through and lateral flow immunoassay formats for visual and smartphone-based multiplex allergen detection. *Biosensors* **2019**, *9*, 143. [CrossRef]

50. Valand, R.; Tanna, S.; Lawson, G.; Bengtström, L. A review of Fourier Transform Infrared (FTIR) spectroscopy used in food adulteration and authenticity investigations. *Food Addit. Contam. Part A Chem. Anal. Control. Expo. Risk Assess.* **2020**, *37*, 19–38. [CrossRef]

51. Danezis, G.P.; Tsagkaris, A.S.; Camin, F.; Brusic, V.; Georgiou, C.A. Food authentication: Techniques, trends & emerging approaches. *TrAC Trends Anal. Chem.* **2016**, *85*, 123–132. [CrossRef]

52. Crocombe, R.A. Portable spectroscopy. *Appl. Spectrosc.* **2018**, *72*, 1701–1751. [CrossRef]

53. SCiO—The World's Only Pocket-Sized NIR Micro Spectrometer. Available online: https://www.consumerphysics.com/ (accessed on 22 March 2021).

54. Lin, Z.; He, L. Recent advance in SERS Techniques for food safety and quality analysis: A brief review. *Curr. Opin. Food Sci.* **2019**, *28*, 82–87. [CrossRef]

55. Hoppmann, E.P.; Yu, W.W.; White, I.M. Highly sensitive and flexible inkjet printed SERS sensors on paper. *Methods* **2013**, *63*, 219–224. [CrossRef] [PubMed]

56. Gong, Z.; Wang, C.; Pu, S.; Wang, C.; Cheng, F.; Wang, Y.; Fan, M. Rapid and direct detection of illicit dyes on tainted fruit peel using a pva hydrogel surface enhanced raman scattering substrate. *Anal. Methods* **2016**, *8*, 4816–4820. [CrossRef]

57. Shi, Q.; Huang, J.; Sun, Y.; Deng, R.; Teng, M.; Li, Q.; Yang, Y.; Hu, X.; Zhang, Z.; Zhang, G. A SERS-based multiple immuno-nanoprobe for ultrasensitive detection of neomycin and quinolone antibiotics via a lateral flow assay. *Microchim. Acta* **2018**, *185*, 3–10. [CrossRef]

58. Fales, A.M.; Vo-Dinh, T. Silver embedded nanostars for SERS with internal reference (SENSIR). *J. Mater. Chem. C* **2015**, *3*, 7319–7324. [CrossRef]

59. Li, X.; Zhang, S.; Yu, Z.; Yang, T. Surface-enhanced raman spectroscopic analysis of phorate and fenthion pesticide in apple skin using silver nanoparticles. *Appl. Spectrosc.* **2014**, *68*, 483–487. [CrossRef] [PubMed]

60. Lin, S.; Lin, X.; Liu, Y.; Zhao, H.; Hasi, W.; Wang, L. Self-assembly of Au@Ag core–shell nanocubes embedded with an internal standard for reliable quantitative SERS measurements. *Anal. Methods* **2018**, *10*, 4201–4208. [CrossRef]

61. Xie, J.; Li, L.; Khan, I.M.; Wang, Z.; Ma, X. Flexible Paper-based SERS substrate strategy for rapid detection of methyl parathion on the surface of fruit. *Spectrochim. Acta Part A Mol. Biomol. Spectrosc.* **2020**, *231*, 118104. [CrossRef] [PubMed]

62. Chen, J.; Huang, M.; Kong, L.; Lin, M. Jellylike flexible nanocellulose SERS substrate for rapid in-situ non-invasive pesticide detection in fruits/vegetables. *Carbohydr. Polym.* **2019**, *205*, 596–600. [CrossRef]

63. Moreno, V.; Adnane, A.; Salghi, R.; Zougagh, M.; Ríos, Á. Nanostructured hybrid surface enhancement raman scattering substrate for the rapid determination of sulfapyridine in milk samples. *Talanta* **2019**, *194*, 357–362. [CrossRef] [PubMed]

64. Sánchez, M.T.; Flores-Rojas, K.; Guerrero, J.E.; Garrido-Varo, A.; Pérez-Marín, D. Measurement of pesticide residues in peppers by near-infrared reflectance spectroscopy. *Pest Manag. Sci.* **2010**, *66*, 580–586. [CrossRef] [PubMed]

65. de Girolamo, A.; Cervellieri, S.; Visconti, A.; Pascale, M. Rapid analysis of deoxynivalenol in durum wheat by FT-NIR spectroscopy. *Toxins* **2014**, *6*, 3129–3143. [CrossRef] [PubMed]

66. Lasch, P.; Stämmler, M.; Zhang, M.; Baranska, M.; Bosch, A.; Majzner, K. FT-IR hyperspectral imaging and artificial neural network analysis for identification of pathogenic bacteria. *Anal. Chem.* **2018**, *90*, 8896–8904. [CrossRef] [PubMed]

67. Skolik, P.; Morais, C.L.M.; Martin, F.L.; McAinsh, M.R. Attenuated total reflection fourier-transform infrared spectroscopy coupled with chemometrics directly detects pre- and post-symptomatic changes in tomato plants infected with botrytis cinerea. *Vib. Spectrosc.* **2020**, *111*, 103171. [CrossRef]

68. Yang, Y.; Zhang, Y.; He, C.; Xie, M.; Luo, H.; Wang, Y.; Zhang, J. Rapid screen of aflatoxin-contaminated peanut oil using fourier transform infrared spectroscopy combined with multivariate decision tree. *Int. J. Food Sci. Technol.* **2018**, *53*, 2386–2393. [CrossRef]

69. Meza-Márquez, O.G.; Gallardo-Velázquez, T.; Osorio-Revilla, G.; Dorantes-Álvarez, L. Detection of clenbuterol in beef meat, liver and kidney by mid-infrared spectroscopy (FT-Mid IR) and multivariate analysis. *Int. J. Food Sci. Technol.* **2012**, *47*, 2342–2351. [CrossRef]

70. Kim, J.; Kumar, R.; Bandodkar, A.J.; Wang, J. Advanced materials for printed wearable electrochemical devices: A review. *Adv. Electron. Mater.* **2017**, *3*, 1600260. [CrossRef]

71. Gao, M.; Li, L.; Song, Y. Inkjet printing wearable electronic devices. *APL Mater.* **2017**, 2971–2993. [CrossRef]

72. Dobbelaere, T.; Vereecken, P.M.; Detavernier, C. HardwareX A USB-controlled potentiostat/galvanostat for thin-film battery characterization. *HardwareX* **2017**, *2*, 34–49. [CrossRef]

73. Dryden, M.D.M.; Wheeler, A.R. DStat: A versatile, open-source potentiostat for electroanalysis and integration. *PLoS ONE* **2015**, e140349. [CrossRef]

74. Ainla, A.; Mousavi, M.P.S.; Tsaloglou, M.; Redston, J.; Bell, G.; Ferna, M.T. Open-source potentiostat for wireless electrochemical detection with smartphones. *Anal. Chem.* **2018**. [CrossRef]

75. Ronkainen, N.J.; Brian, H.; Heineman, W.R. Electrochemical biosensors. *Chem. Soc. Rev.* **2010**, 1747–1763. [CrossRef]

76. Zhao, Y.; Choi, S.Y.; Lou-Franco, J.; Nelis, J.L.D.; Zhou, H.; Cao, C.; Campbell, K.; Elliott, C.; Rafferty, K. Smartphone modulated colorimetric reader with color Subtraction. In Proceedings of the IEEE Sensors, Montreal, QC, Canada, 27–30 October 2019; Volume 2019.

77. Nelis, J.L.D.; Migliorelli, D.; Jafari, S.; Generelli, S.; Lou-Franco, J.; Salvador, J.P.; Marco, M.P.; Cao, C.; Elliott, C.T.; Campbell, K. The benefits of carbon black, gold and magnetic nanomaterials for point-of-harvest electrochemical quantification of domoic acid. *Microchim. Acta* **2020**, *187*, 1–11. [CrossRef]

78. Nelis, J.L.D.; Migliorelli, D.; Mühlebach, L.; Generelli, S.; Stewart, L.; Elliott, C.T.; Campbell, K. Highly sensitive electrochemical detection of the marine toxins okadaic acid and domoic acid with carbon black modified screen printed electrodes. *Talanta* **2021**, *228*, 122215. [CrossRef] [PubMed]

79. Sierra, T.; Crevillen, A.G.; Escarpa, A. Electrochemical detection based on nanomaterials in ce and microfluidic systems. *Electrophoresis* **2019**, *40*, 113–123. [CrossRef] [PubMed]

80. Huang, X.; Xu, D.; Chen, J.; Liu, J.; Li, Y.; Song, J.; Ma, X.; Guo, J. Smartphone-based analytical biosensors. *Analyst* **2018**, *143*, 5339–5351. [CrossRef]

81. Nelis, J.L.D.; Tsagkaris, A.S.; Zhao, Y.; Lou-Franco, J.; Nolan, P.; Zhou, H.; Cao, C.; Rafferty, K.; Hajslova, J.; Elliott, C.T.; et al. The end user sensor tree: An end-user friendly sensor database. *Biosens. Bioelectron.* **2019**, *130*, 245–253. [CrossRef] [PubMed]

82. Dungchai, W.; Chailapakul, O.; Henry, C.S. Electrochemical detection for paper-based microfluidics. *Anal. Chem.* **2009**, *81*, 5821–5826. [CrossRef]

83. Nie, Z.; Nijhuis, C.A.; Gong, J.; Chen, X.; Kumachev, A.; Martinez, A.W.; Narovlyansky, M.; Whitesides, G.M. Electrochemical sensing in paper-based microfluidic devices. *Lab Chip* **2010**, *10*, 477–483. [CrossRef] [PubMed]

84. Ataide, V.N.; Mendes, L.F.; Gama, L.I.L.M.; de Araujo, W.R.; Paixão, T.R.L.C. Electrochemical paper-based analytical devices: Ten years of development. *Anal. Methods* **2020**, *12*, 1030–1054. [CrossRef]

85. Noviana, E.; McCord, C.P.; Clark, K.M.; Jang, I.; Henry, C.S. Electrochemical paper-based devices: Sensing approaches and progress toward practical applications. *Lab Chip* **2020**, *20*, 9–34. [CrossRef]

86. de Araujo, W.R.; Frasson, C.M.R.; Ameku, W.A.; Silva, J.R.; Angnes, L.; Paixão, T.R.L.C. Single-step reagentless laser scribing fabrication of electrochemical paper-based analytical devices. *Angew. Chemie Int. Ed.* **2017**, *56*, 15113–15117. [CrossRef]

87. Sun, G.; Wang, P.; Ge, S.; Ge, L.; Yu, J.; Yan, M. Photoelectrochemical sensor for pentachlorophenol on microfluidic paper-based analytical device based on the molecular imprinting technique. *Biosens. Bioelectron.* **2014**, *56*, 97–103. [CrossRef]

88. Marín-Barroso, E.; Messina, G.A.; Bertolino, F.A.; Raba, J.; Pereira, S. V Electrochemical immunosensor modified with carbon nanofibers coupled to a paper platform for the determination of gliadins in food samples. *Anal. Methods* **2019**, *11*, 2170–2178. [CrossRef]

89. Vojdani, A. Cross-reaction between gliadin and different food and tissue antigens. *Food Nutr. Sci.* **2013**, *4*, 20–32. [CrossRef]

90. Wang, Y.; Ping, J.; Ye, Z.; Wu, J.; Ying, Y. Impedimetric immunosensor based on gold nanoparticles modified graphene paper for label-free detection of escherichia Coli O157:H. *Biosens. Bioelectron.* **2013**, *49*, 492–498. [CrossRef] [PubMed]

91. Rattanarat, P.; Dungchai, W.; Cate, D.; Volckens, J.; Chailapakul, O.; Henry, C.S. Multilayer paper-based device for colorimetric and electrochemical quantification of metals. *Anal. Chem.* **2014**, *86*, 3555–3562. [CrossRef]

92. Bhardwaj, J.; Devarakonda, S.; Kumar, S.; Jang, J. Development of a paper-based electrochemical immunosensor using an antibody-single walled carbon nanotubes bio-conjugate modified electrode for label-free detection of foodborne pathogens. *Sens. Actuators B Chem.* **2017**, *253*, 115–123. [CrossRef]

93. Ge, S.; Liu, W.; Ge, L.; Yan, M.; Yan, J.; Huang, J.; Yu, J. In situ assembly of porous au-paper electrode and functionalization of magnetic silica nanoparticles with HRP via click chemistry for microcystin-LR immunoassay. *Biosens. Bioelectron.* **2013**, *49*, 111–117. [CrossRef] [PubMed]

94. Arduini, F.; Cinti, S.; Caratelli, V.; Amendola, L.; Palleschi, G.; Moscone, D. Origami multiple paper-based electrochemical biosensors for pesticide detection. *Biosens. Bioelectron.* **2019**, *126*, 346–354. [CrossRef]
95. Cinti, S.; Basso, M.; Moscone, D.; Arduini, F. A paper-based nanomodified electrochemical biosensor for ethanol detection in beers. *Anal. Chim. Acta* **2017**, *960*, 123–130. [CrossRef] [PubMed]
96. Cinti, S.; Moscone, D.; Arduini, F. Preparation of paper-based devices for reagentless electrochemical (bio)sensor strips. *Nat. Protoc.* **2019**, *14*, 2437–2451. [CrossRef]
97. Kant, K.; Shahbazi, M.A.; Dave, V.P.; Ngo, T.A.; Chidambara, V.A.; Than, L.Q.; Bang, D.D.; Wolff, A. Microfluidic devices for sample preparation and rapid detection of foodborne pathogens. *Biotechnol. Adv.* **2018**, *36*, 1003–1024. [CrossRef]
98. Puiu, M.; Bala, C. Microfluidics-integrated biosensing platforms as emergency tools for on-site field detection of foodborne pathogens. *TrAC Trends Anal. Chem.* **2020**, *125*, 115831. [CrossRef]
99. Rackus, D.G.; Shamsi, M.H.; Wheeler, A.R. Electrochemistry, biosensors and microfluidics: A convergence of fields. *Chem. Soc. Rev.* **2015**, *44*, 5320–5340. [CrossRef] [PubMed]
100. Lafleur, J.P.; Jönsson, A.; Senkbeil, S.; Kutter, J.P. Recent advances in lab-on-a-chip for biosensing applications. *Biosens. Bioelectron.* **2016**, *76*, 213–233. [CrossRef]
101. Park, Y.M.; Lim, S.Y.; Shin, S.J.; Kim, C.H.; Jeong, S.W.; Shin, S.Y.; Bae, N.H.; Lee, S.J.; Na, J.; Jung, G.Y.; et al. A film-based integrated chip for gene amplification and electrochemical detection of pathogens causing foodborne illnesses. *Anal. Chim. Acta* **2018**, *1027*, 57–66. [CrossRef] [PubMed]
102. Drechsel, L.; Schulz, M.; Von Stetten, F.; Moldovan, C.; Zengerle, R.; Paust, N. Electrochemical pesticide detection with autodip—A portable platform for automation of crude sample analyses. *Lab Chip* **2015**, *15*, 704–710. [CrossRef]
103. EFSA. The 2010 European Union Report on Pesticide Residues in Food. *EFSA J.* **2013**, *11*. [CrossRef]
104. Uludag, Y.; Esen, E.; Kokturk, G.; Ozer, H.; Muhammad, T.; Olcer, Z.; Basegmez, H.I.O.; Simsek, S.; Barut, S.; Gok, M.Y.; et al. Lab-on-a-chip based biosensor for the real-time detection of aflatoxin. *Talanta* **2016**, *160*, 381–388. [CrossRef] [PubMed]
105. European Commission. *Commission Regulation (EC) No 1881/2006 of 19 December 2006 Setting Maximum Levels for Certain Contaminants in Foodstuffs*; European Commission: Brussels, Belgium, 2006.
106. Medina-Sánchez, M.; Mayorga-Martinez, C.C.; Watanabe, T.; Ivandini, T.A.; Honda, Y.; Pino, F.; Nakata, A.; Fujishima, A.; Einaga, Y.; Merkoçi, A. Microfluidic platform for environmental contaminants sensing and degradation based on boron-doped diamond electrodes. *Biosens. Bioelectron.* **2016**, *75*, 365–374. [CrossRef]
107. Panini, N.V.; Salinas, E.; Messina, G.A.; Raba, J. Modified paramagnetic beads in a microfluidic system for the determination of zearalenone in feedstuffs samples. *Food Chem.* **2011**, *125*, 791–796. [CrossRef]
108. Lin, Y.; Zhou, Q.; Tang, D.; Niessner, R.; Knopp, D. Signal-on photoelectrochemical immunoassay for aflatoxin b1 based on enzymatic product-etching MnO2 nanosheets for dissociation of carbon dots. *Anal. Chem.* **2017**, *89*, 5637–5645. [CrossRef]
109. Hervás, M.; López, M.A.; Escarpa, A. Integrated electrokinetic magnetic bead-based electrochemical immunoassay on microfluidic chips for reliable control of permitted levels of zearalenone in infant foods. *Analyst* **2011**, *136*, 2131–2138. [CrossRef] [PubMed]
110. Chen, Q.; Wang, D.; Cai, G.; Xiong, Y.; Li, Y.; Wang, M.; Huo, H.; Lin, J. Fast and sensitive detection of foodborne pathogen using electrochemical impedance analysis, urease catalysis and microfluidics. *Biosens. Bioelectron.* **2016**, *86*, 770–776. [CrossRef] [PubMed]
111. Olcer, Z.; Esen, E.; Muhammad, T.; Ersoy, A.; Budak, S.; Uludag, Y. Fast and sensitive detection of mycotoxins in wheat using microfluidics based real-time electrochemical profiling. *Biosens. Bioelectron.* **2014**, *62*, 163–169. [CrossRef] [PubMed]
112. Freitas, T.A.; Proença, C.A.; Baldo, T.A.; Materón, E.M.; Wong, A.; Magnani, R.F.; Faria, R.C. Ultrasensitive immunoassay for detection of citrus tristeza virus in citrus sample using disposable microfluidic electrochemical device. *Talanta* **2019**, *205*, 120110. [CrossRef] [PubMed]
113. De Oliveira, T.R.; Martucci, D.H.; Faria, R.C. Simple disposable microfluidic device for salmonella typhimurium detection by magneto-immunoassay. *Sens. Actuators B Chem.* **2018**, *255*, 684–691. [CrossRef]
114. Lu, L.; Gunasekaran, S. Dual-Channel ITO-microfluidic electrochemical immunosensor for simultaneous detection of two mycotoxins. *Talanta* **2019**, *194*, 709–716. [CrossRef]
115. Chiriacò, M.S.; De Feo, F.; Primiceri, E.; Monteduro, A.G.; De Benedetto, G.E.; Pennetta, A.; Rinaldi, R.; Maruccio, G. Portable gliadin-immunochip for contamination control on the food production chain. *Talanta* **2015**, *142*, 57–63. [CrossRef]
116. Crew, A.; Lonsdale, D.; Byrd, N.; Pittson, R.; Hart, J.P. A screen-printed, amperometric biosensor array incorporated into a novel automated system for the simultaneous determination of organophosphate pesticides. *Biosens. Bioelectron.* **2011**, *26*, 2847–2851. [CrossRef]
117. Chen, J.; Zhou, Y.; Wang, D.; He, F.; Rotello, V.M.; Carter, K.R.; Watkins, J.J.; Nugen, S.R. UV-nanoimprint lithography as a tool to develop flexible microfluidic devices for electrochemical detection. *Lab Chip* **2015**, *15*, 3086–3094. [CrossRef]
118. Lillehoj, P.B.; Wei, F.; Ho, C.M. A Self-pumping lab-on-a-chip for rapid detection of botulinum toxin. *Lab Chip* **2010**, *10*, 2265–2270. [CrossRef]
119. Singh, C.; Ali, M.A.; Kumar, V.; Ahmad, R.; Sumana, G. Functionalized MoS2 nanosheets assembled microfluidic immunosensor for highly sensitive detection of food pathogen. *Sens. Actuators B Chem.* **2018**, *259*, 1090–1098. [CrossRef]
120. Ramalingam, S.; Chand, R.; Singh, C.B.; Singh, A. Phosphorene-gold nanocomposite based microfluidic aptasensor for the detection of okadaic acid. *Biosens. Bioelectron.* **2019**, *135*, 14–21. [CrossRef]

121. Li, S.; Li, J.; Luo, J.; Xu, Z.; Ma, X. A Microfluidic chip containing a molecularly imprinted polymer and a DNA aptamer for voltammetric determination of carbofuran. *Microchim. Acta* **2018**, *185*, 1–8. [CrossRef] [PubMed]
122. Tan, F.; Leung, P.H.M.; Liu, Z.B.; Zhang, Y.; Xiao, L.; Ye, W.; Zhang, X.; Yi, L.; Yang, M. A PDMS Microfluidic Impedance Immunosensor for E. Coli O157:H7 and staphylococcus aureus detection via antibody-immobilized nanoporous membrane. *Sens. Actuators B Chem.* **2011**, *159*, 328–335. [CrossRef]
123. Mondal, K.; Ali, A.; Srivastava, S.; Malhotra, B.D.; Sharma, A. Sensors and Actuators B: Chemical electrospun functional micro/nanochannels embedded in porous carbon electrodes for microfluidic biosensing. *Sens. Actuators B Chem.* **2016**, *229*, 82–91. [CrossRef]
124. Tian, F.; Lyu, J.; Shi, J.; Tan, F.; Yang, M. A polymeric microfluidic device integrated with nanoporous alumina membranes for simultaneous detection of multiple foodborne pathogens. *Sens. Actuators B Chem.* **2016**, *225*, 312–318. [CrossRef]
125. Thiha, A.; Ibrahim, F.; Muniandy, S.; Julian, I.; Jyan, S.; Lin, K.; Fen, B.; Madou, M. All-carbon suspended nanowire sensors as a rapid highly-sensitive label-free chemiresistive biosensing platform. *Biosens. Bioelectron.* **2018**, *107*, 145–152. [CrossRef]
126. Rezazadeh, M.; Seidi, S.; Lid, M.; Pedersen-bjergaard, S. Trends in analytical chemistry the modern role of smartphones in analytical chemistry. *Trends Anal. Chem.* **2019**, *118*, 548–555. [CrossRef]
127. Fan, Y.; Liu, J.; Wang, Y.; Luo, J.; Xu, H.; Xu, S.; Cai, X. A wireless point-of-care testing system for the detection of neuron-specific enolase with microfluidic paper-based analytical devices. *Biosens. Bioelectron.* **2017**, *95*, 60–66. [CrossRef]
128. Xu, K.; Chen, Q.; Zhao, Y.; Ge, C.; Lin, S.; Liao, J. Cost-effective, wireless, and portable smartphone-based electrochemical system for on-site monitoring and spatial mapping of the nitrite contamination in water. *Sens. Actuators B Chem.* **2020**, *319*. [CrossRef]
129. Liao, J.; Chang, F.; Han, X.; Ge, C.; Lin, S. Wireless water quality monitoring and spatial mapping with disposable whole-copper electrochemical sensors and a smartphone. *Sens. Actuators B Chem.* **2020**, *306*, 127557. [CrossRef]
130. Guan, T.; Huang, W.; Xu, N.; Xu, Z.; Jiang, L.; Li, M.; Wei, X.; Liu, Y.; Shen, X.; Li, X.; et al. Point-of-need detection of microcystin-LR using a smartphone-controlled electrochemical analyzer. *Sens. Actuators B Chem.* **2019**, *294*, 132–140. [CrossRef]
131. Ji, D.; Liu, L.; Li, S.; Chen, C.; Lu, Y.; Wu, J.; Liu, Q. Smartphone-based cyclic voltammetry system with graphene modified screen printed electrodes for glucose detection. *Biosens. Bioelectron.* **2017**, *98*, 449–456. [CrossRef] [PubMed]
132. Lin, H.-Y.; Huang, C.-H.; Park, J.; Pathania, D.; Castro, C.M.; Fasano, A.; Weissleder, R.; Lee, H. Integrated magneto-chemical sensor for on-site food allergen detection. *ACS Nano* **2017**, *11*, 10062–10069. [CrossRef] [PubMed]
133. Mishra, R.K.; Hubble, L.J.; Martín, A.; Kumar, R.; Barfidokht, A.; Kim, J.; Musameh, M.M.; Kyratzis, I.L.; Wang, J. Wearable flexible and stretchable glove biosensor for on-site detection of organophosphorus chemical threats. *ACS Sens.* **2017**, *2*, 553–561. [CrossRef] [PubMed]
134. Barfidokht, A.; Mishra, R.K.; Seenivasan, R.; Liu, S.; Hubble, L.J.; Wang, J.; Hall, D.A. Wearable electrochemical glove-based sensor for rapid and on-site detection of fentanyl. *Sens. Actuators B Chem.* **2019**, *296*, 126422. [CrossRef] [PubMed]
135. Hipple, J.A., Jr. Portable mass spectrometer. *Nature* **1942**, *150*, 111–112. [CrossRef]
136. Yang, M.; Kim, T.-Y.; Hwang, H.-C.; Yi, S.-K.; Kim, D.-H. Developmen of a palm portable mass spectrometer. *J. Am. Soc. Mass Spectrom.* **2008**, *19*, 1442–1448. [CrossRef] [PubMed]
137. Ouyang, Z.; Cooks, R.G. Miniature mass spectrometers. *Annu. Rev. Anal. Chem.* **2009**, *2*, 187–214. [CrossRef]
138. Mielczarek, P.; Silberring, J.; Smoluch, M. Miniaturization in mass spectrometry. *Mass Spectrom. Rev.* **2019**. [CrossRef]
139. Snyder, D.T.; Pulliam, C.J.; Ouyang, Z.; Cooks, R.G. Miniature and fieldable mass spectrometers: Recent advances. *Anal. Chem.* **2016**, *88*, 2–29. [CrossRef]
140. Da Silva, L.C.; Pereira, I.; De Carvalho, T.C.; Allochio Filho, J.F.; Romão, W.; Vaz, B.G. Paper Spray ionization and portable mass spectrometers: A review. *Anal. Methods* **2019**, *11*, 999–1013. [CrossRef]
141. Diaz, J.A.; Giese, C.F.; Gentry, W.R. Portable double-focusing mass-spectrometer system for field gas monitoring. *Field Anal. Chem. Technol.* **2001**, *5*, 156–167. [CrossRef]
142. Bryden, W.A.; Benson, R.C.; Eeelberger, S.A.; Phillips, T.E.; Cotter, R.J.; Fenselau, C. The Tiny-TOF mass spectrometer for chemical and biological Sensing. *Johns Hopkins APL Tech. Dig.* **1995**, *16*, 296–310. [CrossRef]
143. Black, C.; Chevallier, O.P.; Elliott, C.T. The current and potential applications of ambient mass spectrometry in detecting food fraud. *TrAC Trends Anal. Chem.* **2016**, *82*, 268–278. [CrossRef]
144. Guo, X.-Y.; Huang, X.-M.; Zhai, J.-F.; Bai, H.; Li, X.-X.; Ma, X.-X.; Ma, Q. Research advances in ambient ionization and miniature mass spectrometry. *Chin. J. Anal. Chem.* **2019**, *47*, 335–346. [CrossRef]
145. Huang, G.; Xu, W.; Visbal-Onufrak, M.A.; Ouyang, Z.; Cooks, R.G. Direct analysis of melamine in complex matrices using a handheld mass spectrometer. *Analyst* **2010**, *135*, 705–711. [CrossRef]
146. European Commission. Commission Regulation (EU) No 594/2012 of 5 July 2012 Amending Regulation (EC) 1881/2006 as regards the maximum levels of the contaminants Ochratoxin A, non dioxin-like PCBs and melamine in foodstuffs. *Off. J. Eur. Union* **2012**, *176*, 43–45.
147. Soparawalla, S.; Tadjimukhamedov, F.K.; Wiley, J.S.; Ouyang, Z.; Cooks, R.G. In situ analysis of agrochemical residues on fruit using ambient ionization on a handheld mass spectrometer. *Analyst* **2011**, *136*, 4392–4396. [CrossRef]
148. Wiley, J.S.; Shelley, J.T.; Cooks, R.G. Handheld low-temperature plasma probe for portable "point-and- shoot" ambient ionization mass spectrometry. *Anal. Chem.* **2013**, *85*, 6545–6552. [CrossRef]
149. Janfelt, C.; Græsbøll, R.; Lauritsen, F.R. Portable electrospray ionization mass spectrometry (ESI-MS) for analysis of contaminants in the Field. *Int. J. Environ. Anal. Chem.* **2012**, *92*, 397–404. [CrossRef]

150. Gerbig, S.; Neese, S.; Penner, A.; Spengler, B.; Schulz, S. Real-time food authentication using a miniature mass spectrometer. *Anal. Chem.* **2017**, *89*, 10717–10725. [CrossRef]
151. Li, L.; Chen, T.C.; Ren, Y.; Hendricks, P.I.; Cooks, R.G.; Ouyang, Z. Mini 12, miniature mass spectrometer for clinical and other applications—Introduction and characterization. *Anal. Chem.* **2014**, *86*, 2909–2916. [CrossRef] [PubMed]
152. Pulliam, C.J.; Bain, R.M.; Wiley, J.S.; Ouyang, Z.; Cooks, R.G. Mass spectrometry in the home and garden. *J. Am. Soc. Mass Spectrom.* **2015**, *26*, 224–230. [CrossRef] [PubMed]
153. Ma, Q.; Bai, H.; Li, W.; Wang, C.; Li, X.; Cooks, R.G.; Ouyang, Z. Direct identification of prohibited substances in cosmetics and foodstuffs using ambient ionization on a miniature mass spectrometry system. *Anal. Chim. Acta* **2016**, *912*, 65–73. [CrossRef]
154. Mini β Miniature Mass Spectrometer. Available online: http://www.purspec.us/params (accessed on 16 June 2021).
155. Kang, M.; Lian, R.; Zhang, X.; Li, Y.; Zhang, Y.; Zhang, Y.; Zhang, W.; Ouyang, Z. Rapid and on-site detection of multiple fentanyl compounds by dual-ion trap miniature mass spectrometry system. *Talanta* **2020**, *217*, 121057. [CrossRef]
156. Zhang, X.; Zhang, H.; Yu, K.; Li, N.; Liu, Y.; Liu, X.; Zhang, H.; Yang, B.; Wu, W.; Gao, J.; et al. Rapid monitoring approach for microplastics using portable pyrolysis-mass spectrometry. *Anal. Chem.* **2020**, *92*, 4656–4662. [CrossRef] [PubMed]
157. Gomez-Rios, G.A.; Vasiljevic, T.; Gionfriddo, E.; Yu, M.; Pawliszyn, J. Towards on-site analysis of complex matrices by solid-phase microextraction-transmission mode coupled to a portable mass spectrometer via direct analysis in real time. *Analyst* **2017**, *142*, 2928–2935. [CrossRef] [PubMed]
158. Tomkins, B.A.; Ilgner, R.H. Determination of atrazine and four organophosphorus pesticides in ground water using Solid Phase Microextraction (SPME) followed by gas chromatography with selected-ion monitoring. *J. Chromatogr. A* **2002**, *972*, 183–194. [CrossRef]
159. Hrbek, V.; Vaclavik, L.; Elich, O.; Hajslova, J. Authentication of milk and milk-based foods by Direct Analysis in Real Time Ionization–High Resolution Mass Spectrometry (DART–HRMS) Technique: A critical assessment. *Food Control* **2014**, *36*, 138–145. [CrossRef]
160. Blokland, M.H.; Gerssen, A.; Zoontjes, P.W.; Pawliszyn, J.; Nielen, M.W.F. Potential of recent ambient ionization techniques for future food contaminant analysis using (trans)portable mass spectrometry. *Food Anal. Methods* **2019**, *13*, 706–717. [CrossRef]
161. Wright, S.; Malcolm, A.; Wright, C.; O'Prey, S.; Crichton, E.; Dash, N.; Moseley, R.W.; Zaczek, W.; Edwards, P.; Fussell, R.J.; et al. A Microelectromechanical systems-enabled, miniature triple quadrupole mass spectrometer. *Anal. Chem.* **2015**, *87*, 3115–3122. [CrossRef] [PubMed]
162. Jager, J.; Gerssen, A.; Pawliszyn, J.; Sterk, S.S.; Nielen, M.W.F.; Blokland, M.H. USB-powered coated blade spray ion source for on-site testing using transportable mass spectrometry. *J. Am. Soc. Mass Spectrom.* **2020**, *31*, 2243–2249. [CrossRef] [PubMed]
163. Gallo, M.; Ferrara, L.; Calogero, A.; Montesano, D.; Naviglio, D. Relationships between Food and Diseases: What to Know to Ensure Food Safety. *Food Res. Int.* **2020**, *137*. [CrossRef]
164. European Commission. Commission Regulation (EC) No 149/2008 of 29 January 2008 Amending Regulation (EC) No 396/2005 of the European Parliament and of the Council by Establishing Annexes II, III and IV Setting Maximum Residue Levels for Products Covered by Annex I Thereto. *Off. J. Eur. Union* **2008**, *58*, 1–398.
165. Geballa-Koukoula, A.; Gerssen, A.; Nielen, M.W.F. From smartphone lateral flow immunoassay screening to direct MS Analysis: Development and validation of a semi-quantitative Direct Analysis in Real-Time Mass Spectrometric (DART-MS) approach to the analysis of deoxynivalenol. *Sensors* **2021**, *21*, 1861. [CrossRef] [PubMed]
166. Geballa-Koukoula, A.; Gerssen, A.; Nielen, M.W.F. Direct analysis of lateral flow immunoassays for deoxynivalenol using electrospray ionization mass spectrometry. *Anal. Bioanal. Chem.* **2020**. [CrossRef]

Article

Extending the Lipidome Coverage by Combining Different Mass Spectrometric Platforms: An Innovative Strategy to Answer Chemical Food Safety Issues

Jérémy Marchand [1,2], Yann Guitton [1], Estelle Martineau [2,3], Anne-Lise Royer [1], David Balgoma [1], Bruno Le Bizec [1], Patrick Giraudeau [2,*] and Gaud Dervilly [1,*]

[1] LABERCA, Oniris, INRAE, 44307 Nantes, France; jeremy.marchand@hotmail.fr (J.M.); yann.guitton@oniris-nantes.fr (Y.G.); anne-lise.royer@oniris-nantes.fr (A.-L.R.); david.balgoma@ilk.uu.se (D.B.); bruno.lebizec@oniris-nantes.fr (B.L.B.)
[2] CEISAM UMR 6230, Université de Nantes, CNRS, 44000 Nantes, France; estelle.martineau@univ-nantes.fr
[3] SpectroMaîtrise, CAPACITES SAS, 26 Bd Vincent Gâche, 44200 Nantes, France
* Correspondence: patrick.giraudeau@univ-nantes.fr (P.G.); gaud.dervilly@oniris-nantes.fr (G.D.); Tel.: +33-251125709 (P.G.); +33-240687880 (G.D.)

Citation: Marchand, J.; Guitton, Y.; Martineau, E.; Royer, A.-L.; Balgoma, D.; Le Bizec, B.; Giraudeau, P.; Dervilly, G. Extending the Lipidome Coverage by Combining Different Mass Spectrometric Platforms: An Innovative Strategy to Answer Chemical Food Safety Issues. *Foods* **2021**, *10*, 1218. https://doi.org/10.3390/foods10061218

Academic Editors: Philippe Delahaut and Riccardo Marega

Received: 3 March 2021
Accepted: 22 May 2021
Published: 28 May 2021

Abstract: From a general public health perspective, a strategy combining non-targeted and targeted lipidomics MS-based approaches is proposed to identify disrupted patterns in serum lipidome upon growth promoter treatment in pigs. Evaluating the relative contributions of the platforms involved, the study aims at investigating the potential of innovative analytical approaches to highlight potential chemical food safety threats. Serum samples collected during an animal experiment involving control and treated pigs, whose food had been supplemented with ractopamine, were extracted and characterised using three MS strategies: Non-targeted RP LC-HRMS; the targeted Lipidyzer™ platform (differential ion mobility associated with shotgun lipidomics) and a homemade LC-HRMS triglyceride platform. The strategy enabled highlighting specific lipid profile patterns involving various lipid classes, mainly in relation to cholesterol esters, sphingomyelins, lactosylceramide, phosphatidylcholines and triglycerides. Thanks to the combination of non-targeted and targeted MS approaches, various compartments of the pig serum lipidome could be explored, including commonly characterised lipids (Lipidyzer™), triglyceride isomers (Triglyceride platform) and unique lipid features (non-targeted LC-HRMS). Thanks to their respective characteristics, the complementarity of the three tools could be demonstrated for public health purposes, with enhanced coverage, level of characterization and applicability.

Keywords: serum; lipidomics; Lipidyzer™; LC-HRMS; ractopamine; β-agonist

1. Introduction

While the use of anabolic compounds has been banned in livestock for more than 30 years [1], the recently updated regulatory scheme confirms such provision [2]. From a public health perspective related to the chemical safety of food from animal origin, it reaffirms the European commitment to the performance of the associated controls. In this firmly reaffirmed context, the search for ever more innovative control strategies is even more topical. In particular, continuing and extending the promising work initiated 15 years ago on the investigation of the physiological effects induced as a consequence of illegal practices through metabolomics approaches appears to be a priority [3,4]. Although main proofs of concepts have been obtained focusing on the polar metabolome [5–11], the apolar and lipidic fraction was also shown to be relevant while highlighting the disruption of phosphatidylglycerols (PG), phosphatidylethanolamine (PE), phosphatidylcholine (PC) and phosphatidic acid (PA) in bovine serum upon trenbolone/estradiol administration [12] and the disruption of PE, phosphatidylinositol (PI) and sphingomyelin (SM) in muscle tissues collected on ractopamine (RAC) fed pigs [13]. However, these preliminary results did not allow a thorough characterisation of the lipidome since the non-targeted methods applied lacked proper data validation or extensive lipid coverage. Moreover, because the application of these forbidden veterinary drugs in livestock aims at modifying the animal carcass composition for leaner meat promotion, a significant shift of associated lipid profiles is expected, justifying further methodological efforts to be dedicated to allowing robust lipidomics to be applied. Characterising lipidome disruptions as a consequence of growth promoter application would indeed allow generating new knowledge on the mechanism of action of these anabolic agents and especially discovering relevant biomarkers for more efficient screening of such practices.

Lipidomics, which has grown as a major field over the last decade, is recognised as a complex compartment of the metabolome [14–17] with many sample preparation methods [18–21] and mass spectrometric-based analysis methods [22–28]. Across these numerous methods, two main categories can be distinguished and referred to as targeted approaches—where a limited number of specific lipid classes or species are monitored and quantified—and non-targeted approaches—where an open-list of compounds is analysed for subsequent identification using annotation tools [29]. The latter provides rich information on the lipidome, as they theoretically allow the measurement of any detectable lipid signals [30], resulting in thousands of features. However, this requires data cleaning steps to remove noise and redundancies (isotopes, adducts). Moreover, the assignment of these signals remains a challenging step of the workflow. In contrast, targeted approaches are more selective, thus increasing confidence in the results, even if the acquired information is much more limited. Globally, across the diverse methods and strategies, it appears that no single workflow is sufficient for a wide and complete lipidome characterisation. In such a context, the combination of non-targeted and targeted approaches from various complementary techniques is expected to provide an optimal strategy [31] that would further allow discovering unexpected biomarker signals.

The present article describes the implementation of three MS platforms (namely: non-targeted LC-HRMS, Lipidyzer™ and an in-house platform for triglyceride regioisomers) to determine changes in lipidomic profiles in serum of ractopamine treated pigs. Since they differ in technology (ion mobility, LC, HRMS, MS/MS) and approach (targeted, non-targeted), this combination is expected to provide both enhanced lipid coverage and reliability in the obtained results. In comparison with other multi-platform approaches published by other groups [31], this original strategy aims to further enhance TG analysis, using a dedicated platform for quantifying their regioisomeric composition.

2. Materials and Methods

2.1. Animal Experiment

The blood samples used in this study were obtained from a previously described ethically approved experiment [32], specifically designed to evaluate the disruptions induced in pig blood serum metabolite profiles upon ractopamine administration. Two groups constituted of randomly divided 5 healthy 4-month-old female pigs, involved over 4 weeks. After a 3-day acclimatisation, animals from the treated group were exposed to RAC hydrochloride (Sigma Aldrich, Saint Quentin Fallavier, France) through a 10 ppm daily dose in pre-weighted feed (corresponding to 0.45 mg/kg bw/day). The dosage for each animal was verified through complete eating of the daily portion. A total of 6 blood samples were collected, respectively at Day-3 (D3), Day-9 (D9), Day-16 (D16), Day-18 (D18), Day-23 (D23) and Day-29 (D29) for each individual from both groups: control (individuals P1 to P5) and treated (individuals P6 to P10). The samples were then allowed to clot at room temperature in order to obtain serum samples.

QC samples were prepared by pooling the same amount of all collected and carefully homogenised samples.

All samples were prepared into suitable 100 µL aliquots and immediately stored at −20 °C until analysis.

2.2. Analytical Platforms

To characterise the lipidome as widely as possible and evaluate the added value of combining multiple tools, 3 mass spectrometry platforms were involved for the analysis of the samples from the animal experiment, each of them providing a different level of characterisation. A first option was the non-targeted analysis using Reversed-Phase Ultra High-Performance Liquid Chromatography coupled to mass spectrometry (RP UHPLC-HRMS), completed by the targeted platform Lipidyzer™ (differential ion mobility associated with shotgun lipidomics), dedicated to the quantitative analysis of lipids from several classes and finally an in-house developed LC-HRMS platform able to quantify the regioisomeric composition of triglycerides (TG).

For each platform, a dedicated sample preparation protocol was carried out, as described in Table 1. While the non-targeted approach was applied on all samples, the targeted tools were implemented for samples collected at the beginning and end of the animal experiment, based on the results from the former. Each time, specific parameters and processing were used, as well as quality assurance (QA) and quality control procedures (QC), which are summarised in Table 1. The full details of the sample preparation and analysis procedures can be found in Supplementary Materials.

Table 1. Characteristics of the three used platforms and associated experimental details.

Platform	Non-Targeted RPLC-HRMS [12]	Targeted Lipidyzer™ [33,34]	Targeted TG Platform [35]
Extraction type	Bligh and Dyer—like [12]	Two solvent addition/organic phase transfer cycles	Bligh and Dyer—[12]
Samples	D3, D9, D16, D18, D23 and D29 QC	D3, D18 and D23 QC Lipidyzer-specific QC and QC spike samples	D3, D16, D18, D23 and D29 QC
Serum volume	30 μL	30 μL	10 μL, completed with 20 μL H_2O
Solvents	Methanol (MeOH), Chloroform ($CHCl_3$), Water (H_2O)	MeOH, dichloromethane (DCM), H_2O	MeOH, $CHCl_3$, H_2O
Centrifugation	Yes	Yes, two times	Yes
Internal standards	$n = 7$ In $CHCl_3$, 0.5 mg·L^{-1}	Lipidyzer™ standard kit, $n = 54$ 30 μL added at beginning (See Supplementary Materials)	$n = 3$ In $CHCl_3$, 0.132 μmol·L^{-1}
Transfer	200 μL organic phase	Multiple organic phases	200 μL organic phase
Evaporation	Yes	Yes	Yes
Reconstitution solvent	Acetonitrile(AcN):Isopropanol(IPA):H_2O (65:30:5, v:v:v)	DCM:MeOH (50:50, v:v), 10 mM Ammonium Acetate	AcN:IPA (50:50, v:v)
Reconstitution volume	200 μL	300 μL	200 μL
Analysis Technique	LC-HRMS (full-scan + data dependent MS/MS)	DMS-MS/MS (direct introduction)	LC-MS/MS
Quantification	No	Yes	No
Targeted	No	Yes	Yes
Analytical system	LC: Thermo UltiMate® 3000 MS: Thermo Q-Exactive	Sciex QTRAP 5500, with SelexION differential mobility spectrometry (DMS)	LC: Waters Acquity UPLC MS: Waters Acquity-Synapt G2S Q-TOF
Column	Waters CSH C18 (100 × 2.1 mm i.d., 1.7 μm particle size)	None (direct introduction)	Waters BEH C18 (150 × 2.1 mm i.d. 1.7 μm particle size)
Mobile phase	A: ACN:H_2O (60:40, v:v) B: IPA:ACN:H_2O (88:10:2, v:v:v) Both: 10 mM ammonium acetate + 0.1% acetic acid	DCM:MeOH (50:50, v:v) 10 mM Ammonium Acetate	A: MeOH B: MeOH/IPA (50:50, v:v) Both: 2 mM ammonium acetate + 6 mM acetic acid

Table 1. *Cont.*

Platform	Non-Targeted RPLC-HRMS [12]	Targeted Lipidyzer™ [33,34]	Targeted TG Platform [35]
Data processing	MSConvert [36] XCMS [37], CAMERA Batch drift correction [38] Annotation: Lipidsearch (Thermo Fisher Scientific) after additional data dependent MS/MS—Top 15 (Full MS/dd-MS^2-Top 15) acquisitions	Automated Lipidyzer™ framework	MassWolf XCMS [37] In-house R algorithm
Number of features/lipids in analysed samples	ESI−: 1612 features ESI+: 2914 features	873 lipids *	50 TG **
Quality Assurance/Quality Control	Randomisation, QC (pooled samples), Internal standards, Extraction blanks	Randomisation, QC (pooled samples), Control plasma, Spiked samples, Internal standards, Extraction blanks	Cross checking of platform performance [35], calibration, QC (pooled samples), extraction blanks

* 383 individual species + 490 TG, including redundancies (see details in appropriate section) ** 143 regioisomers in total when considering proportion estimates (see details in appropriate section).

2.3. Data Analysis

For non-targeted data, multivariate analysis was performed using SIMCA 13.0.2 (Umetrics AB, Umeå, Sweden), where log transformation, Pareto scaling and centring were applied. Two-component Principal Component Analyses (PCA) provided an overview of the data and checking the quality of the analysis. Results were then analysed by Partial Least Square Discriminant Analysis (PLS-DA) (centred, UV-scaled). Each PLS-DA model was further validated thanks to permutation tests (n = 100 permutations) and CV-ANOVA. For better interpretability, Orthogonal Projection to Latent Structure Discriminant Analyses (OPLS-DA) were also performed.

Univariate analysis was performed on all datasets using a Wilcoxon test in R studio and p-values were calculated using the *coin* package (R studio).

3. Results

In order to investigate changes in the lipidome profiles and the complementarity of different MS fingerprinting strategies, a set of samples from which the lipid profiles were expected to be disrupted was chosen as a proof of concept [39]. Below are described and compared the results obtained from three methods: Non-targeted RP UHPLC-HRMS and two targeted approaches, namely Lipidyzer™ and a platform focused on TG regioisomers.

3.1. Non-Targeted RPLC-HRMS

In the frame of a global lipidomics study, a common method is the non-targeted fingerprinting using LC-HRMS, as it allows studying a large set of lipid species without any a priori hypothesis [40], i.e., theoretically all lipids accessible to the analysis technique. In the present case, the objective was not to develop a new analytical approach but rather evaluate the contribution of an already established workflow [12] in the frame of a multi-platform study.

After acquisition and verification of the fingerprint quality (see details in Supplementary Materials), 1612 and 2914 features were selected in the ESI− and ESI+ datasets, respectively. A PCA allowed highlighting clustering of the QC samples, thus demonstrating the reproducibility of the analysis (see Figure S1). Furthermore, in PCAs score plots, samples from D3 and D9 did not show major differences between groups, probably because of the slow response of the lipidome to such growth-promoting treatment as previously observed [39]. Consequently, these early collection points were removed, and the PCAs generated on the resulting ESI+ and ESI− datasets (D16, D18, D23, D29 samples) exhibited separation trends between groups (see Figure S2). PLS-DA were then performed and a discrimination between groups was observed (Figure 1, left panel) with the following performance: $R^2 = 0.882$ and $Q^2 = 0.444$ for ESI−; $R^2 = 0.697$ and $Q^2 = 0.482$ for ESI+. The models were further assessed with CV-ANOVA (p-value $= 9.5 \times 10^{-4}$ for ESI− and p-value $= 3.1 \times 10^{-4}$ for ESI+) indicating significant statistical models [41]. For both models, high R^2 values demonstrated high descriptive ability, while Q^2 values (<0.5) pointed out limited predictability, as confirmed by permutation tests (Figure S3). This was attributed to the high number of features—generating noise—while better predictive models were expected through refined selection of features. Such selection would also answer our needs in terms of classification model practical implementation. Consequently, the features of interest were determined using a strategy successfully applied in previous works based on OPLS-DA outcomes [42], here through assessment of variable importance for projection of the predictive component (VIPpred) [43], using Workflow4metabolomics 3.3 [44–46]. VIPpred was specifically chosen as it is purely associated with the consequences of the treatment, as opposed to the orthogonal component, associated with the experimental variability and time-related evolution of the individuals. In order to select only robust and discriminating features between the groups studied, the threshold applied to their selection (VIP pred >1.8) was deliberately chosen to be more stringent than the classically reported value of VIPpred >1.5. The consequence of such a choice was the reduction of the number of features thus selected (46 from the ESI+ datasets/94 from the ESI− vs. 374 from ESI+ and 203 from ESI−, respectively), but to the benefit of the quality of these potential biomarkers. All of these features exhibited higher signal intensity in the samples from treated animals. From these features, new PLS-DA models were built (Figure 1, right panel), showing a strong discrimination between groups, with, the following performance for the reduced ESI+ model: $R^2 = 0.544$; $Q^2 = 0.465$, CV-ANOVA p-value $= 5.3 \times 10^{-4}$; and reduced ESI− model: $R^2 = 0.620$; $Q^2 = 0.487$, CV-ANOVA p-value $= 3.3 \times 10^{-4}$. The quality of the reduced models was also confirmed by permutation tests (Figure S4).

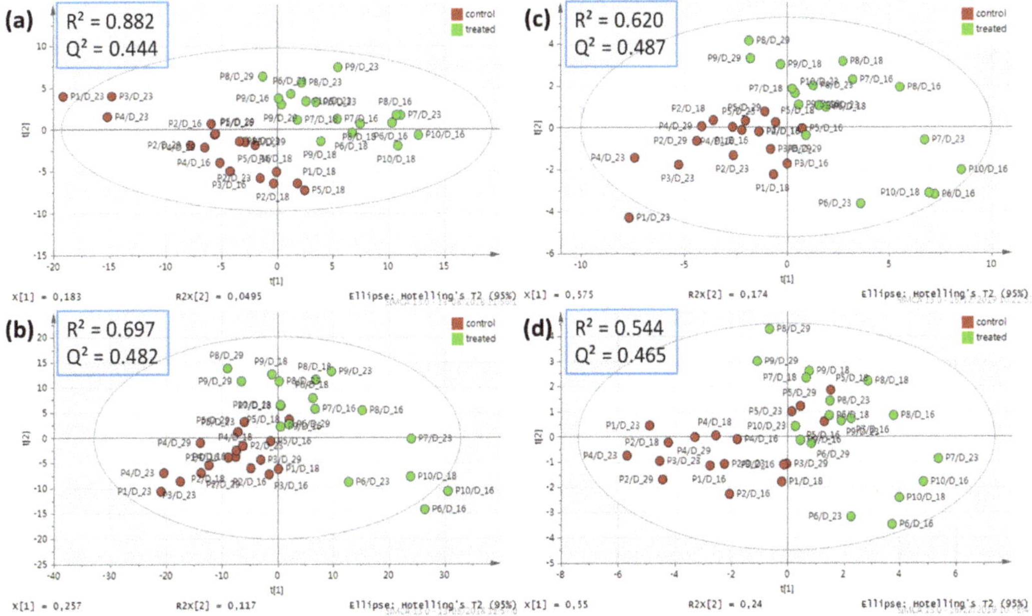

Figure 1. PLS-DA score plots after removing QC, D0, D3, D9 samples from the cleaned ESI− and ESI+ datasets acquired with RP UHPLC-HRMS. Datasets containing 1612 (ESI−) (**a**) and 2914 (ESI+) (**b**) features, $n = 36$. Reduced datasets containing 94 (ESI−) (**c**) and 46 (ESI+) (**d**) features, $n = 36$. Log 10 transformation, Pareto scaling and centering were applied.

The relevance of the selected features was confirmed by "day-by-day" Wilcoxon tests. Thanks to additional data from dependent MS/MS—Top 15 (Full MS/dd-MS^2-Top 15) experiments performed on QC samples and four typical samples (P4 (control) and P8 (treated) at D18 and D23), a few of them could be putatively using the LipidSearch tool. Annotations and statistical results are detailed in Table 2. Detailed results from LipidSearch for these features can be found in Supplementary Materials (Table S1). From the reduced ESI− dataset, 3 PC, 8 PE and 1 phosphatidylserine (PS) could be annotated whereas 1 PC, 2 PE and 9 TG were annotated from the reduced ESI+ dataset, including 1 PE, which was annotated in both ionisation modes (PE(17:0_20:4)). From these preliminary results, it can be noticed that the discrimination between samples from control and treated animals mainly relies on phospholipids and TG, which was consistent with recent literature [13]. PC appears to be mostly discriminant (p-value ≤ 0.05) at D16, D18 and D29, PE at D16 and D23 and TG at D23. The annotated phosphatidylserine (PS(18:2_21:0)) was found to be discriminant at all kinetic points between D16 and D29. However, three annotated TG (TG(16:0_17:0_18:1) and the two adducts of TG(18:0_17:0_18:1)) did not exhibit p-values ≤ 0.05 and thus could be regarded as modestly involved in the discrimination between groups.

Table 2. Putatively annotated features of interest extracted from the reduced the LC-HRMS datasets, with associated VIPpred values from the OPLS-DA used for variable selection and p-values from a Wilcoxon test. **: p-value < 0.01; *: p-value ≤ 0.05. †: VIPpred values from the OPLS-DA model based on the 1612 (ESI−) and 2914 features (ESI+) after removal of QC, D0, D3 and D9.

Variable ID	VIPpred †	Annotation (LipidSearch)	MS^2 Validation (LipidSearch)	p-Value D16	p-Value D18	p-Value D23	p-Value D29
			ESI−				
M791T491	1.81	[PC(18:1_14:0) + CH$_3$COO]$^-$	✓	0.117	* 0.027	0.117	* 0.034
M805T538	1.98	[PC(15:0_18:1) + CH$_3$COO]$^-$	✓	** 0.009	* 0.014	* 0.028	* 0.034
M833T633	1.84	[PC(17:0_18:1) + CH$_3$COO]$^-$	✓	* 0.028	* 0.014	0.076	* 0.034
M715T534	1.92	[PE(16:0_18:2)-H]$^-$	✓	0.117	0.221	** 0.009	0.480
M717T611	2.18	[PE(16:0_18:1)-H]$^-$	✓	* 0.028	0.806	** 0.009	0.480
M739T518	1.97	[PE(16:0_20:4)-H]$^-$	✓	* 0.047	0.086	* 0.016	0.480
M745T705	2.05	[PE(18:0_18:1)-H]$^-$	✓	0.076	0.142	* 0.047	0.289
M753T566	2.02	[PE(17:0_20:4)-H]$^-$	✓	* 0.028	* 0.014	0.076	0.480
M765T524	2.17	[PE(18:1_20:4)-H]$^-$	✓	0.076	0.142	* 0.016	0.157
M723T563	1.90	[PE(16:0p_20:4)-H]$^-$	✓	** 0.009	0.221	** 0.009	0.077
M751T659	1.84	[PE(16:0p_22:4)-H]$^-$	✓	** 0.009	0.327	* 0.028	0.157
M829T472	1.80	[PS(18:2_21:0)-H]$^-$	✓	* 0.016	* 0.050	* 0.028	* 0.034
			ESI+				
M777T719	1.95	[PC(16:0_19:0) + H]$^+$	X	* 0.047	* 0.027	0.175	0.289
M755T566	1.81	[PE(17:0_20:4) + H]$^+$	✓	** 0.009	* 0.014	* 0.047	0.077
M759T836	1.84	[PE(20:0p_18:1) + H]$^+$	✓	* 0.028	* 0.050	* 0.047	0.077
M865T1051	1.87	[TG(16:0_17:0_18:1) + NH$_4$]$^+$	✓	0.117	0.086	0.076	0.157
M879T1059	1.92	[TG(18:0_16:0_18:1) + NH4]$^+$	✓	0.076	0.142	* 0.028	0.157
M891T1051	1.84	[TG(17:0_18:1_18:1) + NH$_4$]$^+$	✓	0.117	0.086	* 0.047	0.157
M893T1066	1.91	[TG(18:0_17:0_18:1) + NH$_4$]$^+$	✓	0.076	0.086	0.076	0.289
M898T1065	1.84	[TG(18:0_17:0_18:1) + Na]$^+$	✓	0.117	0.086	0.076	0.157
M921T1080	2.35	[TG(18:0_18:1_19:0) + NH$_4$]$^+$	✓	0.117	0.086	* 0.047	0.157
M926T1080	1.99	[TG(18:0_18:1_19:0) + Na]$^+$	✓	* 0.047	0.086	0.076	0.157
M919T1066	1.88	[TG(19:1_18:0_18:1) + NH$_4$]$^+$	✓	* 0.047	0.142	* 0.016	0.077
M924T1066	1.84	[TG(19:0_18:1_18:1) + Na]$^+$	✓	0.076	0.142	* 0.047	0.157

3.2. Lipidyzer™ Platform

In order to provide additional insight into lipids involved in the sample group separation observed above, an alternative MS lipidomics approach was applied. Lipidyzer™ is a commercial lipid quantification tool based on shotgun lipidomics and benefiting from ion mobility, coming with its own workflow and dedicated framework. As it is based on targeted analysis and differs in the separation mode, complementary results from the non-targeted method presented above are expected. Lipidyzer™ was originally designed for human blood serum and plasma studies, but its applicability may be tested for other species. However, since it was used here for porcine serum samples, the associated results cannot be considered as absolute concentrations. Because of the lack of validation for pig samples, the Lipidyzer results detailed in this work were, therefore, considered as "estimated" concentration. Here, this experiment required a limited number of samples, hence samples collected at D3 (as a basis for comparison), D18 and D23 were characterised with Lipidyzer™, as a result of RPLC-HRMS outcomes described above. From the analysed samples, 795 lipid species were actually measured (i.e., above limit of quantification in at least one sample), namely: 26 Cholesterol Esters (CE), 10 Ceramides (CER), 7 Dihydroceramides (DCER), 11 Hexosyl ceramides (HCER), 10 Lactosyl ceramides (LCER), 54 Diacylglycerols (DAG), 26 Free Fatty Acids (FFA), 18 Lysophosphatidylcholine (LPC), 89 PC, 13 Lysophosphatidylethanolamines (LPE), 107 PE, 12 SM and 490 TG. Univariate statistical tests (Wilcoxon, day by day) were performed, showing significant shifts upon RAC treatment for 22 CE, 1 CER, 11 DAG, 1 DCER, 1 FFA, 3 HCER, 3 LCER, 1 LPE, 26 PC, 12 PE, 5 SM and 152 TG (see Table 3). Details about these species can be found in Table S2.

Table 3. Lipid class analysis results from Lipidyzer™, with associated *p*-values from a Wilcoxon test. **: *p*-value $\leq$ 0.01; *: *p*-value $\leq$ 0.05.

Lipid Class	*p*-Value D3	*p*-Value D18	*p*-Value D23
CE	0.55	*0.03	0.10
CER	0.22	0.11	0.31
DAG	0.42	0.20	** 0.01
DCER	0.42	1.00	0.22
FFA	1.00	0.20	0.42
HCER	0.15	* 0.03	0.69
LCER	0.69	* 0.03	** 0.01
LPC	0.06	0.34	0.84
LPE	0.15	0.11	0.55
PC	0.69	0.06	0.06
PE	0.84	* 0.03	** 0.01
SM	0.22	* 0.03	1.00
TG	0.55	0.20	0.06

When looking at the differences of concentration between samples from control and treated animals at D3 for all measured lipids, only 1 HCER (HCER(24:1)) and 1 PE (PE(O-18:0_18:1)) were shown as significant (*p*-value $\leq$ 0.05), while 1 PC (PC(16:0_18:0)) and 1 TG (TG42:1-FA14:0) were marginally significantly affected (*p*-value $\leq$ 0.06). This correlates non-targeted results, where no significant patterns could be observed so early in the experiment. Interestingly, CE, CER, DCER, HCER, LCER, LPE and SM species appeared as significant in the discrimination almost exclusively at D18, while DAG exhibited a significant shift in lipid profiles mainly at D23. The significant shift of species from other classes was distributed evenly between D18 and D23. All the lipids were observed to be more concentrated in the serum of treated animals, except for 1 HCER measured in lower concentration in the serum of treated animals at D3 (HCER(24:1)). Globally, when the number of significant lipid species in either D3, D18, or D23 samples was proportionated to the number of analysed species per class, the most altered classes were CE, (85% of analysed species deemed as significant), SM, (42%), TAG (31%), LCER (30%), PC (29%) and HCER (27%). Examples of boxplots illustrating differences in concentration levels between samples from control and treated groups for four particular species are presented in Figure 2.

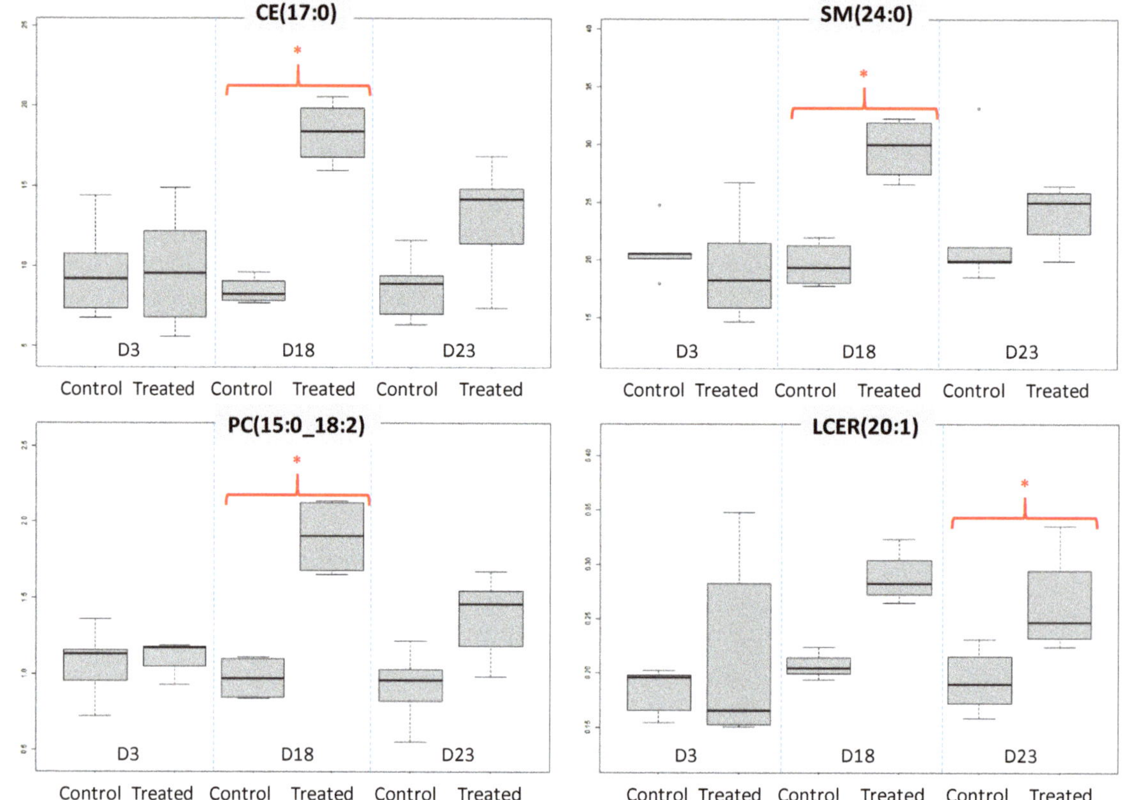

Figure 2. Comparison of estimated concentration ($nmol \cdot g^{-1}$) from four lipid species analysed with Lipidyzer™ between the two animal groups of interest, and for different serum collection points. Here, the quantification cannot be considered as accurate (hence "estimated") since it is has not been validated on pig serum, as opposed to human. *: p-value ≤ 0.05.

3.3. TG Platform

The characterisation of the different TG isomers is an issue that was not completely addressed by Lipidyzer™, which justified resorting to a dedicated TG platform, originally developed for the annotation and semi-quantification of TG isomers in vegetable oils [35].

Through modelling of the fragmentation patterns in TG containing common fatty acids, using multivariate constrained regression, this TG platform was able to determine their regioisomeric composition. This analytical method is semi-quantitative and aimed at highlighting TG patterns, together with their fatty acid composition. Relative proportions for each regioisomer (TG(rac-A/B/C); A, B and C corresponding to the constituting fatty acyl chains) can also be determined. The analysis was performed on a limited number of relevant samples: D3 as a reference and samples from D16 to D29, corresponding to time points for which most important TG shifts had been observed using both previous platforms.

From univariate statistical tests (Table 4), five TG (TG(52:5), two TG(54:6), TG(54:5) and TG(54:7)) were detected as significant (Wilcoxon test) in the context of the study for the discrimination between control and treated sample groups at D23, with higher concentrations upon RAC treatment. Two of them, namely TG(54:6) at the retention time (Rt) 555.9 s and TG(54:7), were also found as significant at D16, but with a limit p-value (0.05) and slightly lower concentrations in treated individuals. For detected TGs, the proportions of the corresponding regioisomers can also be estimated. For instance, the significant variable TG(54:7), detected at Rt 476.03 s was mainly constituted by TG(rac-18:2/18:2/18:3) (around 60%) but also TG(rac-18:2/18:3/18:2) (around 40%).

Table 4. Results from the TG platform, with associated *p*-values from a Wilcoxon test. *: *p*-value $\leq$ 0.05. For each TG signal, the corresponding regioisomers and associated estimated proportions are detailed. The main regioisomers are in bold.

TG_Rt	Corresponding Regioisomers with Estimated Proportions	*p*-Values				
		D3	D16	D18	D23	D29
TG(52:5)_553.44s	TG(rac-18:3/16:0/18:2)∼15% **TG(rac-16:0/18:2/18:3)∼50%** TG(rac-16:0/18:3/18:2)∼35%	0.44	0.77	0.64	* 0.03	0.06
TG(54:6)_555.9s	**TG(18:2/18:2/18:2)**	0.17	* 0.05	0.39	* 0.03	0.72
TG(54:6)_566.5s	**TG(rac-18:3/18:1/18:2)∼60%** TG(rac-18:1/18:2/18:3)∼10% TG(rac-18:1/18:3/18:2)∼30%	0.17	0.18	0.25	*0.03	1.00
TG(54:5)_685.8s	**TG(rac-18:3/18:0/18:2)∼60%** TG(rac-18:0/18:2/18:3)∼20% TG(rac-18:0/18:3/18:2)∼20%	1.00	0.65	0.15	* 0.05	0.51
TG(54:7)_476.03s	**TG(rac-18:2/18:2/18:3)∼60%** TG(rac-18:2/18:3/18:2)∼40%	0.65	* 0.05	0.64	* 0.03	1.00

4. Discussion

4.1. Assessment of the Complementarity between Platforms

Three platforms differently addressing the lipidome were involved in the characterisation of a set of serum samples in which specific lipid patterns are expected to be observed. The results have been carefully compared for assessing their respective contributions and complementarity in lipidomics in general and for the proposed application. As a preliminary step, reproducibility was compared between the platforms, which were assessed by CV(QC)% on common lipid targets (n = 30), resulting in median values below 8%, which were considered to fit our requirements.

Whatever the platform used, the disruption of various lipid classes could be highlighted in pig serum after several weeks of RAC treatment, as illustrated in Table 5. The same trends could be observed with the three tools, as higher lipid levels were observed in the serum of treated individuals, e.g., for TG (non-targeted, Lipidyzer™ and TG platforms) but also for PC and PE (non-targeted and Lipidyzer™ platforms). A graphical illustration of these shared trends can be found in Figure S5.

To check the consistency between these results, the annotated lipids highlighted by the reduced models in the non-targeted analysis were searched in Lipidyzer™ outcomes. Most of them could easily be retrieved and were also found to be significant (p-value <0.05) with the same variations towards RAC treatment, highlighting good consistency, in particular for PC(15:0_18:1), PC(17:0_18:1), PC(18:1_14:0) PE(16:0_20:4); PE(16:0_18:2) and PE(16:0p_20:4). The collection dates when these lipids were found to be significant were generally in accordance, although minor differences were observed. For instance, PC(18:1_14:0) was only highlighted at D18 with the non-targeted analysis, whereas it was also found to be marginally significant at D23 (p-value = 0.056), using Lipidyzer™ (as "PC(14:0_18:1)"). Still, some lipids that were highlighted with the non-targeted approach were not observed as significant with Lipidyzer™, usually due to a corresponding signal below the limit of quantification with the latter, as observed for PE(17:0_20:4). In other cases, the reason for this difference was less clear; e.g., PE(16:0_18:1) and PE(16:0p_22:4), which were retained from ESI− non-targeted results were not found as significant with Lipidyzer™. This could be explained by different measurement biases or by erroneous annotation, even if no obvious inconsistency was observed. Conversely, significant Lipidyzer™ features were curated in the non-targeted datasets. Even though some lipid classes from which the lipid species were deemed as significant by Lipidyzer™ (p-value ≤ 0.05) were annotated in the non-targeted dataset, some did not belong to the set of features selected for the reduced model. Indeed, CE and DAG were detected and annotated in the ESI+ dataset, LPE and FFA were observed in the ESI− datasets, whereas CER, DCER and SM were characterised in both.

Table 5. Comparison of the results from various MS platform. The analysed lipid classes are mentioned with the level of significance, determined from a univariate Wilcoxon test.

	Non-Targeted RP LC-HRMS		Lipidyzer™		TG Platform	
Class of the Relevant Lipids	Analysed and Annotated?	Variation (If Significant)	Analysed and Annotated?	Variation (If Significant)	Analysed and Annotated?	Variation (If Significant)
CE	Yes [†]		Yes	↗ **D18** * ↗ D23 *	No	-
CER	Yes [†]		Yes	↗ D18 *	No	-
DAG	Yes [†]		Yes	↗ D18 * ↗ **D23** *	No	-
DCER	Yes [†]		Yes	↗ D18 *	No	-
FFA	Yes [†]		Yes	↗ D18 * ↗ D23 *	No	-
HCER	Yes [†]		Yes	↘ D3 * ↗ **D18** *	No	-
LCER	No		Yes	↗ **D18** * ↗ D23 *	No	-
LPC	Yes [†]		Yes	-	No	-
LPE	Yes [†]		Yes	↗ D18 *	No	-
PC	Yes	↗ D16 *, ↗ D18 *, ↗ D23 *, ↗ D29 *	Yes	↗ D18 * ↗ D23 *	No	-
PE	Yes	↗ D16 *, ↗ D18 *, ↗ D23 *	Yes	↗ D3 * ↗ **D18** * ↗ **D23** *	No	-
PS	Yes	↗ D16 *, ↗ D18 *, ↗ D23 *, ↗ D29 *	No	-	No	-
SM	Yes [†]		Yes	↗ D18 *	No	-
TG	Yes	↗ D16 *, ↗ D23 *	Yes	↗ **D18** * ↗ **D23** *	Yes	↘ D16 *, ↗ D23 *

Level of significance after Wilcoxon test is indicated with asterisks: *: p-value ≤ 0.05. [†]: Lipid class analysed and annotated by non-targeted RP UPLC-HRMS but not observed in the set of selected features from OPLS-DA (VIPpred > 1.8). ↘: More concentrated in control samples. ↗: More concentrated in samples from treated animals. In bold: Days where main disruptions are observed.

An important matter to consider when comparing the results between platforms is their relative capability for lipid annotation, which as a consequence, directly influences the biological interpretation.

With the non-targeted strategy proposed, the annotation is only putative (level 2 or 3 of identification), and a small portion (<20% for both datasets) of the original features could be annotated, thus demonstrating the challenge of this step. Using targeted approaches, such an issue is less likely to happen as their workflows were optimised to target specific lipids of interest. Implementing Lipidyzer™ and the TG platforms thus enabled confident lipid assignment.

While comparing the platform's outcomes and lipid annotation, a particular case is the one of TG, where the assignment of the fatty acyl chains (*sn*-1(3) versus *sn*-2) is recognised as a serious analytical challenge, leading to multiple dedicated research studies [47–49].

- From the non-targeted method, TGs were annotated from their three FA chains (e.g., "TG(16:0_17:0_18:1)"), based on the annotation results from LipidSearch after data-dependent MS/MS. Although allowing confident assignment, the results of such an approach may in some particular cases be considered with caution as illustrated hereafter. Among the selected features, for instance, some lipids (M926T1080 and M921T1080; highlighted in light grey as well as M898T1065 and M893T1066 highlighted in dark grey in Table 2) were annotated as adducts of the same TG. These features were initially not discarded during the data processing step because of an inconsistency between the adduct annotation between the CAMERA package and LipidSearch. In addition, two other features (M919T1066 and M924T1066; highlighted in blue in Table 2) were annotated as two different TG when they could potentially be two adducts of the same lipid as they are isomers of TG(55:2).

- In Lipidyzer™, TG results were expressed with the shorthand annotation nomenclature (total number of carbons and unsaturations among the three FA chains and the precision on one of them), such as "TG51:1-FA16:0". While technically correct, this leads to an overestimation of the TG, as previously highlighted in the literature [31]. Moreover, several Lipidyzer™ candidates (e.g., TG51:1-FA18:1 and TG51:1-FA16:0) can correspond to a single TG feature in RP LC-HRMS (e.g., TG(16:0_17:0_18:1)), and vice-versa, thus complicating result comparison.

- Because of previous issues in TG assignment, a dedicated platform for the determination of TG regioisomeric composition was used [35]. It is interesting to note that the TGs highlighted with the dedicated tool were not those annotated in non-targeted data. Moreover, after conversion to the corresponding shorthand annotation to allow such a comparison, none of them was deemed as significant with Lipidyzer™, which could be due to the overestimation of TG with the latter. Conversely, none of the discriminant TG highlighted within the RPLC-HRMS results were monitored with the TG platform since it is designed for the analysis of even FA chains TG only. It is interesting to note that this specific platform allowed obtaining confident results on TG and the position (*sn*-1(3) versus *sn*-2) of their constituting FA chains. Thus, it yielded finer results than the combined use of non-targeted and Lipidyzer platforms—an approach that was already explored by Contrepois et al. [31].

Between all evaluated platforms, Lipidyzer™ offered the most detailed analysis for a large number of lipids, providing a large amount of biologically interpretable data. Yet, interpretation issues were observed when considering the TG because of the overestimated occurrence of this class, whereas the TG platform could bring information on the regioisomers of interest without doubt. However, the latter was designed for this class only, and the number of followed species and regioisomers is limited.

Nonetheless, targeted platforms focus on a limited number of lipids, originally selected for a particular application, i.e., the human serum/plasma studies for Lipidyzer™ and vegetable oils for the TG platform. Hence, the relevance of the monitored compounds is not guaranteed when applied to a different research question, and species of interest are also likely to be overlooked, as opposed to the non-targeted strategy. For instance, applying the latter enabled highlighting PS(18:2_21:0) in ESI− as well as PC(15:0_16:0) and PC(16:0_19:0) in ESI+ as relevant upon RAC treatment.

Regarding practical considerations, Lipidyzer can be performed in an easy manner, thanks to the entirely software-guided workflow, from instrument calibration to processing. Comparatively, the non-targeted platform requires more expertise, in particular for data processing, even though tools are available to make this step more accessible, such as Workflow4Metabolomics (W4M) [44,45]. Since it is still recent, the TG platform still requires a high level of expertise for using the dedicated in-house R algorithm.

Among the three platforms, Lipidyzer™ can be considered as the quickest since the analysis (two 15-min injections, comparable with the 30-min of the non-targeted method and 18-min of TG platform) is compensated by the assisted data processing, allowing a direct interpretation of the results. Nevertheless, since it entails the purchase of dedicated instrument/software/kits, Lipidyzer™ implies a substantial financial investment, whereas the other two can be adapted to various instrument types, although buying pure standards is still required for the confirmation of lipid assignment or the calibration of TG regioisomers.

The investigation of the serum lipidome disruptions upon RAC administration to pigs showed the added value of the three tools. Rather than heaving up one particular platform above the others, these results clearly demonstrate how comprehensive lipidome characterisation is a challenging task, requiring several tools for both enhanced lipidic coverage and increased confidence in the observations.

4.2. Biological Interpretation

The results enabled further investigating ractopamine effects on pig blood lipids profile. Although a full biological interpretation of the metabolic pathways was out of the scope of this work, our observations are discussed below in light of the current knowledge regarding the impact of RAC on metabolism.

RAC is a synthetic drug belonging to the β-agonist family, widely used as a growth promoter in several countries, as it has been shown to improve growth performance such as average daily gain [50] in pigs. However, as such, it is banned in the European Union [1,51], and robust screening methods are required to detect any potential abuse. In such context, metabolomics has been successfully applied for screening β-agonists treatment in bovine, thus highlighting the signature of administration, enabling the construction of new robust models based on these biomarkers [42]. That is why RAC effects have been similarly studied in porcine, using non-targeted tissue screening [13] and serum metabolomics [32]. As the lipids are known to be disrupted by the use of this compound [52–54], the lipidome appears as a promising compartment for inspecting the effects of RAC, prompting their study by NMR lipidomics [39] and the currently presented work. The mechanism of action of RAC as a growth promoter is relatively well-known; it stimulates β_2-adrenergic receptors, linked with the relaxation of smooth muscles. They enhance the synthesis and decrease the degradation of proteins [55]. A reduction of adipose tissues as an effect of RAC treatment is commonly reported through two pathways: reduction of lipogenesis and/or increase in lipolysis, as reported by Ferreira et al. [54]. This review concluded on a predominance of the former, as a treatment generally does not induce an increase of serum non-esterified fatty acids (NEFA), which is characteristic of the latter.

As observed above, the blood serum levels of various lipid classes appear to be affected by the RAC treatment, starting on the third week of the experiment. The disrupted phospholipid profiles observed in the present study are in accordance with the NMR study [39] and previous observations on muscle where modified diacylglycerophosphoethanolamine and phosphatidylinositol profiles have been associated with RAC administration to pigs [13]. Among the highlighted classes, the disruption of SM is in accordance with reported observations in tissue, associating changes in sphingomyelin profiles with RAC administration to pigs [13]. For all involved lipid classes, a delay in the action of RAC can be observed. Further, a limited effect is observed at D29, although the animals were still exposed to the drug. Such observation could be hypothesised to be linked with a de-sensitisation regarding the RAC treatment, which occurs from 21 to 28 days, according to Ferreira et al. [54]. Interestingly, some odd-numbered fatty acids such as C17-cholesteryl esters and C-15 containing phosphatidyl choline were highlighted as modified upon ractopamine treatment, which is quite unexpected as almost all natural occurring fatty acids are even-numbered, although some odd-numbered fatty acids also exist. The metabolism of odd-numbered fatty acids is, however, specific in that they are reported not to be favorable substrates for beta-oxidation-related enzymes, thus leading to accumulation in the tissues [56].

These observations could form the basis for a better understanding of the mechanism underlying β-agonist treatment on lipid metabolism. Here, no particular effect of RAC could be observed on the free FA profiles (covered by Lipidyzer™). Hence, even if a deeper biological interpretation is necessary before drawing definitive conclusions, this seems in accordance with Ferreira's review [54], suggesting inhibition of lipogenesis as a preferential mechanism of the effect of RAC, rather than an increase in lipolysis, which would have conducted to higher free FAs levels in the blood.

5. Conclusions

This work describes the combination of three different fingerprinting approaches in order to join their forces for one single study dedicated to food safety. Serum samples from an animal experiment involving a repartitioning agent of interest were characterised. This combination allowed a fine characterisation of the lipid profiles, showing particular lipid classes and species disruptions in pig blood serum following RAC treatment. Specific benefits could be highlighted from the three described platforms in terms of lipidome coverage, level of characterisation or applicability. Although these platforms enabled reaching complementary information, further work should be conducted to validate the proposed workflows.

For optimising lipidome characterisation, the next refinements of the strategy will be directed towards the improvement of lipid annotation from non-targeted RP UPLC-HRMS. Many tools have been reported in the recent literature such as LOBSTAHS [57] or LipidMatch [58], and their evaluation/implementation would ensure higher confidence in results and facilitated link with other platforms. The selection of relevant features could also be improved, through the use of sparse methods [59,60] or the recent *biosigner* algorithm [61], precisely aiming at building reduced models. Moreover, the TG platform could be extended in order to include more lipid species, thus requiring further developments in order to increase its suitability to a wider range of lipidomics applications. Improvements could also be made for the development of a more user-friendly data processing interface, which would make this platform accessible to less-experienced analysts and accelerate the time dedicated for such data handling.

Regarding the study of the effects of RAC on pig's lipidome, further work is still necessary to fully understand the biological implications underlined by the presented results. Additional animal experiments could also be performed involving, for instance, different dosages or individuals with different characteristics, for confirming these outcomes and validate candidate biomarkers. From a public health perspective, it is expected that the outcomes of the present study may serve risk analysis, either at the risk assessment level while proposing new insight on the mode of action and associated effects or at the risk management steps, as the basis for an alternative screening method based on lipid biomarkers.

Supplementary Materials: The following are available online at https://www.mdpi.com/article/10.3390/foods10061218/s1, Figure S1: PCA Score plot (with QC) from the non-targeted RP LC-HRMS datasets, Figure S2: PCA Score plot (Without QC, Day-3 and Day-9 samples) from the non-targeted RP LC-HRMS datasets, Figure S3: Permutation tests ($n = 100$ permutations) associated with the PLS-DA models (Without QC, Day-3 and Day-9 samples) from the non-targeted RP LC-HRMS datasets, Figure S4: Permutation tests ($n = 100$ permutations) associated with the reduced PLS-DA models (Without QC, Day-3 and Day-9 samples) from the non-targeted RP LC-HRMS datasets, Figure S5: Intensity trend comparison of two lipid species analysed with RPLC-HRMS (ESI−) and Lipidyzer in control and treated samples from D23, Figure S6: m/z measurement error (ppm) on the internal standards signals from QC samples injections, Table S1: Results from the LipidSearch annotation after MS^2 analysis for the features of interest, Table S2: Statistical results from the Lipidyzer platform.

Author Contributions: Conceptualization, G.D., B.L.B., and P.G.; investigation, J.M.; methodology, Y.G., A.-L.R., and D.B.; formal analysis, J.M., Y.G., and D.B.; validation, J.M., E.M., Y.G., G.D., and P.G.; writing—original draft preparation, J.M.; writing—review and editing, P.G. and G.D.; supervision, G.D., B.L.B., and P.G. All authors have read and agreed to the published version of the manuscript.

Funding: This work was supported by Région Pays de la Loire, FRANCE, through the "Recherche-Formation-Innovation: Food 4.2" program [grant LipidoTool].

Institutional Review Board Statement: The study was conducted according to the guidelines of the Declaration of Helsinki, and approved by the Ethics Committee of Oniris (protocol code 2.015.092.516.084.715/APAFIS 1914 (CRIP-2015-054) 2015).

Informed Consent Statement: Not applicable.

Data Availability Statement: Not applicable.

Acknowledgments: The authors would like to thank the Corsaire metabolomics core facility (Biogenouest). Sciex is also acknowledged for providing access to the Lipidyzer™ platform.

Conflicts of Interest: The authors declare no conflict of interest.

References

1. Council Directive 88/146/EEC. Council Directive 88/146/EEC Prohibiting the Use Livestock Farming of Certain Substances Having a Hormonal Action. Available online: http://data.europa.eu/eli/dir/1988/146/oj (accessed on 25 May 2021).
2. European Parliament and Council. Regulation (EU) 2017/625 on Official Controls and Other Official Activities Performed to Ensure the Application of Food and Feed Law, Rules on Animal Health and Welfare, Plant Health and Plant Protection Products, Amending Regulations (EC) No 999/2001, (EC) No 396/2005, (EC) No 1069/2009, (EC) No 1107/2009, (EU) No 1151/2012, (EU) No 652/2014, (EU) 2016/429 and (EU) 2016/2031 of the European Parliament and of the Council, Council Regulations (EC) No 1/2005 and (EC) No 1099/2009 and Council Directives 98/58/EC, 1999/74/EC, 2007/43/EC, 2008/119/EC and 2008/120/EC, and repealing Regulations (EC) No 854/2004 and (EC) No 882/2004 of the European Parliament and of the Council, Council Directives 89/608/EEC, 89/662/EEC, 90/425/EEC, 91/496/EEC, 96/23/EC, 96/93/EC and 97/78/EC and Council Decision 92/438/EEC (Official Controls Regulation). OJ L 95, 7.4. 2017, pp. 1–142. Available online: http://data.europa.eu/eli/reg/2017/625/oj (accessed on 25 May 2021).
3. Pinel, G.; Weigel, S.; Antignac, J.-P.; Mooney, M.; Elliott, C.; Nielen, M.; Le Bizec, B. Targeted and untargeted profiling of biological fluids to screen for anabolic practices in cattle. *TrAC Trends Anal. Chem.* **2010**, *29*, 1269–1280. [CrossRef]
4. Gallart-Ayala, H.; Chéreau, S.; Dervilly-Pinel, G.; Le Bizec, B. Potential of mass spectrometry metabolomics for chemical food safety. *Bioanalysis* **2015**, *7*, 133–146. [CrossRef] [PubMed]
5. Courant, F.; Pinel, G.; Bichon, E.; Monteau, F.; Antignac, J.-P.; Le Bizec, B. Development of a metabolomic approach based on liquid chromatography-high resolution mass spectrometry to screen for clenbuterol abuse in calves. *Analyst* **2009**, *134*, 1637–1646. [CrossRef] [PubMed]
6. Stella, R.; Dervilly-Pinel, G.; Bovo, D.; Mastrorilli, E.; Royer, A.-L.; Angeletti, R.; Le Bizec, B.; Biancotto, G. Metabolomics analysis of liver reveals profile disruption in bovines upon steroid treatment. *Metabolomics* **2017**, *13*, 80. [CrossRef]
7. Lu, H.; Zhang, H.; Zhu, T.; Xiao, Y.; Xie, S.; Gu, H.; Cui, M.; Luo, L. Metabolic Effects of Clenbuterol and Salbutamol on Pork Meat Studied Using Internal Extractive Electrospray Ionization Mass Spectrometry. *Sci. Rep.* **2017**, *7*, 5136. [CrossRef] [PubMed]
8. Dervilly-Pinel, G.; Courant, F.; Chéreau, S.; Royer, A.-L.; Boyard-Kieken, F.; Antignac, J.-P.; Monteau, F.; Le Bizec, B. Metabolomics in food analysis: Application to the control of forbidden substances. *Drug Test. Anal.* **2012**, *4*, 59–69. [CrossRef]
9. Dervilly-Pinel, G.; Weigel, S.; Lommen, A.; Chereau, S.; Rambaud, L.; Essers, M.; Antignac, J.-P.; Nielen, M.W.; Le Bizec, B. Assessment of two complementary liquid chromatography coupled to high resolution mass spectrometry metabolomics strategies for the screening of anabolic steroid treatment in calves. *Anal. Chim. Acta* **2011**, *700*, 144–154. [CrossRef]
10. Jacob, C.C.; Dervilly-Pinel, G.; Biancotto, G.; Monteau, F.; Le Bizec, B. Global urine fingerprinting by LC-ESI(+)-HRMS for better characterization of metabolic pathway disruption upon anabolic practices in bovine. *Metabolomics* **2014**, *11*, 184–197. [CrossRef]
11. Nzoughet, J.J.K.; Dervilly-Pinel, G.; Chéreau, S.; Biancotto, G.; Monteau, F.; Elliott, C.T.; Le Bizec, B. First insights into serum metabolomics of trenbolone/estradiol implanted bovines; screening model to predict hormone-treated and control animals' status. *Metabolomics* **2015**, *11*, 1184–1196. [CrossRef]
12. Nzoughet, J.K.; Gallart-Ayala, H.; Biancotto, G.; Hennig, K.; Dervilly-Pinel, G.; Le Bizec, B. Hydrophilic interaction (HILIC) and reverse phase liquid chromatography (RPLC)–high resolution MS for characterizing lipids profile disruption in serum of anabolic implanted bovines. *Metabolomics* **2015**, *11*, 1884–1895. [CrossRef]
13. Guitton, Y.; Dervilly-Pinel, G.; Jandova, R.; Stead, S.; Takats, Z.; Le Bizec, B. Rapid evaporative ionisation mass spectrometry and chemometrics for high-throughput screening of growth promoters in meat producing animals. *Food Addit. Contam. Part A* **2018**, *35*, 900–910. [CrossRef] [PubMed]
14. Wenk, M.R. Lipidomics: New Tools and Applications. *Cell* **2010**, *143*, 888–895. [CrossRef]
15. Ryan, E.; Reid, G.E. Chemical Derivatization and Ultrahigh Resolution and Accurate Mass Spectrometry Strategies for "Shotgun" Lipidome Analysis. *Accounts Chem. Res.* **2016**, *49*, 1596–1604. [CrossRef] [PubMed]
16. Zhao, Y.-Y.; Wu, S.-P.; Liu, S.; Zhang, Y.; Lin, R.-C. Ultra-performance liquid chromatography–mass spectrometry as a sensitive and powerful technology in lipidomic applications. *Chem. Interact.* **2014**, *220*, 181–192. [CrossRef] [PubMed]
17. Li, M.; Yang, L.; Bai, Y.; Liu, H. Analytical Methods in Lipidomics and Their Applications. *Anal. Chem.* **2014**, *86*, 161–175. [CrossRef] [PubMed]
18. Bligh, E.G.; Dyer, W.J. A rapid method of total lipid extraction and purification. *Can. J. Biochem. Physiol.* **1959**, *37*, 911–917. [CrossRef]

19. Folch, J.; Ascoli, I.; Lees, M.; Meath, J.; LeBaron, F. Preparation of lipide extracts from brain tissue. *J. Biol. Chem.* **1951**, *191*, 833–841. [CrossRef]

20. Tumanov, S.; Kamphorst, J.J. Recent advances in expanding the coverage of the lipidome. *Curr. Opin. Biotechnol.* **2017**, *43*, 127–133. [CrossRef]

21. Yang, K.; Han, X. Lipidomics: Techniques, Applications, and Outcomes Related to Biomedical Sciences. *Trends Biochem. Sci.* **2016**, *41*, 954–969. [CrossRef]

22. Hu, C.; van der Heijden, R.; Wang, M.; van der Greef, J.; Hankemeier, T.; Xu, G. Analytical strategies in lipidomics and applications in disease biomarker discovery. *J. Chromatogr. B* **2009**, *877*, 2836–2846. [CrossRef]

23. Triebl, A.; Hartler, J.; Trötzmüller, M.; Köfeler, H.C. Lipidomics: Prospects from a technological perspective. *Biochim. Biophys. Acta (BBA) Mol. Cell Biol. Lipids* **2017**, *1862*, 740–746. [CrossRef] [PubMed]

24. Han, X.; Gross, R.W. Global analyses of cellular lipidomes directly from crude extracts of biological samples by ESI mass spectrometry: A bridge to lipidomics. *J. Lipid Res.* **2003**, *44*, 1071–1079. [CrossRef] [PubMed]

25. Schwudke, D.; Oegema, J.; Burton, L.; Entchev, E.; Hannich, J.T.; Ejsing, C.S.; Kurzchalia, T.; Shevchenko, A. Lipid Profiling by Multiple Precursor and Neutral Loss Scanning Driven by the Data-Dependent Acquisition. *Anal. Chem.* **2006**, *78*, 585–595. [CrossRef]

26. Schwudke, D.; Schuhmann, K.; Herzog, R.; Bornstein, S.R.; Shevchenko, A. Shotgun Lipidomics on High Resolution Mass Spectrometers. *Cold Spring Harb. Perspect. Biol.* **2011**, *3*, a004614. [CrossRef]

27. Almeida, R.; Pauling, J.K.; Sokol, E.; Hannibal-Bach, H.K.; Ejsing, C.S. Comprehensive Lipidome Analysis by Shotgun Lipidomics on a Hybrid Quadrupole-Orbitrap-Linear Ion Trap Mass Spectrometer. *J. Am. Soc. Mass Spectrom.* **2014**, *26*, 133–148. [CrossRef] [PubMed]

28. Hinz, C.; Liggi, S.; Griffin, J.L. The potential of Ion Mobility Mass Spectrometry for high-throughput and high-resolution lipidomics. *Curr. Opin. Chem. Biol.* **2018**, *42*, 42–50. [CrossRef]

29. Lee, H.-C.; Yokomizo, T. Applications of mass spectrometry-based targeted and non-targeted lipidomics. *Biochem. Biophys. Res. Commun.* **2018**, *504*, 576–581. [CrossRef]

30. Cajka, T.; Fiehn, O. Toward Merging Untargeted and Targeted Methods in Mass Spectrometry-Based Metabolomics and Lipidomics. *Anal. Chem.* **2016**, *88*, 524–545. [CrossRef]

31. Contrepois, K.; Mahmoudi, S.; Ubhi, B.K.; Papsdorf, K.; Hornburg, D.; Brunet, A.; Snyder, M. Cross-Platform Comparison of Untargeted and Targeted Lipidomics Approaches on Aging Mouse Plasma. *Sci. Rep.* **2018**, *8*, 17747. [CrossRef]

32. Peng, T.; Royer, A.-L.; Guitton, Y.; Le Bizec, B.; Dervilly-Pinel, G. Serum-based metabolomics characterization of pigs treated with ractopamine. *Metabolomics* **2017**, *13*, 77. [CrossRef]

33. Lintonen, T.P.I.; Baker, P.R.S.; Suoniemi, M.; Ubhi, B.K.; Koistinen, K.M.; Duchoslav, E.; Campbell, J.L.; Ekroos, K. Differential Mobility Spectrometry-Driven Shotgun Lipidomics. *Anal. Chem.* **2014**, *86*, 9662–9669. [CrossRef]

34. Ubhi, B.K.; Conner, A.; Duchoslav, E.; Evans, A.; Robinson, R.; Wang, L.; Baker, P.R.; Watkins, S. *A Novel Lipid Screening Platform that Provides a Complete Solution for Lipidomics Research*; Sciex: Vaughan, ON, Canada, 2016; pp. 1–4.

35. Balgoma, D.; Guitton, Y.; Evans, J.J.; Le Bizec, B.; Dervilly-Pinel, G.; Meynier, A.; David, B.; Yann, G.; Jason, J.E.; Bruno, L.B.; et al. Modeling the fragmentation patterns of triacylglycerides in mass spectrometry allows the quantification of the regioisomers with a minimal number of standards. *Anal. Chim. Acta* **2019**, *1057*, 60–69. [CrossRef]

36. Kessner, D.; Chambers, M.; Burke, R.; Agus, D.; Mallick, P. ProteoWizard: Open source software for rapid proteomics tools development. *Bioinformatics* **2008**, *24*, 2534–2536. [CrossRef] [PubMed]

37. Smith, C.A.; Want, E.J.; O'Maille, G.; Abagyan, R.; Siuzdak, G. XCMS: Processing Mass Spectrometry Data for Metabolite Profiling Using Nonlinear Peak Alignment, Matching, and Identification. *Anal. Chem.* **2006**, *78*, 779–787. [CrossRef] [PubMed]

38. Van Der Kloet, F.M.; Bobeldijk, I.; Verheij, E.R.; Jellema, R. Analytical Error Reduction Using Single Point Calibration for Accurate and Precise Metabolomic Phenotyping. *J. Proteome Res.* **2009**, *8*, 5132–5141. [CrossRef] [PubMed]

39. Marchand, J.; Martineau, E.; Guitton, Y.; Le Bizec, B.; Dervilly-Pinel, G.; Giraudeau, P. A multidimensional 1H NMR lipidomics workflow to address chemical food safety issues. *Metabolomics* **2018**, *14*, 60. [CrossRef] [PubMed]

40. Navas-Iglesias, N.; Carrasco-Pancorbo, A.; Cuadros-Rodríguez, L. From lipids analysis towards lipidomics, a new challenge for the analytical chemistry of the 21st century. Part II: Analytical lipidomics. *TrAC Trends Anal. Chem.* **2009**, *28*, 393–403. [CrossRef]

41. Eriksson, L.; Trygg, J.; Wold, S. CV-ANOVA for significance testing of PLS and OPLS® models. *J. Chemom.* **2008**, *22*, 594–600. [CrossRef]

42. Dervilly-Pinel, G.; Chereau, S.; Cesbron, N.; Monteau, F.; Le Bizec, B. LC-HRMS based metabolomics screening model to detect various β-agonists treatments in bovines. *Metabolomics* **2015**, *11*, 403–411. [CrossRef]

43. Galindo-Prieto, B.; Eriksson, L.; Trygg, J. Variable influence on projection (VIP) for orthogonal projections to latent structures (OPLS). *J. Chemom.* **2014**, *28*, 623–632. [CrossRef]

44. Giacomoni, F.; Le Corguillé, G.; Monsoor, M.; Landi, M.; Pericard, P.; Pétéra, M.; Duperier, C.; Tremblay-Franco, M.; Martin, J.-F.; Jacob, D.; et al. Workflow4Metabolomics: A collaborative research infrastructure for computational metabolomics. *Bioinformatics* **2015**, *31*, 1493–1495. [CrossRef]

45. Guitton, Y.; Tremblay-Franco, M.; Le Corguillé, G.; Martin, J.-F.; Pétéra, M.; Roger-Mele, P.; Delabrière, A.; Goulitquer, S.; Monsoor, M.; Duperier, C.; et al. Create, run, share, publish, and reference your LC–MS, FIA–MS, GC–MS, and NMR data analysis workflows with the Workflow4Metabolomics 3.0 Galaxy online infrastructure for metabolomics. *Int. J. Biochem. Cell Biol.* **2017**, *93*, 89–101. [CrossRef] [PubMed]

46. Thévenot, E.A.; Roux, A.; Xu, Y.; Ezan, E.; Junot, C. Analysis of the Human Adult Urinary Metabolome Variations with Age, Body Mass Index, and Gender by Implementing a Comprehensive Workflow for Univariate and OPLS Statistical Analyses. *J. Proteome Res.* **2015**, *14*, 3322–3335. [CrossRef] [PubMed]

47. Baba, T.; Campbell, J.L.; Le Blanc, J.C.Y.; Baker, P.R. Structural identification of triacylglycerol isomers using electron impact excitation of ions from organics (EIEIO). *J. Lipid Res.* **2016**, *57*, 2015–2027. [CrossRef]

48. Nagy, K.; Sandoz, L.; Destaillats, F.; Schafer, O. Mapping the regioisomeric distribution of fatty acids in triacylglycerols by hybrid mass spectrometry. *J. Lipid Res.* **2013**, *54*, 290–305. [CrossRef]

49. Bird, S.S.; Marur, V.R.; Sniatynski, M.J.; Greenberg, H.K.; Kristal, B.S. Serum Lipidomics Profiling Using LC–MS and High-Energy Collisional Dissociation Fragmentation: Focus on Triglyceride Detection and Characterization. *Anal. Chem.* **2011**, *83*, 6648–6657. [CrossRef] [PubMed]

50. E Watkins, L.; Jones, D.J.; Mowrey, D.H.; Anderson, D.B.; Veenhuizen, E.L. The effect of various levels of ractopamine hydrochloride on the performance and carcass characteristics of finishing swine. *J. Anim. Sci.* **1990**, *68*, 3588–3595. [CrossRef]

51. Council Directive 96/22/EC. Council Directive 96/22/EC of 29 April 1996 Concerning the Prohibition on the Use in Stockfarming of Certain Substances Having a Hormonal or Thyrostatic Action and of Beta-Agonists, and Repealing Directives 81/602/EEC, 88/146/EEC and 88/299/EEC. 1996. Available online: http://data.europa.eu/eli/dir/1996/22/oj (accessed on 25 May 2021).

52. Dunshea, F.R. Effect of metabolism modifiers on lipid metabolism in the pig. *J. Anim. Sci.* **1993**, *71*, 1966–1977. [CrossRef] [PubMed]

53. Dunshea, F.R.; Leur, B.J.; Tilbrook, A.J.; King, R.H. Ractopamine increases glucose turnover without affecting lipogenesis in the pig. *Aust. J. Agric. Res.* **1998**, *49*, 1147. [CrossRef]

54. Ferreira, M.S.D.S.; Garbossa, C.A.P.; Oberlender, G.; Pereira, L.J.; Zangeronimo, M.G.; De Sousa, R.V.; Cantarelli, V.D.S. Effect of ractopamine on lipid metabolism in vivo—A systematic review. *Braz. Arch. Biol. Technol.* **2013**, *56*, 35–43. [CrossRef]

55. Paris, A.; André, F.; Antignac, J.P.; Bonneau, M.; Briant, C.; Caraty, A.; Chilliard, Y.; Cognié, Y.; Combarnous, Y.; Cravedi, J.P.; et al. L'utilisation des Hormones en Elevage: Les Développements Zootechniques et les Préoccupations de Santé Publique. 2008. Available online: https://hal.archives-ouvertes.fr/hal-01173447 (accessed on 25 May 2021).

56. Gotoh, N.; Moroda, K.; Watanabe, H.; Yoshinaga, K.; Tanaka, M.; Mizobe, H.; Ichioka, K.; Tokairin, S.; Wada, S. Metabolism of odd-numbered fatty acids and even-numbered fatty acids in mouse. *J. Oleo Sci.* **2008**, *57*, 293–299. [CrossRef]

57. Collins, J.R.; Edwards, B.R.; Fredricks, H.F.; Van Mooy, B.A.S. LOBSTAHS: An Adduct-Based Lipidomics Strategy for Discovery and Identification of Oxidative Stress Biomarkers. *Anal. Chem.* **2016**, *88*, 7154–7162. [CrossRef]

58. Koelmel, J.P.; Kroeger, N.M.; Ulmer, C.Z.; Bowden, J.A.; Patterson, R.E.; Cochran, J.A.; Beecher, C.W.W.; Garrett, T.J.; Yost, R.A. LipidMatch: An automated workflow for rule-based lipid identification using untargeted high-resolution tandem mass spectrometry data. *BMC Bioinform.* **2017**, *18*, 1–11. [CrossRef] [PubMed]

59. Shen, H.; Huang, J.Z. Sparse principal component analysis via regularized low rank matrix approximation. *J. Multivar. Anal.* **2008**, *99*, 1015–1034. [CrossRef]

60. Cao, K.-A.L.; Boitard, S.; Besse, P. Sparse PLS discriminant analysis: Biologically relevant feature selection and graphical displays for multiclass problems. *BMC Bioinform.* **2011**, *12*, 253. [CrossRef]

61. Rinaudo, P.; Boudah, S.; Junot, C.; Thévenot, E.A. biosigner: A New Method for the Discovery of Significant Molecular Signatures from Omics Data. *Front. Mol. Biosci.* **2016**, *3*, 26. [CrossRef] [PubMed]

Review

Modernization of Control of Pathogenic Micro-Organisms in the Food-Chain Requires a Durable Role for Immunoaffinity-Based Detection Methodology—A Review

Aldert A. Bergwerff * and Sylvia B. Debast

Laboratory of Clinical Microbiology and Infectious Diseases, Isala Hospital, Dr. van Heesweg 2, NL-8025 AB Zwolle, The Netherlands; s.b.debast@isala.nl
* Correspondence: aldert.bergwerff@freedom.nl

Abstract: Food microbiology is deluged by a vastly growing plethora of analytical methods. This review endeavors to color the context into which methodology has to fit and underlines the importance of sampling and sample treatment. The context is that the highest risk of food contamination is through the animal and human fecal route with a majority of foodborne infections originating from sources in mass and domestic kitchens at the end of the food-chain. Containment requires easy-to-use, failsafe, single-use tests giving an overall risk score in situ. Conversely, progressive food-safety systems are relying increasingly on early assessment of batches and groups involving risk-based sampling, monitoring environment and herd/flock health status, and (historic) food-chain information. Accordingly, responsible field laboratories prefer specificity, multi-analyte, and high-throughput procedures. Under certain etiological and epidemiological circumstances, indirect antigen immunoaffinity assays outperform the diagnostic sensitivity and diagnostic specificity of e.g., nucleic acid sequence-based assays. The current bulk of testing involves therefore *ante-* and *post-mortem* probing of humoral response to several pathogens. In this review, the inclusion of immunoglobulins against additional invasive micro-organisms indicating the level of hygiene and *ergo* public health risks in tests is advocated. Immunomagnetic separation, immunochromatography, immunosensor, microsphere array, lab-on-a-chip/disc platforms increasingly in combination with nanotechnologies, are discussed. The heuristic development of portable and ambulant microfluidic devices is intriguing and promising. *Tant pis*, many new platforms seem unattainable as the industry standard. Comparability of results with those of reference methods hinders the implementation of new technologies. Whatever the scientific and technological excellence and incentives, the decision-maker determines this implementation after weighing mainly costs and business risks.

Keywords: food microbiology; immunoaffinity assays; immunoagglutination; immunosensors; immunochromatographic testing; immunomagnetic separation; one health; pathogenic micro-organisms; responsive monitoring; review

Citation: Bergwerff, A.A.; Debast, S.B. Modernization of Control of Pathogenic Micro-Organisms in the Food-Chain Requires a Durable Role for Immunoaffinity-Based Detection Methodology—A Review. *Foods* **2021**, *10*, 832. https://doi.org/10.3390/foods10040832

Academic Editors: Philippe Delahaut and Riccardo Marega

Received: 17 February 2021
Accepted: 2 April 2021
Published: 11 April 2021

1. Introduction

The ever-increasing number of studies on the determination of micro-organisms (further abbreviated as MOs) gives the impression that the development of analytical platforms is solely technology-driven. Many of these developments seem not to connect to or seem not to fit in the daily practice of systems that should warrant the safety of food. Several inventions stick in testing relative pure reference materials or test spiked field samples only. Such samples do not reflect naturally infected animals, plants, or food products.

This paper will discuss which categories of disease-causing MOs need most and immediate attention and control (Sections 1.1 and 1.2). Examination of failing hygiene, pest control, and extrinsic and extramural factors will clarify the need for surveillance of the environment and rationalize the type of testing needed, *viz.* tests delivering multiple data from a series of samples timely and accurately (Section 1.3). Subsequently, this paper will show that consumers and persons preparing meals are responsible for the majority of foodborne infections and wonders whether there is a more prominent role for tests in the kitchen and domestic situations (Section 1.3). To understand how immunoaffinity (IA) testing may fit in all of this, in particular in intervention and control programs, this review presents first the basic concepts and historical perspectives of MOs analysis (Sections 2.2 and 2.3).

The conventional, emerging, and novel alternative approaches lined up in this paper are compared for their tradeoffs, merits, demerits, and usefulness in the light of a demand for more confidence (Sections 2.6, 2.7 and 3). The authors recognize that their views are based upon Western standards with the comfort of advanced technology at one's disposal. Colleagues from other regions may have other opinions on the utility of techniques and validity of some conclusions.

The authors realize also the numerous reviews and overviews which appeared on the subject. They attempted to look at developments in food microbiology analysis in another way by combining views on monitoring, reliability of results, the efficacy of the intervention, etc. It is for this reason that an example of *Salmonella* intervention in pig production is presented as a showcase of the analytical challenges in practice (Section 4).

This review does not desire to discuss specific technical and (bio)chemical details of innovative IA platforms. The reader interested in these details is given many (recent) references and other starting points to find this information. In addition, the article is written with routine, high-volume analyses of the food-chain in mind. Nevertheless, the authors are also familiar with specialistic methodology to determine bioagents that are non-cultivable or fastidious to determine. These methods are commonly Herculean for screening purposes and are therefore not discussed or only touched. Finally, the paper will elaborate modestly on the future of IA-based platforms, methodology, and the microbes of concern (Sections 5 and 6). After all, the SARS-CoV-2 pandemic has made us frustratingly and unmistakably clear: not macro-organisms but micro-organisms rule the world!

1.1. Food and Micro-Organisms

Eating and digesting food will maximize the entropy of the consumed plant or animal [1,2]. Food stabilizes the entropy of the consumer and facilitates the renewal of life. Food is thus essential to life but comes with the risk of other forms of entropy maximization. For humans, these (physical injury) risks do not arise from catching, harvesting, processing, or preparing food *per se*. Generally, food contains invisible micro-organisms; a term suggesting tiny living material with an autonomous nature. It can be debated whether this is completely true. As defined by the European Commission, MOs compromise bacteria, viruses, yeasts, molds, algae, parasitic protozoa, microscopic parasitic helminths, and their toxins and metabolites [3]. So, not all MOs are living materials and autonomous.

MOs are part of biogeochemical cycles, are highly adaptable, can survive anywhere and in all facets of life. Most MOs are "friendly" and many have closed a pact with plants, animals, and humans. In particular symbiotic bacteria protect and support our skin and gut, and help to convert certain food components while also producing e.g., vitamin K. In reality, microbes in the gut are essential for a healthy person stimulating the immune system. Beneficial MOs ferment (unpalatable) matters into favorable and enjoyable food products. Besides adding taste and quality, MOs can spoil food products and/or form a health threat for food workers and consumers. Following the invasion of a host, some MOs can replicate and cause infectious diseases. Pathogenic MOs can be harmful even without invasion of the eater by leaving toxins in the food and causing so-called toxic infections. In other cases, not only the injurious action of the MO but also the consumer's immune response to the MO leads to health damage. For example, although rare and affecting only approximately in 1 out of 1,000 infected individuals, different reactive (autoimmune) diseases are possible consequences of a *Campylobacter* infection [4,5]. Annually, 1,400 cases of reactive arthritis, 60 cases of Guillain-Barré syndrome, and 10 cases of inflammatory bowel disease as a result of campylobacteriosis are estimated in The Netherlands alone [6]. Here, *Campylobacter* is an example of bacterial zoonosis, i.e., a pathogen transmitted from animal to human.

Microbial hazards, like zoonoses, can enter the food-chain at any point: water, soil, in the growing, weaning, fattening phase up to the preparation of a meal. Water and soil are listed not without reason. They may contain harmful bacteria, parasites, and viruses which infect or contaminate plant or animal. For example, cows may contract zoonotic cysticercosis from feeding on tainted grass. Pet food is a part of that food-chain as well and also requires high hygienic standards. In several surveys, disease-causing agents, such as *Salmonella, Listeria,* toxin-producing *Escherichia coli, Mycobacterium* spp., *Sarcocystis* spp. and many other parasites were found in, predominantly raw, cat and dog food, not only diseasing companion animals but also transmitting the pathogens to (juvenile) humans [7–9]. Besides the indirect route, remind that pet food is consumed occasionally by people who cannot afford other food [10].

A subject often forgotten, is that food safety does not only concern the final consumer, but also the professionals coming into contact with infected or contaminated animals, plants, and food products, namely the butcher, (spouse and children of) farmers, fishermen, inspector, processor, retailer, slaughterer, veterinarian, etc. [11], and residents close to farms and slaughterhouses. Many risks, such as hepatitis E virus (HEV), *Salmonella, Streptococcus suis, Vibrio* sp. (fish products), have been underestimated and underreported for food-workers for a long time [12,13]. As an inquisitiveness, *vice versa,* professionals can infect plants (hepatitis A virus (HAV)), animals (*Streptococcus* Lancefield, *Taenia saginata* or *T. solium*), also referred to as anthropozoonosis, or food products (*Cryptospora, Cyclospora, Giarda,* HAV, norovirus, *Salmonella, Streptococcus aureus, Shigella*) as well.

1.2. Need to Contain Foodborne Pathogenic Micro-Organisms

Food contaminated with chemical substances and harmful MOs cause over 200 diseases [14]. An estimated number of 600 million persons worldwide fall ill from foodborne infections every year [14]. A total of 420,000 fatalities and 33 million disability-adjusted life years (DALYs), i.e., lost healthy years, are ascribed to food-transmitted diseases. Others report almost one million deaths due to water- and foodborne gastroenteritis by thirteen MOs in children aged less than five years in Africa, Asia, and Latin America [15]. When only zoonotic risks of food are considered, the most commonly detected causative agents in 2017 in Europe were bacteria (33.9%) followed by bacterial toxins (17.7%), viruses (9.8%), other causative agents (2.2%), and parasites (0.4%) [16]. The foods involved were eggs (23.0%), poultry meat (18.5%), fishery products including crustaceans, bivalves (22.4%), meat products other than poultry (21.7%), and dairy (14.4%) [16].

1.2.1. Bacteria

Most bacterial foodborne illnesses worldwide are caused by *Campylobacter,* namely an estimated 96 million cases in 2010 [17]. However, on a world-scale, the highest health burden of foodborne MOs comes from non-typhoidal *Salmonella enterica* infections, namely 4.07 million DALYs in 2010 [17]. In most confirmed cases in EU member states and associated countries, food was the carrier of pathogens with a zoonotic origin (Table 1) [18,19]. Despite its much lower incidence, the highest case fatality was caused by *Listeria* (17.6%), whereas the top two, campylobacteriosis and salmonellosis, caused 0.03% and 0.22% mortality, respectively [19].

When studying food-caused outbreaks exclusively, *viz.* cases in which two or more persons fall ill from the same foodborne sickness after eating or drinking the same food, *Salmonella* is causing substantially more instances than *Campylobacter* (Figure 1). It seems that certain strains, such as *Salmonella enterica* Enteritidis, evolved in such a way that it efficiently transmits to humans [20]. Although their incidence is low, opportunistic bacterial contaminants are causing, some very serious, hospitalizations, such as *Enterobacter sakazakii, Klebsiella* spp., *Citrobacter* spp., *Serratia* spp. [21,22].

Table 1. Confirmed human cases associated with foodborne pathogens in EU member states and associated countries in 2019 [19].

Pathogenic Micro-Organism	Confirmed Human Cases (Number)	Case Fatality (%)
Campylobacter	220,682	0.03
Salmonella	87,923	0.22
Shiga toxin-producing *E. coli* (STEC)	7775	0.21
Yersinia	6961	0.05
Listeria	2621	17.6

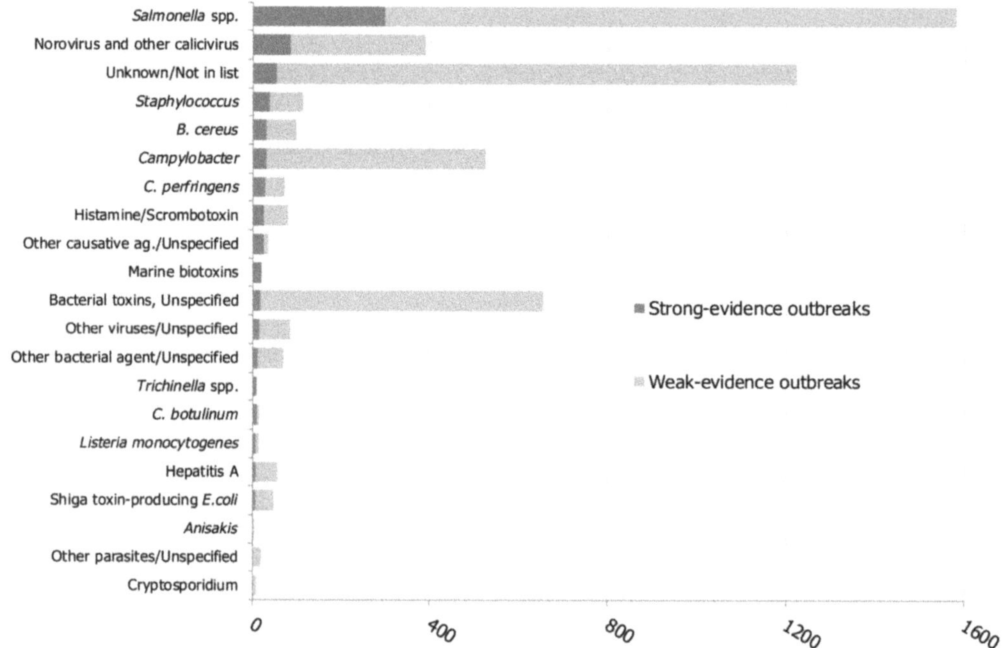

Figure 1. Food and water-borne outbreaks, i.e., cases in which ≥2 persons fell ill from the same food and the same agent, per causative pathogen in 2018 in the EU and associated countries. The explanation of "other" relevant for this paper is as follows. Other bacteria include *Aeromonas hydrophila*, (enteroinvasive (EIEC) or enterotoxigenic (ETEC)) *Escherichia coli*, *Enterococcus*, *Leptospira* spp., *Shigella* spp., *Shigella flexneri*, *Shigella sonnei*, *Yersinia enterocolitica*, and unspecified bacteria. Other viruses include adenovirus, flavivirus, hepatitis E virus, rotavirus, and unspecified viruses. Other parasites include *Giardia intestinalis* (*lamblia*), *Giardia* spp., *Taenia saginata*, and unspecified parasites. Figure courtesy of EFSA from [18].

1.2.2. Parasites

From Figure 1 it is evident that bacteria play an important role, but that there are microbial pathogens from other phylogenic domains to reckon with. Besides biotoxins, such as histamine and phycotoxins, viruses and parasites should not be neglected. Parasites fall into two main groups: protozoa and helminths (worms) and they are characterized by, sometimes very complicated, life cycles which can take years to complete. The fecal-oral route is predominantly contaminating the food-chain. Parasitic eggs, oocytes, of some species can easily survive years in the environment. Fruits or salads tainted by contaminated water may be the source of an infection among which e.g., tapeworm *Echinococcus multilocularis* and *Echinococcus granulosis* are feared. Large outbreaks through insufficiently decontaminated surface water prepared for drinking water is frequently the cause of human protozoan infections in the Eastern-Europe, UK, and the USA, such as *Giardia*, *Cryptosporidium*, and *Cyclospora* [23].

The potential implications of foodborne worms and protozoans, such as *Anisakis* spp., *Taenia* spp., *Sarcocystis* (*sui*) *hominis*, *Toxoplasma gondii*, or *Trichinella* spp., should not be underestimated. To demonstrate the impact of one of these parasites, approximately 45 million persons worldwide, of which 11 million persons in Europe, suffer from taeniosis caused by *Taenia saginata* [24]. A *T. saginata*, or beef tapeworm, infection affects annually up to 30,000 persons in The Netherlands alone [23]. This mild infection runs exclusively through cattle with humans as the final host. For cattle to become infected it has to come in contact with human sewage. As a consequence, prevalence is higher in countries with poor sewage control [25]. Taeniosis is thus also a sign of other pathogenic risks associated with improper hygiene. The infection is one of the blind spots in our surveillance systems; the sensitivity of the obligatory visual inspection for the causative worm is only 10% to 30% [26]. There is thus a world to win for test developers, routine test laboratories, and authorities. An available IA method is used for research purposes only and not suitable for routine analyses [27].

1.2.3. Viruses

Nearly all food-transmitted viruses originate from the human gut. Infections are predominantly caused by fecal contamination of (prepared) food as a result of insufficient personal hygiene of people handling the food. An important risk factor is the pre-symptomatic spreading of a virus by an e.g., infected cook or worker in the food industry. Consequently, several important health-threatening virions become foodborne at the end of the food-chain. Potential food-contaminating human fecal viruses belong to the adenoviruses, astroviruses, caliciviruses (norovirus), enteroviruses, hepatoviruses, parvoviruses, and rotaviruses. In this list is the all-time highest-scoring food-poisoning organism worldwide causing an estimated 125 million cases of gastro-enteritis in 2010: norovirus [17]. To make clear that norovirus is not an issue of developing regions alone, 60% and 75% of all foodborne illnesses in the European region [28] and in the USA, respectively, are caused by norovirus. Norovirus is highly contagious: the infectious dose can be only a few virus particles [29].

An indirect human fecal-to-food route is the contamination of bivalves cultured in estuaries, coastal regions, or other places with (brackish) water tainted by human excreta. In these cases, sewage can come from inefficiently working water purification systems dumping its pathogen-containing product in open water systems. Not to underestimate are also the sources of (a)symptomatic carriers defecating in open waters and fields of which many cases have been described in developed countries. In this way, bivalves were a vehicle for poliovirus, HAV, HEV, norovirus (and several dangerous bacteria such as *Shigella*, *Vibrio cholera*) [30]. Only 10–100 HAV particles are sufficient to infect a person and similar to norovirus, it is predominantly poor hygiene that the virus is transferred via fruit and (undercooked) meat. Fortunately, in none of the 1248 samples of fruit, bivalves and meat HAV were detected in the EU in 2018 [18]. Concordantly, none of the 535 fruit and vegetable samples were positive in 2019 either [19].

An exception to the human fecal route, although its etiology and epidemiology are not yet completely clear, is the risk of an HEV infection through consumption of raw or mildly prepared pig, wild boar, and deer meat products [31]. Fermented/dried sausages provide the highest HEV risk following raw *porcine* liver. The cause is liver-contaminated diaphragm muscle used to produce sausages.

1.2.4. Other Types of Pathogens

Among bacteria, viruses, parasites, the reader may miss foodborne infections caused by algae, fungi, yeasts, and infectious proteins (prions). The last category does not initiate (adaptive) immune responses and although detectable using IA techniques (e.g., immuno-PCR [32]), it is considered to be beyond the subject of this paper. Whereas immunocompromised patients have to be careful, foodborne algae, fungi, yeasts are seldomly harmful to healthy people. Although these infections are rare, toxicoinfections through the production of, some extremely harmful, toxins by fungi growing in/on plants or (raw) food products are not rare at all [33]. Toxicoinfections with a bacterial origin are caused by food-transmitted toxins produced by for example *Bacillus cereus*, *Clostridium botulinum*, *Clostridium perfringens*, and *Staphylococcus aureus* [34]. Their exercised menace is responsible for 16% of all food- and water-borne outbreaks in the EU in 2017 [18]. Immunoaffinity testing methodology to assess these biotoxin-caused risks is beyond the scope of this review and the subject is only touched.

1.3. Failing Containment of Pathogenic Micro-Organisms

With appropriate measures and intervention, the risks of MOs in food can be mitigated greatly if not completely. Their chance of survival in the chain is determined by extrinsic (such as humidity, temperature, gas composition), intrinsic (such as water activity, pH, structure), and implicit (such as the ability of the MO to adapt to its environment) factors which can be controlled. Humankind has discovered and invented many food handling and preparation methods, including cooling, salting, fermentation, drying, acidifying, marinating, to influence the risk of MOs. In some cases, new microbial risks were however introduced unintentionally by these methods, such as the very dreaded botulinum toxin-producing *Clostridium botulinum* bacteria in e.g., air-dried sausage (*botulus* in Latin), dried fish or oil-submerged protein-rich food products (like mushrooms) (see ProMed mailings for case descriptions [35]).

Even when the cooling chain is closed and maintained at low temperatures as a common and effective intervention method should, bacteria, like pathogenic *Yesinia enterocolitica* and *Listeria monocytogenes*, can still grow to risky amounts. Actually, it is hypothesized that *Listeria monocytogenes* has adapted itself to processing plant conditions, such as osmotic, detergent, acid, and oxidative stress [36], and became psychrotrophic, i.e., growing even at low temperatures [37], since the introduction of food-processing plants and home refrigerators, respectively [38].

Under optimal conditions, bacteria can propagate very fast. For example, *Clostridium perfringens* can produce a new generation every ten minutes. MOs can be contained in food by obeying strict (environmental) hygiene and by maintaining a closed cooling chain at low temperatures. Apparently, as a relative number of cases are dropping over the years [39,40], humankind succeeds increasingly well, but, as entropy strives for maximization [2], hygiene and other measures can never be neglected and the risk will never be zero. Therefore, continuous intervention is crucial and food microbiology plays an important role in control.

Notably, over 50% of the registered foodborne infections in Europe and North America were contracted at home [41]. Following domestic causes, food from mass catering, such as fast-food outlets, in-flight caterings, hotels, hospital restaurants, takeaways, causes most salmonellosis cases [39]. Persons preparing food make many food-handling errors, including inadequate handwashing, poor surface cleaning, undercooking, etc. [42]. Knowledge of safe food storage and handling is poorly developed under young consumers [43]. Household sources of infection are mainly contaminated food and water, people, pests, and pets greatly by a fecal-oral route. The reader notices that the bioagents causing a large part of the foodborne illnesses are thus not originating from the original food product upstream in the chain.

Sensible domestic hygiene principles seem partly lost. We learn about correct and incorrect hygiene in the kitchen effectively from watching cooking shows [44]. When asked, people generally are not aware to prepare their salad and vegetables before that of their meat, are not conscious of knife hygiene, hand washing, or refrigerator cleanliness (authors' outcomes of inquiries, unpublished). It is therefore not a surprise that fecal coliforms originating from intestines of warm-blooded animals, including from ourselves, and a warning for the possible presence of pathogens, were found in all 250 sampled home kitchens [41]. Within the scope of this review, affordable, convenient and reliable microbiological assays for household testing to give a consumer a means to see imperfect cleaning and risks lurking, are hardly available.

The way we prepare food has evolved hugely. A trend is the increasing use of locally produced products that may evade (not maliciously *per se*) usual food-safety inspection regimes. In addition, we seem to appreciate serving raw or barely cooked dishes to emphasize the freshness and excellence of the ingredients used. Along with tidbits of for example uncooked salmon tartare, raw milk directly from the dairy farm, cheese from non-pasteurized milk, and thin slices of beef carpaccio, so fresh are also the parasites, bacteria, and viruses that can reside in/on them. We also increasingly prefer ready-to-eat food and expect that these are pathogen-free. Well, a third of all strong-evidence outbreaks involved buffet meals, mixed food, and unspecified foods [16,39].

2. Analytical Microbiology

There is no future without a past. This chapter will therefore first clarify why analysis in the food-chain is indispensable (Section 2.1). Following a short description of the discovery of bacteria, parasites, and viruses (Section 2.2) and their biochemical characteristics (Section 2.3.1), this chapter will throw a light on two important available measuring principles (Section 2.3.2) and suitable sample types for these measuring principles. These insights will help to explain where, what, when, for which purpose, and at what costs measurements are performed (Sections 2.4 and 2.5) before specific analysis immuno-affinity assay techniques (Section 2.6) are described. This chapter will close with a short description of a few important analysis principles other than those based on a specific reaction of an immunoglobulin with an antigen and their comparison with immunoaffinity assays (Section 2.7).

2.1. Why Need to Measure?

Pathogenic MOs in the food-chain are not only a threat to human health, and cause of DALYs and lost lives, they have a socio-economic impact and influence world history [45,46]. The economic losses due to lost labor capacity, health-care costs, damage to product brand, producer's standing, etc. can be devastating. Although foodborne illnesses are known for thousands of years [47], foodborne infections were considered long not life-threatening. Some pathogens were considered a problem of for example canners, such as botulism, while eggs were regarded sterile as long as the shell was closed [48]. The first serious consumer concern with respect to foodborne pathogens with a significant societal impact was at the end of the 1980s of the last century. These public concerns on infections contracted from contaminated food were surprisingly much later than public concerns on chemical contaminants in food, such as anabolic hormones in particular diethylstilbestrol (DES) [49]. Still, the consumer appears to accept pathogens in food more than regulated and tolerated chemical food contaminants, such as residues of veterinary pharmaceuticals. Nonetheless, in comparison, morbidity and mortality clearly comes prevalently from foodborne MOs, including bacterial and fungal toxins, not from chemical contaminants and additives in food. Colleagues and authors assume chemophobia stronger than fear for 'natural' biological phenomena like pathogens, but, apparently, this has not been investigated scientifically [50].

The first big food-scare involving pathogens playing havoc with the food-production industry was namely in 1988 and went into history as the "egg affair" [46]. It started with the impromptu announcement in the UK that salmonellosis in humans was linked with the consumption of *Salmonella*-contaminated eggs and poultry. This declaration was hyped by the press and caused a collapse of the British egg industry. It marked the birth of (supra)national *Salmonella* screening and control programs by organized private and by governmental/NGO parties.

Hygiene and control measures are important to stop a bioagent in the food-chain and to prevent harm to the consumer (Figure 2). Whether these general actions are sufficient, and whether additional, specific interventions are necessary, is revealed by continuous, appropriate, and efficacious analysis of (indicators of) bioagents at critical moments in the chain.

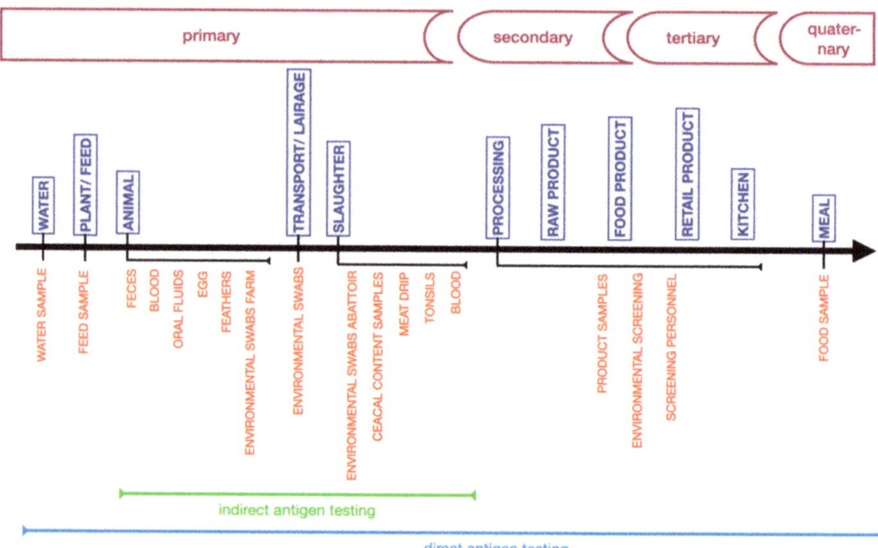

Figure 2. When, what, and how to screen in the animal food production chain (dark blue color) to secure food safety with respect to microbial hazards. Hazard critical point analysis will reveal optimal sampling moments. The sectors (purple color) which determine the type of sampling (red color) and testing (green and light blue colors) possible are indicated. The primary sector includes the pre-harvest stage until slaughter and comprises reproduction, egg and milk, fattening, transport, and slaughter phases. The secondary sector includes all food-processing steps converting milk, eggs, and meat into complex products. The tertiary and quaternary sectors include wholesale, street vendors, catering, institutional kitchens, and finally private kitchens and (domestic) consumption.

2.2. Historical Overview of the Discovery of Micro-Organisms

In the awareness of infectious diseases transmitted through food, there is since the discovery of MOs an increasing need to escape agnostic situations and to pin and destroy harmful bioagents in their habitats. However, it took almost two centuries after Antoni van Leeuwenhoek (1632–1723) observed bacteria in 1675 as the first human till Rudolf Virchow (1821–1902) formulated his cellular pathology concept in 1858. Virchow formulated the third dictum of cell theory *Omnis cellula e cellula* ("All cells come from cells") and coined the term "zoonosis" as he had noticed a link of diseases between man and animals. The new cellular pathogen concept formed the basis of modern (analytical) bacteriology funded by Louis Pasteur (1822–1895) and Robert Koch (1843–1910). In addition, August Gärtner (1848–1934) showed that several bacteria were able to produce thermostable toxins giving cholera and typhus-like diseases after consumption of contaminated food. This finding showed that bacteriology does not end with the absence of a pathogenic bacterium and that food is an important vehicle for diseases.

In parallel, the propagation cycles of several parasites were elucidated mid-19th century and paved the way to modern parasitology. In addition, in 1892, the first clue for the existence of viruses was presented by Dmitri Ivanovsky (1864–1920). His famous experiments with porcelain filters showed retention of bacteria in the residue above the filter but the filtrate was as infectious nevertheless. The name virus (*contagium vivium fluidum*) was coined in 1908 by Martinus Beijerinck (1851–1931).

These new etiological insights involving bacteria, parasites, and viruses, gave an impulse to preventive control of pathogenic MOs in animals, plants, and in or on their derived food products. These insights also fueled further technical development of a methodology to investigate, detect, and screen MOs of which many reviews report, for example [51,52].

2.3. Immunoaffinity Principle

2.3.1. Microbial Handles for Analysis

Unique morphological, chemico-structural, biochemical, and genetical characteristics of MOs are opportunities for analytical methods to spot a pathogen sensitively and specifically. The antigenicity of some elements of MOs triggers the immune system of vertebrates in several ways. One of these ways is a humoral response, i.e., the production of antigen-reacting immunoglobulins (IGs) or antibodies (ABs). This reaction is specific between a certain antigen and a particular antibody, and therefore informative. The unraveling of antigenic, but also of biochemical and genetic, differences are the basis for the taxonomy of MOs.

In the case of bacteria, ABs generated by a vertebrate are directed against motifs at the outer cell wall (Figure 3), but, in principle, the host can also raise IGs against excreted bacterial moieties. As produced ABs react specifically with certain structures, they can be used for analytical purposes to determine the presence of antigens and thus of a disease-causing agent. Originally, difference was made between heat-labile (proteinaceous) and heat-stable (involving e.g., polysaccharides) antigens. Characterization is now performed using multiple specific antisera against various outer cell envelope structures, predominantly capsular polysaccharides (K-antigens), fimbriae (F-antigen), flagella (H-antigen), and lipopolysaccharides (O-antigens) (Figure 3). *Cave*, the formation of capsular polysaccharides may obscure antigenic cell wall structures for detection as observed for e.g., *Staphylococcus aureus* [53].

Over 150 O-antigens and approximately 50 and 60 K- and H-antigens, respectively, are described for *Escherichia coli*. Of the over 2400 known O: K:H-*E. coli* variants, most are not pathogenic. Identification of *Salmonella* is primarily through its somatic O-antigens (LPS) and has revealed the occurrence of many (sub-)subspecies, so-called serovars. Through shared O-antigens, *Salmonella* serovars are categorized into serogroups indicated by letters. Of interest to food microbiologists are for example zoonotic *Salmonella* serovars belonging to serogroups B, C, and D.

Immunodetermination of *Salmonella* using specifically reacting ABs should be performed carefully as identical O-antigens are also found on other potential food-contaminants such as *Citrobacter freundii* and *Escherichia coli* O157 [54]. For this and other reasons, *Salmonella* subspecies are occasionally further specified by supplementary profiling of the flagellar (H) antigens, which are more specific than O-antigens. Other antigenic *Salmonella* membrane proteins are generally not used for serological analysis as these proteins show cross-reactivity with other Enterobacteriaceae *genera*. In this way, serovars of *Salmonella* are described by a unique combination of O- and H-antigens. This classification is known as the Kauffman-White (-Le Minor) Antigenic Scheme. The list of *Salmonella* serovars is growing with new sub-subspecies verified formally by the WHO Collaborating Centre for Reference and Research on *Salmonella* (WHOCC-Salm). Hitherto, over 2600 *Salmonella* serovars have been described [55] of which less than 100 serovars account for infections in humans. Genomic analyses have shown that the 2600 *Salmonella* strains belong actually to two species, namely *Salmonella bongori* and *Salmonella enterica*.

Some bacteria show relatively fast genetic shifts in their antigenic structures and manifest in many different variants and serotypes. This phenomenon hampers the development of specific binders to facilitate the detection of for example *Campylobacter jejuni* and *Listeria monocytogenes* at the species level [56].

Campylobacter spp. does not show a one-to-one serovar-genotype relation, and an isolate can change its serovar over time [57]. This makes categorization and analysis complicated. Nevertheless, the Penner-typing scheme based on antisera against capsular polysaccharides and LPS is used to identify *Campylobacter jejuni* ssp. Jejuni which is the most frequent cause of campylobacteriosis worldwide [57]. It demonstrates that some MOs need specialistic knowledge before one can start to develop, validate or exploit immunological methods.

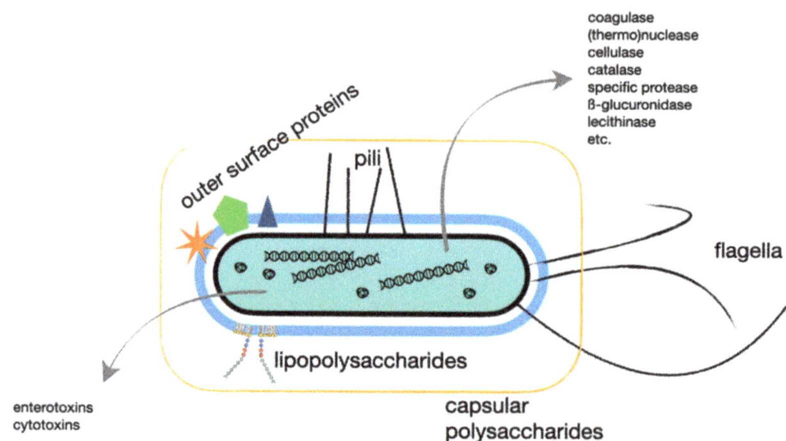

Figure 3. A simplified overview of "handles" of a Gram-negative bacterium analytically available to find a bacterial cell among all other food components. Flagella, H-antigen, are only present on motile cells and when present, it can be a single flagellum or multiple flagella organized mono-/lopho-/amphi-/peritrichously [57]. Lipopolysaccharides, O-antigens, form the outside of the outer membrane of the bacterial cell. Pili or fimbriae, F-antigens, are of little relevance to food microbiologists. Capsular polysaccharides (CPS), K-antigen, can be formed in both Gram-negative and Gram-positive bacteria. A special subtype of the K-antigen is the Vi-antigen in Salmonella. The activity of excreted enzymes is partly specific and used as a marker to identify a bacterium. Analysis of excreted toxins can also be used to trace and identify a pathogenic bacterial cell. The depicted lists of enzymes and toxins are not exhaustive.

Detection and identification of parasites rely for a great part on visual inspection (cysts) and microscopic techniques. In general, molecular, i.e., nucleic acid sequence-based (NASB), techniques have poor sensitivity due to a low parasite burden (see also below). Protozoans and helminths are very diverse and do not have generic (morphologic) structures shared between *genera*, family, order, and classes such as bacteria do. Nevertheless, parasites give themselves away by triggering an immune response not only as an intact entity but also by releasing excretory-secretory antigens into e.g., the circulation [58–60].

Viruses are actually obligate intracellular parasites. Like protozoans and helminths, viruses are extremely diverse in their structures, genetic compositions and in their ability to infect, persist and initiate disease in a host [61,62]. Any viral protein may provoke the generation of antibodies. However, many viruses have evolved mechanisms to sabotage the arms of the immune system, including those of a humoral response. They do this for example through antigenic drift and shifts resulting in mutation of protein regions that are normally targeted by ABs.

Despite an antigen-antibody reaction is specific, the primary or stereochemical structure of epitopes may be related so that the paratopic loci on antibodies are not able to discriminate different analytes sufficiently, in particular when MOs are phylogenetically closely related. This cross-reaction is a frequently occurring feature of ABs. The developer, producer and end-user deploying an IA method should be conscious of false-positive outcomes caused by this phenomenon.

2.3.2. Antigens and Antibodies as Potential Analytical Tools

Following infection, the immune system can react in various ways, such as through cellular (by releasing e.g., cytokines) and/or by the production of IGs (humoral response) [61,62]. Whatever the response, each can be exploited for diagnostic and other analytical purposes. Cytokines regulate and mediate immunity and are commonly less specific than antibodies to trace infections or to determine the success of vaccination. This review focuses on antibodies reacting specifically with an antigenic structure of a pathogenic MO. The antigen can be a unique biomolecule, structural element, a primary or secondary metabolite, e.g., an excretory/secretory product of the MO (*cf.* Figure 3). An example of a secondary metabolite is a bacterial or fungal toxin. An example of an excretory/secretory product is the ES antigen of the parasitic worm *Trichinella spiralis* used in ELISA serology [63].

Specific bio-recognizing antibodies are produced, isolated, and occasionally purified, and used in a plethora of analytical tests for diagnoses, monitoring, screening, and surveillance. A bio-recognizing antibody used as an analytical tool, cannot be designed by a rational template or code. They have to be produced by immunization techniques. Sufficient specificity and affinity are not guaranteed. Traditionally, antibodies, polyclonal antibodies (PABs), are isolated and concentrated from a mammalian animal exposed to an isolated antigen or an attenuated MO, or is infected under controlled conditions. Factually, these approaches are all vaccinations.

On the other hand, monoclonal antibodies (MABs) are produced ex vivo by hybridoma cell technology developed by Nobel laureates Milstein and Köhler in 1975 [64], which was a huge step forward in the history of immunoassays. Immortalized cell clones yielding ABs with favorable characteristics, such as a high avidity, are picked for MAB production.

MABs are generally rather specific through binding a single epitope only. In contrast, PABs are diverse in terms of e.g., subclasses and have an affinity for a variety of epitopes/antigens, which can improve sensitivity. The tradeoff is that PABs are associated with a much higher risk of a-specific binding by "reading errors" and/or non-specific binding events compared to MABs. The choice of the primary AB, either MAB or PAB, can markedly affect the specificity of the final IA assay. Therefore, MABs and PABs have to be exploited strategically and require different binding conditions in a test set-up.

Alternatively, to take advantage of a multi-epitope binding ability as PABs have, a mixture of MABs binding to multiple epitopes, so-called oligoclonal antibodies (oABs), may improve test performance. In the first report in 1983, this approach was coined "cooperative immunoassay" (CIA) and increased the sensitivity of an assay twice [65]. The exploitation of oABs was re-introduced in several recent studies (see [66] for more information).

Besides ex vivo hybridoma cell technology and in vivo production in mammalians, antibodies can be manufactured alternatively through exploiting phage-display banks [67]. In fact, phage-display technology was used to produce single-chain fragment (scFv) binders from different clones to bind a wide range of *Listeria monocytogenes* serotypes and strains where "traditional" methods failed in many attempts [68]. Finally, specific non-antibody binders, such as aptamers and molecular imprinted polymers (MIPs), as alternatives for antibodies are attracting increasing attention as well.

For analytical purposes, a specific interaction between immunoglobulin and an antigen is exploited in vitro in two ways: either in a direct antigen (Section 2.3.2.1) or an indirect antigen (Section 2.3.2.2), also known as an antibody or serological test. Reference [69] gives the interested reader a brilliant overview of the many different direct and indirect IA assays developed since the end of the 19th century illustrated with authentic pictures and figures.

Finally, the following has to be emphasized here. Any immunoaffinity technique starts with the antigen. After all, no immunogenic structures, no immunological response, no immunoaffinity methods. Consequently, the antigen will determine the quality of the direct and the indirect antigen test.

2.3.2.1. Measuring Principle of Direct Antigen Tests

As described above, the ways to detect the MO of interest are an indirect antigen or a direct antigen test. In a direct antigen immunoassay, isolated specific antibodies bind to probed antigens if present in the sample. The types of samples suitable for direct antigen testing are numerous. A sample can be (a swab of) any part of the (culled) animal, fruit, vegetables, salads, environment (of the farm, abattoir, processing plant, truck, bakery, butcher, supermarket, kitchen), packaging material, but also eggshells, feces, feathers, hair, and saliva. In the last matrix, for example, the authors succeeded to detect zoonotic *Streptococcus suis* serotype 2 through its secreted antigenic extracellular factor (EF) in the broth of cultured swine saliva using a surface plasmon resonance (SPR) biosensor [70].

Selection of a sample type and the moment of sampling should be done carefully, and with consideration of circumstances and implicit factors to increase the chance of a hit. Specialistic epidemiological, etiological and pathological knowledge of a MO, including period and routes of shedding, tissue predilection, is essential for effective microbiological screening. As an example, pathogens are not distributed proportionally over a carcass. It is for this reason that the EU Regulation on microbiological criteria has specified that four sites of a carcass have to be sampled with a minimal surface of 20 cm^2 in the case of aerobic colony counts or Enterobacteriaceae, or, when swabs are used, at least 100 cm^2 [3]. In the case of *Salmonella*, an abrasive sponge must sample at least 400 cm^2 of the most likely contaminated areas.

In direct antigen tests, detection is performed in many different ways not only using the antigenic moieties of the pathogen exclusively. In so-called sandwich assays, the immobilized antigen-binders are MABs but more likely PABs for sensitivity purposes, and a secondary AB, often a MAB for specificity aims, provides a signal (through a conjugated label).

2.3.2.2. Measuring Principle of Indirect Antigen Tests

An infection is recognized by increasing levels of immunoglobulins specifically reacting with the invading MO. The antibodies are of the IgA, IgG, IgM, or IgY (IgG homolog in fowl) class and have different binding characteristics, which are seldomly deployed in food analysis unlike in medical microbiology. Antibodies are found in body fluids, including blood, cerebrospinal (liquor) or synovial fluid and saliva/mucosa, but also in egg (fowl), extracellular fluids, milk and colostrum (lactating animals), and tears. There is a correlation between the concentration of antibodies (titer) and disease burden but this correlation is not strong. Immunoglobulins are (stereo)chemically and biochemically relatively stable, including quite resistant towards proteolytic attack. Therefore, the determination of antibodies can be performed rather long after the onset of death and after sampling.

In an indirect antigen test, responsive antibodies will bind to an isolated (purified) and specific antigen and in this manner indirectly demask the presence of an MO. This testing principle is used predominantly to assess the risks of animals and raw animal products before they are released to the secondary sector (Figure 2) because responsive ABs are a good indicator or biomarker of a (past) infection. The detection of specific antibodies is also named serology, but this designation is avoided here as it can be confused with serological confirmation of an isolated bacterial cell. In many instances, the antigen is (semi-)synthesized or produced in cultured immortalized cells by recombinant DNA protein techniques, such as antigens representing the *porcine* reproduction and respiratory syndrome (PRRS) virus [71].

The diagnostic sensitivity, i.e., the ability to correctly determine infected individuals or populations, of this type of assay is hampered by:

(*i*)   Low immunogenic response of the individual animal, and
(*ii*) The so-called seroconversion window.

Campylobacter jejuni and *C. coli*, the most frequently isolated foodborne pathogens from humans suffering from gastro-enteritis, colonize the intestinal tract of most animals [72], but these do not show clinical symptoms of the disease. Animals apparently lack specific factors, such as receptors, and/or have an effective immune mechanism [4]. In fact, the lack of an animal model, except non-human primates or genetically or surgically modified animals, hinders significantly scientific research to understand campylobacteriosis [4]. For example, circulating antibodies neutralizing *Campylobacter* cytolethal distending toxin are not developed in chickens, while these antibodies are elicited in humans [73]. So far, no indirect antigen assays are routinely in use for the detection of *Campylobacter* [72].

The course of an infection, i.e., invasion, colonization of a host by a MO and MO clearance, the onset and development of a humoral immune response, and the duration of detectable IGs against the MO are asynchronous and do not match. In accordance, the results of direct antigen tests do not balance well with those of indirect antigen assays [62]. In other words, a seropositive individual may be free from the tested pathogen; a positive indirect antigen test reports in that case a past but cleared exposure to a pathogen. On the other hand, many pathogens, including viruses, bacteria (e.g., *Salmonella*), and parasites (e.g., *Toxoplasma*), can "hide" themselves, for example intracellularly, leaving them undetectable in a direct antigen test but detectable indirectly by specific IGs. Under certain circumstances, such seropositive animals become contagious again and for this reason, they should be traced to prevent contaminated food.

A seroconversion window refers to the time needed for a vertebrate to respond to an infection, *i.c.* the interval after commencement of the infection to produce detectable amounts of immunoglobulins. This window varies considerably between approximately four days and even weeks depending on the bioagent, animal, and the type of matrix in which the IGs are searched. A negative antibody test on an individual is therefore no assurance that this tested individual is free from the pathogen unless the test is repeated (in some cases multiple times) after an appropriate time. But even following an adequate time-interval, the IG-concentration can be less than the limit of detection of the assay. A notorious example is the failing or weak seroconversion of poultry upon infection of *Salmonella enterica* serovars in the C serogroup (O:6), such as zoonotic *Salmonella enterica* Infantis [74], while it is a commonly isolated serovar in laying hens and broilers. This is a major drawback of the technique and one of the reasons why indirect *Salmonella* testing is not popular in poultry monitoring programs.

In addition, IG-classes have different binding-reactivity properties and different anabolism and catabolism characteristics. One must therefore realize that IG-concentration profiles of different IG-classes in different body fluids are similar but not identical. The concentration of *Salmonella*-binding IGs in meat drip or meat juice (see also below), which is an important analytical matrix in national screening programs in several countries, such as Denmark, is for example lower than found in blood serum [75,76]. Although concerns have arisen from the use of meat drip for *Salmonella* monitoring on several occasions [75,76], an important driver to assay meat drip instead of blood serum is that it is practically easier than collecting blood samples at for example slaughter [77,78]. It is also a matrix with high predictive power for the occurrence of HEV in pork [79]. Its use to screen other pathogens, such as the protozoan *Toxoplasma gondii* in other food animals, including goats and sheep [80,81], has been well demonstrated. However, anti-*Toxoplasma gondii* IgG levels in heart, diaphragm, tongue samples vary significantly [82] and test sensitivity using diaphragm juice is 60–77% compared to serum [83]. For the preparation of meat drip samples, these muscle types are nevertheless preferred as they represent a low economic value.

Meat drip is suitable to screen carcasses, but obviously not suitable to screen animals in the pre-harvest phase. Typically, blood samples for indirect antigen screening are collected in this primary phase. However, unlike collecting feces, sampling blood in pigs at a farm gives labor and animal welfare issues. It is therefore that other body fluids, like oral fluids, are investigated as an alternative [84,85]. The use of oral fluids for agglutination tests goes back as early as 1909 [85]. In saliva samples, IgG, IgM but predominantly IgA is found. The antibody concentrations are lower than in serum and methods should be adapted for oral fluids to let sensitivity in line with those suitable for serum or meat drip [84,86]. The choice of the matrix will thus influence the diagnostic sensitivity of the tests. For example, screening of oral fluids collected on cotton ropes hanging in pens for antibodies against the PRRS virus is becoming increasingly accepted. The pigs bite playfully in the rope leaving saliva. A factor to be aware of is that ill animals bite less frequently (thus an indicator by itself) and can be missed by the "rope test". Although PRRS is a widespread disease affecting swine, not humans, and thus not a food-safety issue, the disease can be spread through careless wasting of food leftovers and using food spills in feed.

For clarity, analytical and diagnostic sensitivity (or specificity for that matter) are fundamentally different terms. Analytical sensitivity refers to the smallest detectable amount of analyte (e.g., antibody in serum), i.e., the detection limit of the assay. It can be assessed by dilutions of commercially available reference samples in samples of negative animals to define the penultimate dilution in which the analyte is no longer detectable or indistinguishable from negative samples. Similarly, analytical specificity is the degree to which the assay does not cross-react with non-targeted analytes in a standard sample. Diagnostic sensitivity and diagnostic specificity determine the probability that a given test result reflects the true infection status of the animal. The values are derived from testing a series of samples from reference animals (which are assessed by the reference method as well).

A way to overcome a disappointing diagnostic sensitivity, even when the analytical sensitivity of the assay is satisfactory, is to screen the seroprevalence of a group instead of assessing the serostatus of an individual animal or product. In this approach, the flock, group, herd, or (products of) carcasses are considered as a single batch within which infection kinetics and dynamics are actually averaged. In other words, assessment of a cluster will normalize results and blot out the impact of false negatives and false positives. As thought fit, this does not imply that samples are pooled before analysis; it means that each individual sample is analyzed separately followed by evaluation of each outcome within the sample series resulting in an overall score for the batch/group.

The chosen sample size is then of principal importance to yield statistically significant results which allow educated decisions on condemning or releasing a batch, i.e., considering it non-compliant or compliant, respectively. Sample size depends on many parameters, circumstances, and decisions, such as prevalence of infection (in a region), sensitivity of the test, (non-)hypergeometric sampling, desired confidence level *et cetera* [87].

Another matter to consider is that, in principle, indirect antigen analyses do not discriminate a seropositive signal coming from a naturally wild-strain-infected animal or a vaccinated animal. It is for this reason that countries having acquired a specific disease-free status, prohibit vaccination against the respective animal disease unless a so-called marker vaccine is used. An example is an immunization against non-zoonotic Aujezký́s disease or pseudorabies virus (PRV) using vaccines containing (attenuated) modified viruses devoid of certain otherwise antigenic proteins [88]. This technique is known as DIVA: differentiating vaccinated from infected animals. Similarly, investigators using NASB methods may miss modified vaccines when they use non-adjusted primers, such as the Riems C-strain of the CSF virus [89].

Finally, some pathogens are typically, not exclusively, introduced *postmortem* in the processing plant or the kitchen and will therefore not provoke an immune response that can be picked up by an indirect antigen test. Examples of such pathogens are *Listeria monocytogenes* or norovirus, but also *Salmonella* by e.g., *postmortem* cross-contamination.

These mismatches between results provided by direct and indirect antigen assays give the impression of false negatives and false positives. This is a needless debate as the purpose of testing, in general, this is risk management, dictates the choice of test approach. The indirect antigen analysis strategy fits well in food quality assurance programs as samples are relatively easily collected and processed with high-throughput and laboratory robotics possibilities. These possibilities facilitate cost-effective monitoring, three months intervals are usual, of pens, herds, and farms providing a risk status.

Specific biomolecules also reveal the presence of MOs indirectly in other ways. An example is the major virulence factor p60 of *Listeria* spp. which it excretes in relatively large quantities. This surface protein biomarker was detected in milk using an electrochemical immunosensor [90]. Among other biomarkers are volatile metabolites which can be profiled at ppb levels using so-called electronic noses (E-noses). *Salmonella enterica* Typhimurium, for example, is recognized by primary and secondary alcohols, and methyl ketones [91]. *Escherichia coli* produces typically large amounts of indoles not produced by other pathogenic or spoilage bacteria [91]. The specific production of indoles by *Escherichia coli* is not surprising, as the bacterium is a prominent intestinal inhabitant and indoles, as degradation products of tryptophan, are causing the characteristic noxious smell of (human) feces. A major drawback of E-noses is that they easily interfere when the composition of the surrounding air changes. Analysis should therefore be performed in rooms with low effluent from other sources. The use of, for example, alcohol to disinfect hands, or a nearby car with a running engine makes E-nose analyses unreliable.

2.3.3. Sample Type and Preparation

2.3.3.1. Sample Preparation for Direct Antigen Tests

Bacterial cells may be stressed, damaged, or otherwise sublethally injured by intrinsic factors such as organic acids, the activity of preservatives, and high salt concentration, or extrinsic factors such as cold or heat. Notwithstanding their bad condition, cells may revive and proliferate in/on a food product under more favorable circumstances, like meat presented improperly at the market lying at elevated temperatures in its protein-rich drip. As tests used for inspection purposes fail to detect the target MO straight in the animal or food product (elaborated on below), bacteria are multiplied first by inoculation of a broth and culturing. This so-called (pre-)enrichment step also offers the possibility to improve the selectivity of the overall method.

There is another reason to execute this culture step. Assuming sufficient sensitivity, a test method could produce false-positive test results, i.e., correct identification but of a non-viable organism which is therefore no longer a public health threat. The culturing step thus not only improves selectivity, but it also prevents wrong conclusions as per definition only viable cells are multiplied.

Overnight incubations in a liquid nutrient at a specific temperature optimal for growth of the target organism(s) and suppression of that of other bacteria are therefore common. In the culturing step, an initial incubation time has to be respected to allow injured cells to resuscitate. It is for this resuscitation interval that bacterial methods are hindered to obtain a fast time-to-result (TTR). In addition, quite some pathogens, such as *Salmonella enterica* group B strains [92], are known to grow slowly. Most food- and water-borne bacterial pathogens require at least 18 h, even up to 72 h, incubation. So-called "same-day" assays, comprising sample preparation, (pre-)enrichment, and detection in a working day (8 h), have therefore to be applied carefully to avoid false-negative outcomes. Notwithstanding, "same-day" methods involving real-time PCR (qPCR) for *Salmonella enterica* determination in carcass swabs and pork, veal, and poultry meat samples show favorable results [93,94].

A means to isolate and concentrate (intact and viable) MOs is immunomagnetic separation (IMS). The enrichment technique involves paramagnetic beads coated with MO-binding immunoglobulins. The microspheres are suspended with the sample, such as incubated broth or food slurry, allowed to capture the target and then collected from the suspension using a magnet. Meanwhile, the particles are washed to remove interfering food particles, (supra)molecules, and any other (competing) MOs. In other words, IMS does not only isolate and concentrate but also purifies the target. The technique is easily combined in automated systems to bring down labor and time and it is a popular step in many rapid methods. In quite some examples, analyte-microspheres complexes are suitable for analysis without further treatment, including traditional culturing, ELISA, molecular biology methods, MALDI-TOF mass spectrometry, and affinity assays. An example is a combination of IMS-beads with a paper-based carrier for color development and colorimetric evaluation using a smartphone. This approach has been demonstrated for *Salmonella enterica* Typhimurium in Starling bird feces and whole milk at 10^5 CFU/g and 10^3 CFU/mL, respectively [95]. As an alternative, bacteria can also be isolated and concentrated up to 500 times using anti-*Salmonella* antibodies immobilized on Affi-Prep, which is cross-linked acrylamide support, or on beads in immunoaffinity chromatography [96].

There is little that is perfect *a posteriori* that appeared perfect *a priori*. The IMS technology is bound by the quality of the binder which limits the range of application. It is a challenge to bind e.g., all food-safety relevant *Salmonella* serovars with similar affinities while not binding MO with related antigens. Specificity of the detection method after IMS must compensate for this failure. However, abundant competing non-targeted MOs can occupy the binding sites and prevent the target to attach causing false-negative results in this way. Non-specific binding causes another problem and may cause false-negative as false-positive results as well.

The IMS efficiency is highly dependent on the sample matrix and on the result of the pre-enrichment culture with recoveries, which are disappointing when sampling complexity increases (Author's experience with *Salmonella* in Pathatrix system; [97,98]). Besides food matrix interference, the costs of IMS beads are an issue in routine food microbiological screening. Therefore, there is a continuous exploration for IMS alternatives for the isolation and concentration of target MOs from large volumes.

As an alternative for classic IMS beads, the magnetic microspheres are coated with (synthesized) binding molecules other than immunoglobulins. For example, bead-linked aptamers have been used to purify *Salmonella enterica* serovar Typhimurium from food samples [99]. Another alternative for IMS, with a long history, is continuous flow separation on metal wires [100].

Whatever steps are needed to prepare a laboratory sample into a test portion, it is without further explanation that treatments have to preserve the phenotypical of genotypical characteristics of the aimed MOs to allow correct determination.

2.3.3.2. Sample Preparation for Indirect Antigen Tests

Indirect antigen tests require samples containing reporting reactive immunoglobulins making the technology not applicable for plant products. Qualified sample types are blood, colostrum, eggs, feces, liquor, meat juice, milk, mucus or saliva, and tears, which can be collected *ante* and/or *postmortem*. Some of the listed matrices contain detectable levels of IgA, such as colostrum, mucus, and feces. However, blood serum and meat drip are used most frequently for indirect antigen (AG) investigations. These materials are superfluously available and relatively easily prepared, handled, and stored.

Whole blood is used seldomly and is treated to yield plasma or, in prevalent cases, serum. Meat drip, also called meat juice, is usually collected from the diaphragm, a muscle with low economic value. It is prepared commonly by placing a piece of muscle tissue (1 to 5 g) in a holder on top of a simple filter in a labeled tube. After freezing the complete set, including holder with tissue, filter, and tube, while, for example, transported to the laboratory, it is thawed just before analysis. Besides cytosol, the drip is composed of extracellular fluid mirroring serum proteins, including immunoglobulins. The matrix is suitable for testing as long as it is stored properly, and denaturing and inactivation of IG by freeze/thaw cycles and heat is prevented (see e.g., [101]). Furthermore, the fat content of the sample should be low as fat and other lipid-like materials can affect the analytical performances of the indirect antigen assay.

A clear advantage of indirect antigen testing is that multiple analytes can be screened in the matrix. Multiplexing bacteria, parasites, viruses, and even residues of pharmaceuticals in a single test is possible [102]. This versatile use and range of operation of indirect antigen tests are difficult to match by NASB techniques. These techniques have a blind spot for MOs giving a low burden, such as parasites and many viruses as well (further elaborated on below). In addition, histochemistry and microscopy in some cases of parasite monitoring (e.g., *Trichinella*) are labor-intensive, time-consuming reference methods for foodborne parasites making indirect antigen assaying a more attractive choice.

2.4. Where and How to Measure?

Despite quality and safety control mechanisms, such as HACCP and the good practices of GAP, GHP, GMP, GVP, effective blocking of contamination cannot be realized without analytical tests at critical points in the chain. Sampling at critical points is essential but is not trivial. The food-chain is complex and highly (internationally) organized with extended complexity (Figure 2). Sample type and sampling moment, but also apparent obvious matters like sample handling, quality of sample transport and repository, autolysis, contamination (with detergents, antiseptic materials), temperature fluctuations, UV exposure, influence the (quality of the) outcome of the result. Taking the wrong sample at the wrong moment, and/or label, store and/or transport it inappropriately and the analytical result of the most sensitive, most specific and most accurate microbiological analytical method is worthless (Table 2). Sample keeping and treatment should be optimal for the best result. In addition, analyses are credible when at least:

(*i*)   The investigator is adequately educated and trained to perform the analysis.

(*ii*)   The place where the test is performed is appropriate and hygienic measures are adequate and not a source for false positive (or false-negative) results.

(*iii*) The test is fit for its intended purpose.

Typically, testing is hindered by:

(1)   Asymptomatic carriers of a pathogen that remain unnoticed and are not excluded from the food-chain or not further investigated following a visual sanitary inspection.

(2)   The sometimes extremely low microbial dose causing disease in humans which therefore needs very sensitive analytical methods, and

(3)   the overwhelming presence of many other, non-harmful, entities obscuring the detection of a disease-causative agent.

These hurdles are relevant to understand to make educated decisions on the samples to be taken at which point in the food-chain, and on the most appropriate analytical method to be applied.

With reference to the first hurdle, *Salmonella enterica* spp. seem to be a ubiquitous part of the chicken's environment. Among the 2600 known *Salmonella* serovars, only two, *Salmonella enterica* serovars Gallinarum and Pollorum, disease chickens. All other *enterica* serovar infections have only minor, subclinical effects for a short period of time and only when the chickens are young [103], after which they become chronic asymptomatic carriers [104]. The chicken is thus prevalently *Salmonella* tolerant [18]. At the abattoir, an inspector is unable to suspect *ante* or *postmortem Salmonella* in a flock of broilers just by visual inspection unless samples are taken for laboratory investigations. The bird is not or little immune responding to *Salmonella enteritidis* serogroup C either and for this reason indirect antigen testing is not suitable for full *Salmonella* assessment of chickens.

In a similar way, animals infested by parasites may be without symptoms as well. In fact, a parasite giving the least disease burden while producing as much as off-spring is usually very successful in biological terms. In contrast to bacteria, some parasite species can be detected visually by a trained inspector or butcher in particular when they are incapsulated. However, most of these parasites are missed visually as especially viable macroscopic parasites are camouflaged by their meat-like color with little contrast.

Concerning the second critical hindrance, the infectious dose in a meal for healthy persons to get ill is as low as 10 CFU bacteria/g food, 10^3 bacterial spores, 10 virus particles, or 10 protozoan cysts [105–107]. Imminent foodborne pathogens such as *Escherichia coli* O157:H7 or *Campylobacter jejuni* require only an estimated number of 15 cells per gram [50] or 500 bacteria [105], respectively, to disease a person. Compare this with the 1,000,000 *Campylobacter* cells that fit on the tip of a pin. Although it must be said that there is great uncertainty over these numbers.

Legislative norms are for a part based on these low infectious doses. For example, *Salmonella* spp. and *Listeria monocytogenes* should be absent in a 25 g sample of most food products [3]. This sensitivity requirement is a significant and important difference with clinical microbiology. Proving the absence of a single bacterial cell in 25 g material (of a symptomless animal) is of a complete other analytical order and challenge than finding the causative pathogen in an appropriate sample of a diseased patient with clear and informative symptoms. In fact, proving the absence of an analyte is fundamental scientifically impossible. Nevertheless, absence in 25 g sample is a regulatory obligation and comes with problems in the poultry sector for the mentioned pathogens. Scientists worldwide have published a scientific pamphlet criticizing this zero-tolerance [108].

Table 2. Steps and phases in an analytical procedure determining a bacterial pathogen which can cause foodborne infections (adapted from [109]). These steps involve prevalently those for direct antigen analyses.

Step/Phase	Note
To investigate	Water: environment, processing water, drinking water (for animals or to prepare a meal). Environment: farm (including wild animals and insects in its surroundings), processing plant, abattoir, butcher, greengrocer, kitchen, etc. Pre-products: carcass, ingredients (herbs), etc. Products: fruit, meat, sliced vegetables, ready-to-eat, salads, etc.
Sampling	Sample quality and size should reflect what is investigated (food, flock, farm, herd, retailer, kitchen, etc.)
Transport	Identifiable at all times, properly cooled, and with no risk of cross-contamination
Sample treatment	Acclimatization Homogenize if required (as a matter of fact, homogenization is a specialism in itself)
Pre-enrichment	Resuscitate and proliferate bacteria to determine even low numbers
Selective enrichment	Proliferate the aimed pathogen exclusively
Culture evaluation	Gauge selective culture by assessing color, smell, turbidity and microscopical investigation, Gram staining, etc.
Analysis	Traditional plating, agglutination, enzymic assays, LFD, (IMS) ELISA, apta-/geno-/immune-/phagosensors, LoaC, LoaD, nucleic acid sequence assays
Verification	Confirmation and identification

Many reviews and overviews of available techniques for tracing health-threatening MOs in the food-chain see a schism between culture-dependent and culture-independent methods. Such partition is surprising. Suitable detection methods, which can verify the mentioned nadirs must be not only extremely sensitive but at these low levels tremendously selective and specific as well. In routine analyses, they have to deliver this performance on a high-throughput and a low-cost basis without rigorous sample treatment, such as inoculation of a culture broth or intensive sample cleanup in the case of e.g., not cultivable MOs. There is not much indication yet that these so-called culture-independent methods are becoming an industry-standard in the short term.

MOs are usually not dispersed homogeneously in or over a food sample. Hence, the compulsory 25 g sample material increases the likelihood to include pathogens in a laboratory sample. This amount is colossal for modern, advanced (instrumental) methods and needs sample processing to bring it back to the volume of a test portion. IMS is a method used often for this purpose (see Section 2.3.3.1). One of the challenges is that the final test portion has to reflect reliably the contamination status of the originally sampled animal, plant, or product. Occasionally, samples are pooled and a test portion from the pool has then to reflect a whole herd, farm, or batch in a direct antigen assay. It is without further explanation that this way of surveillance or monitoring is very critical for failures. Besides the required sensitivity (SE) and ability to process large sample quantities (10–25 g), this second hindrance reveals another analytical requirement related to the third hurdle as well, namely inclusivity or specificity (SP), i.e., the degree to which the assay does not cross-react with other MOs in the large sample matrix.

The third analytical challenge refers to relatively small amounts of the target analyte amidst large amounts of numerous biomolecules, such as proteins, lipids, and many other (large amounts of) MOs. This means that the analytical method used to trace pathogens should not only be very sensitive but be highly exclusive or selective as well. That is, the degree of the assay to distinguish the target from all other components in the sample should be very high. Usually, this cannot be obtained without sample treatment although modern (instrumental) techniques have introduced tremendous resolving power, such as mass spectrometry and several NASB methods.

2.5. Test What, When, for Which Purpose and at What Costs?

A test delivering an apparent simple "yes" or "no" answer to the question "is it safe?", usually does not provide an answer to questions like: "how safe?" and/or "what is making it not safe?". More specific questions in the scope of this review are:

(1)   Which microbes need to be analyzed and intervened?
(2)   Where in the food-chain can these MOs best analyzed and with how many of which (type of) samples? (see also above)
(3)   What are the test quality requirements?
(4)   What is the most effective test methodology?

With respect to question (1), preferably all hazards are screened simultaneously and continuously. However, a perfect world does not exist and for understandable reasons, choices have to be made. These choices should be primarily risk-based. In a simple approach, risks are assessed as:

$$R = E \times P \ (+H)$$

where R is risk, E is effect, P is probability and, optionally, H is the "hype" factor.

This formula exemplifies which MOs need to be intervened upstream in the food-chain. The factors R, E and P are obvious and can be illustrated by *Listeria monocytogenes*: the chance to contract listeriosis is low, but the potential effect of death is substantial (Table 1) and therefore the risk of foodborne *Listeria* is relatively high.

The "hype" factor is an addition to the formula by Prof. Rainer Stephany (Utrecht University, Utrecht, The Netherlands). With this factor, he gave expression to the whether or not justified pressure of media, politics, and/or irrational reactions of laypersons. An example of the H-factor was given here before with the "egg affair".

Typically, risks are region-dependent. As once pointed out by a South-African delegate to the authors, the supranational screening program applying to products exported from Africa to the EU obligates analyses of typical European risks not being African risks per se, while omitting the screening for typical African hazards. On the other hand, sophisticated legislation providing opportunities for risk-based methodology, risk-based sample sizes, and continuous mutual briefing of scientists and policy makers, can support an effective and efficient surveillance system focusing on all relevant health threats [110,111].

In many cases other than toxigenic *Escherichia coli*, *Listeria*, and *Salmonella* (Table 3), a method cannot deliver simply a yes or no answer. It requires a threshold, not the limit-of-detection (LOD) *per se*, which is preferably scientifically determined. A result below the threshold implies acceptable risks. A result over this level indicates risks that are considered not acceptable. However, such determination of a threshold can also be based on whether the risk can be avoided and how much effort (usually this means "costs") it takes to reduce the risk. It is for this reason that some thresholds are determined pragmatically as for *Campylobacter*.

The prevalence of *Campylobacter* spp. in broilers is high, especially in organic and free-range flocks [112], and modern production methods are not able (yet) to deliver the degree of biosecurity to prevent disease in consumer at an acceptable level. In addition, the EU allows only limited use of a few decontamination treatments of carcasses which will not completely abolish this microbial risk of poultry [113]. Therefore, elimination of the hazard would mean the prohibition of the production and sale of poultry meat. As this is not acceptable either, so-called food-safety objectives (FSO) were introduced [114,115]. In the case of *Campylobacter* spp. this means that out of a predefined number of samples per batch, a certain number of samples must be less than an established load of the bacterium.

In fact, the norm of 1000 colony forming units (CFU) per gram broiler carcass relies on this reasoning (Table 3) [114].

Table 3. A selection of norms for pathogenic MOs in some food products [3]. The criteria apply at different stages in the food-chain and are associated with a sampling plan and with certified ISO methods not indicated or included in this table.

Pathogen/Predictor	Food Product/Matrix	Norm
Listeria monocytogenes	Ready-to-eat [a]	Absent in 25 g
Salmonella	Cheese, butter, cream from raw or non-pasteurized milk	Absent in 25 g
Salmonella	Meat products intended to be eaten raw	Absent in 25 g
Salmonella	Meat preparations intended to be eaten cooked [b]	Absent in 10 g
Staphylococcal enterotoxins	Cheeses, milk powder, and whey powder	not detected in 25 g [c]
Enterobacteriaceae	Egg products	10 or 100 CFU/mL or CFU/g
Campylobacter spp.	Carcasses of broilers	1000 CFU/g
STEC O157, O26, O111, O103, O145 and O104:H4	sprouts	not detected in 25 g [c]

[a] Exceptions of the norm if ready-to-eat food is not intended for infants or special medical purposes then 100 CFU/g. [b] If made from poultry, then *Salmonella* should absent in 25 g. [c] "not detected in 25 g" using a specified ISO method: implying an LoD not necessarily meaning "absent in 25 g".

With respect to the second question (2), involving the best position in the food-chain to take samples, how many and which: incidental food contamination must be prevented at all stages of food production, processing, distribution, retailing, and preparation. Although a majority of contamination occurs in the tertiary and quaternary sectors, i.e., mass and home kitchens (see Figure 2), still a significant number of the foodborne infections involve germs originating from a reservoir earlier in the food-chain. These reservoirs can be found in animals used for the production of food and/or in wild animals, including (rodent) pests, living in or near a farm. Occasionally, diseases (*Salmonella*, *Campylobacter*) are transmitted through vectors as flies (*Musca domestica*) [116] and other insects as well. It tells us to screen and tackle the sources of these microbes upstream as early as possible and in the environment close to their source(s).

There is another reason to trace and mitigate risks as early as possible: unpolluted products may be cross-contaminated by a tainted animal, plant, carcass, or food product. Cross-contamination can be prevented when the MO is eliminated or isolated as soon as it is detected. In addition, MOs can multiply themselves in the food-chain when not controlled appropriately. A notable, remarkable and frequently observed phenomenon is that reduction of *Salmonella* prevalence through intervention programs gives opportunity to other *Salmonella enterica* serovars to fill the void created after sanitizing other serovars ([20] and unpublished). Although of another order, but a similar phenomenon nonetheless, de-contamination of food products as an intervention step, destroys the colonization resistance of non-pathogenic bacteria and thereby provides an opportunity for new (pathogenic) bacteria (and fungi and yeasts as a matter of fact) to re-colonize the product and introduce new food-safety risks. After all, biological laws and entropy cannot be fooled. Notably, risks, such as that of *Salmonella enterica*, may also come from pet food [9,117]. It demonstrates that the complete human food-chain, including the pet food branch, needs continuous attention, inspection, and control.

With respect to the third question (3), it should be clear for whom food is produced. Analytical quantity and quality requirements for an analytical procedure to examine food safety are stricter when the consumer is a so-called YOPI, *i.c.* belonging to the group of

Young (<5 years of age), Old (>64 years of age), Pregnant or Ill (immunodeficient) people. Furthermore, quality requirements also depend on the type of investigation (*cf.* Table 4). Different test requirements are needed for example in a case of (legal) dispute or claim settlement, of a surveillance (aimed, close and short investigation) or of a monitoring (continuous assessment) program.

With respect to question four (4), "what is the most effective test methodology?", the spectrum of available analytical strategies is tremendously broad (Figure 4). It is fairly impossible to weigh and compare all these techniques against each other. To assist a user to decide on a method, Table 4 summarizes the characteristics of an ideal analytical procedure. Needless to write that no single existing assay satisfies all listed criteria. The user must weigh a method of choice based on the type of the MO, assessment point in the food-chain, quality of result, throughput, specificity, TTR, etc., not necessarily in this order of importance.

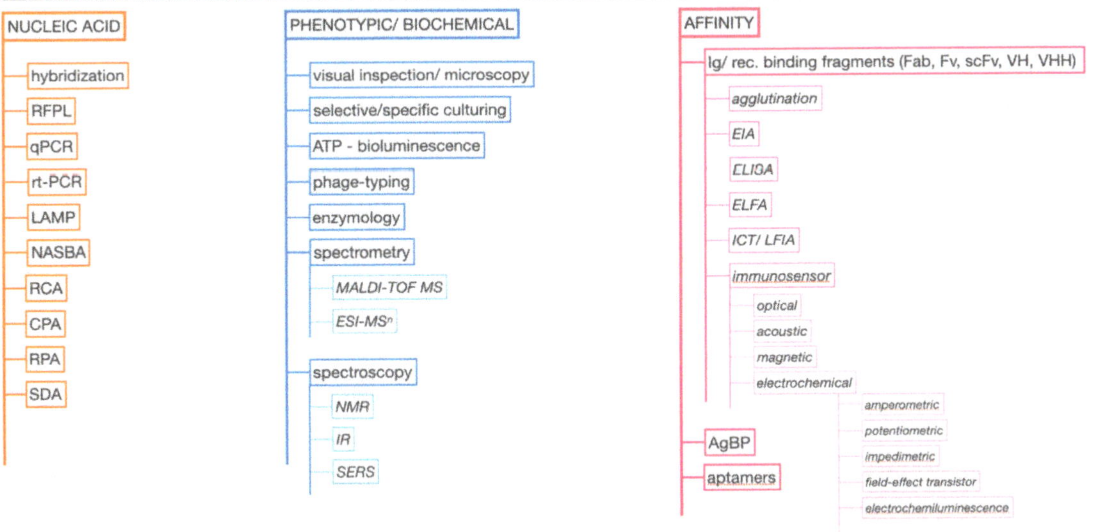

Figure 4. Impression of analytical methods and approaches to find and determine (pathogenic) micro-organisms in the food-chain. It should be noted that electrochemical immunosensors exploit many different detection principles, including amperometry, impedimetry, field-effect transistor (FET), potentiometry. This overview is far from complete. ATP, adenosine triphosphate; AgBP, antigen-binding proteins (not immunoglobulin related); CPA, cross-priming amplification; EIA, enzyme immunoassay; ELFA, enzyme-linked immunofluorescent assay; ELISA, enzyme-linked immunosorbent assay; ESI, electro-spray ionization; Ig, immunoglobulin; IR, infra-red; GC, gas-chromatography; ICT, immunochromatographic test; (RT-)LAMP, (reverse transcription) loop-mediated isothermal amplification; LFIA, lateral flow immunoassays; MALDI-TOF, matrix-assisted laser-desorption ionization time-of-flight; MS, mass-spectrometry; MS ^{n}, multi-stage MS; NASBA, nucleic acid sequence-based amplification; NMR, nuclear magnetic resonance; qPCR, quantitative (real-time) polymerase chain reaction; RCA, rolling circle amplification; rec. binding fragments, parts of immunoglobulins obtained by recombinant DNA protein engineering; RPA, recombinase polymerase amplification; rt-PCR, reverse transcription-polymerase chain reaction; SDA, strand displacement amplification; SERS, surface-enhanced Raman-spectroscopy.

Table 4. Typical end-user requirements for assays that determine MOs in the food-chain. The order of listed requirements is arbitrary.

Requirement	Note
Low cost	Test outcome should provide sufficient information to make a contemplated decision at costs that balances investments and consequences. Costs include overhead, maintenance, personnel, laboratory footprint, disposables, etc.
One-pot reaction	Complete preparation and processing in a single vial/tube. No addition of reagents required (~ easy-to-use).
Range of application/ operation	Versatile: *sensu lato* applicable to fresh and processed food matrices, and/or for samples from feed, plants, and animals. Suitable for all relevant bioagents.
Multi-analyte/multiplex	Able to determine multiple targets in a single analysis run.
High-throughput	Multiple samples run simultaneously.
(relative) accuracy	Free from systematic and random errors.
Reproducible (precise)	No or insignificant variation (in-)between runs, days, machines, analysts, etc.
Repeatable	No or insignificant variation between laboratories, batches, lots.
(Relative) specificity	100% specific compared to the reference method (no false positives); able to distinguish at the strain level.
(Relative) sensitivity	100% sensitive compared to the reference method (no false negatives); able to detect (viable) MO at required sensitivity [a]. Excellent signal-to-noise ratio.
Confirmation of analyte	Juridical problems arise when the identity of the analyte is not confirmed. There should be a reliable answer to the question: what is the degree of certainty of the identity of the aimed target giving the result?
Automable	A stand-alone instrument with limited hands-on time. The frequency of personnel attention is low.
Portable/Point-of-care (PoC)	Ambulant, performing analysis in/at/on-line of the food-chain with interpretable results available on site.
User-friendly and fail-safe	Easy-to-use, performed by non-skilled personnel, preventing facultative and inherent errors. In other words, the assay is rugged and gives the same result even when test conditions such as incubation times, operator, pH, reagent concentrations, temperature, etc. vary. This also includes safe to use.
Easy to interpret	Analysis data give a transparent, unambiguous, and understandable result
Instantaneous result	Results available real- or near-time. Although not meaning the same, time needed for the whole analytical process from sample collection to result (TTR) should be short. In practice, TTR can imply a result within a working day/before the next phase in the food-chain.
No sample treatment	The test portion is obtained directly from the laboratory sample.
Robust	Reliable operation under e.g., humid, varying temperatures, bumping conditions. No drift/long-term stability.

[a] Required sensitivity: Detection level at which the MO does not pose a health risk or at the legislative norm.

Available analytical approaches can be cataloged roughly and not exhaustively into NASB techniques, immunoglobulin-based methods and biosensor-based technologies, and physicochemical techniques, such as NMR and MS. These approaches are more or less discussed here below.

2.6. Immunoaffinity Assays

Immunological methods rely on a specific interaction between immunoglobulin and an antigen to selectively capture, label, and detect antigens associated with target organisms. Besides detection, showing a specific antibody-antigen binding reaction is accepted as a way to confirm the identity of an isolated MO as well. Until the end of the last century, agglutination, EIA, and ELISAs were apparently the only existing IA formats [118,119], but much has changed since then. However, this review does not elaborate on all developments, such as heterogeneous versus homogeneous immunoassays, nor on all possible configurations, such as a sandwich, open sandwich, competitive versus immunometric assays, etc. The interested reader may learn more about these configurations and test settings in [120].

2.6.1. Traditional Methods

2.6.1.1. Direct Antigen Approaches

The first direct antigen methods were developed since the awareness of cellular pathology (see Section 2.2). Agglutination, i.e., macroscopic clumping when agglutinating antibodies in sera react specifically with (bacterial) antigens, was first observed in 1896 by various bacteriologists [121]. This specific antibody-generated clumping of bacteria became known as Gruber–Durham reaction. In the food microbiology laboratory, the agglutination test is hitherto performed by mixing latex spheres coated with target pathogen-binding immunoglobulins with a test cell suspension, which is usually obtained after enrichment. In a positive reaction, conglomerates are visible after some minutes [122]. Typically, tube tests are more sensitive than slide agglutination tests. Since the early bacterial agglutination tests in tubes, on glass slides, and, much later, in microtiter tray wells, the method has evolved into many configurations. For example, gold particles are attached to the agglutinating antibodies to improve sensitivity. In addition, a broader range of applications, including the detection of viruses and parasites, have become available. Besides antibodies, variations were developed using lectins, which bind specific (surface) carbohydrates, and deploying advanced fluorescent emission (see for an example [123]).

Another variation, known as reversed passive latex agglutination (RPLA), is the immobilization of probing immunoglobulins on latex particles which are incubated with food extracts to detect bacterial toxins (the target antigen) to form lattice structures. Examples are the detection of staphylococcal enterotoxins SEA, SEB, SEC, and SED [124] or *Bacillus cereus* enterotoxin [125]. Although RPLA is more sensitive than PCR methods, this method can be less sensitive and less specific than the less easy-to-use cell cytotoxicity tests applied for the same goal [126,127].

In conclusion, immuno-agglutination tests are simple to use, inexpensive, require minimal equipment, and are relatively rapid. One of their disadvantages though is that they consume relatively much antiserum, which can be scarce and costly for some specific (serovars of) MOs.

Lateral flow immunoassays (LFIA) or immunochromatographic tests (ICT), which were sprouting from the continued development of the agglutination test in the 1950s [128], are user-friendly. In ICT tests, IG-binders, including protein A or anti-IG antibodies, are immobilized in a test/capture line on a porous membrane. Other, often gold- or color-dyed particle-conjugated ABs against the antigen of interest are located just under the pad where the sample, such as a portion of (diluted) enriched medium, is put on.

The LFIA tests were one of the first successful attempts with a major impact on fast end-point analysis in, especially, clinical situations. The tests come in different configurations as uncovered immunochromatographic test strips put vertically in a test solution, or "wrapped" in a plastic casing (LFD) which became known to the public as the "pregnancy test". Examples of the use of lateral flow devices (LFD) is the detection of EHEC O157:H7, *Salmonella*, *Listeria* in different food products [129–133]. However, as explained, an enrichment step by culturing is needed before these tests can be executed reliably. Although the development of the test itself takes typically 10–20 min, a result may therefore still take up to 48 h. A clear advantage is that no (expensive) equipment is needed.

The LFD technique also allows multiplexing in different ways and the interested reader can learn more about these developments in [134]. Multiplexing LFDs can characterize different (closely related) bacteria and viruses and have attracted (veterinary) clinical interest. However, they consume relatively much of a sample, show low reproducibility, and do not offer high sample throughput possibilities.

The development of chromophoric or fluorophores enzyme immunoassays (EIA), enzyme-linked immunosorbent assays (ELISA), enzyme-linked immunofluorescent (ELFA), and chemiluminescence (CLIA) assays in various configurations was a consistent next step in the evolution of microbiological antigen tests. Although the original discovery of ELISA is disputed, the technique saw daylight for the first time at the end of the 1960s in the last century. These assays need more hands-on time than LFIA, although steps can be carried out by robotic liquid handers. Development of ELISA(-like) assays takes longer to obtain results (1–2 h), but they have improved sensitivity, offer quantification possibilities, and allow a much higher sample throughput per time unit.

An improvement of up to five orders of magnitude, detecting as less as 10^{-21} moles (zepto moles; closing into the number of Avogadro!), is obtained by replacing the (enzyme-)label-conjugate by an oligonucleotide that can be amplified exponentially by a usual PCR reaction [135,136]. This technique called immuno-PCR (iPCR) exploits the flexibility and versatility of an ELISA and the signal amplification ability of a PCR reaction. As enzyme inhibiting sample components are washed away in the ELISA part of the procedure, the PCR reaction is unhindered. This combination has demonstrated much better sensitivities for *Clostridium botulinum* neurotoxins A and B in different types of milk than the standard mouse bioassay [137]. The iPCR further evolved into real-time, immunoquantitive variants (iqPCR). One of these iqPCR methods was 10^2 to 10^3 times more sensitive than its ELISA counterpart using conventional reporters to detect *Staphylococcus aureus* enterotoxin B (SEB) in different foods, such as coffee creamer, cooked ham, paella [138,139].

The ELFA principle is used, in some cases with better results than traditional methods, to detect *Listeria* spp., *Campylobacter*, *Salmonella*, *Escherichia coli* O157 in various food products [140]. Anno 2020, this measuring principle is available commercially, for example, the VIDAS system (bioMérieux) optionally in combination with IMS to reduce running times significantly.

As most viruses cannot be or otherwise not easily multiplied, their tiny harmful amounts are an arduous analytical challenge to flag products as potentially contaminated. Classic diagnostic methods, including virus isolation, remained principally technically unaltered, or changed and improved only a little. For this reason, but also because viruses are very diverse and change structurally continuously, analytical techniques based upon viral nucleic acid sequences are much more suitable and usually outperform immunoaffinity-based detection approaches. In contrast to traditional tests, NASB methods have advanced dramatically over the last two decades [141].

2.6.1.2. Indirect Antigen Approaches

The first known example of an indirect antigen test is the Widal agglutination test developed by Fernand Widal (1862–1929) using inactivated *Salmonella typhi* as antigen to probe blood serum antibodies in a patient to be able to diagnose typhoid fever [142]. As antigen and serum are minimally prepared, interpretation of a positive result is difficult by the many cross-reactions.

Whatever the indirect antigen assay of choice, the analyst has to be aware of the complex milieu of biological fluids containing components in variable amounts which can affect the assay in a non-specific manner. The matrix may contain soluble receptors and other antigen-binding proteins yielding false-positive outcomes. In other instances, immunoglobulin-ligand interactions may be impaired and may thus produce false-negative results. Moreover, IG-binding proteins, such as complement factors, may occur at unpredictable concentrations and attribute to unreliable results.

For decades, indirect antigen tests were executed prevalently in the well-known 96-wells SBS plates, also known as Microtiter, which is the formal trade name of the plate and spelled like this. Here, SBS refers to the organization of the Society for Biomolecular Sciences which has defined the standards of the plates with 24, 48, 96, and 384 wells. Non-standardized plates with 1536, 3456, and 9600 wells are available as well. The plates were and are used for various indirect antigen ELISA assays, which deployed and deploy still predominantly a chromogenic sandwich configuration. That means, antibodies in a test portion bind to immobilized antigens. In a traditional assay set-up, bound antibodies are then detected by colorimetry after incubation with an enzyme-labeled anti-antibody antibody and chromogenic enzyme substrate. Alternatives for the solid phase of a microtiter plate have brought analytical innovations and some are discussed below.

2.6.2. Advanced Methods

2.6.2.1. Bead-Based Arrays

In 1968, a special sort of microscope was invented in Germany (called "*Impulszytophotometrie*") to count and differentiate cells. This flow cytometer can detect labeled MOs as in a direct antigen test for food analysis (see for an overview [143]). Whereas the flow cytometer is suitable for particles as large as cells, including bacterial cells, viruses are too small for the usual dimensions of the flow cell of the instrument and cannot be detected directly. The detection of *Escherichia coli* O157:H7 in ground beef is an example of a direct bacterial application [144].

Instead of a static two-dimensional solid phase, such as in SBS Microtiter plates, a solid phase when sufficiently small, can also be coated with a ligand and then suspended enabling fluidics through the whole sample column and other handling with the formed immune-bead complexes. Bead-based assays (BBA) deploy microspheres as solid phase and carrier of a ligand [145]. When mixed with a sample, contact between receptor and analyte is intensified substantially compared to ELISA(-like) assays which largely depend on Brownian motion even when the Microtiter plate is agitated. The fluid is namely almost static at the nanometer scale at which the recognition and binding reactions take place. The improved interaction by suspension of the solid phase provides more binding reactions per volume and per time unit allowing shorter incubation time intervals and correspondingly less non-specific binding events [146,147]. In other words, BBA reaction kinetics are typically fast, reproducible with very good signal-to-noise ratios.

The flow cytometer identifies defined beads and measures the signal generated from the binding of an analyte to the ligand immobilized on the microspheres. Actually, the flow cytometer is operated as a flow "beadmeter" performing "beadmetry". Standardized microspheres come with diverse material compositions available from an increasing number of vendors (see for a list [147]). In most applied configurations, microbeads are dyed with a standardized series of varying intensities of a single or mixtures of chromophores or, preferentially, fluorophores. The flow cytometer can not only identify the color of a microsphere but also discriminate physicochemically closely related microspheres (Figure 5). In other words, besides variation of dyes, other chemical and physical variations, such as quantum-dot tagging, different accurate sizes of microspheres, add dimensions resulting in an array. Correspondingly, the BBA becomes a suspension array assay facilitating multiplexing. Systems with up to 500 distinguishable bead-ligand-analyte combinations (multiplexing) are possible. An advantage of BBA over many other multiplex techniques: a bead-ligand combination, a so-called "set", is easily replaced, added, or removed from an array of beads.

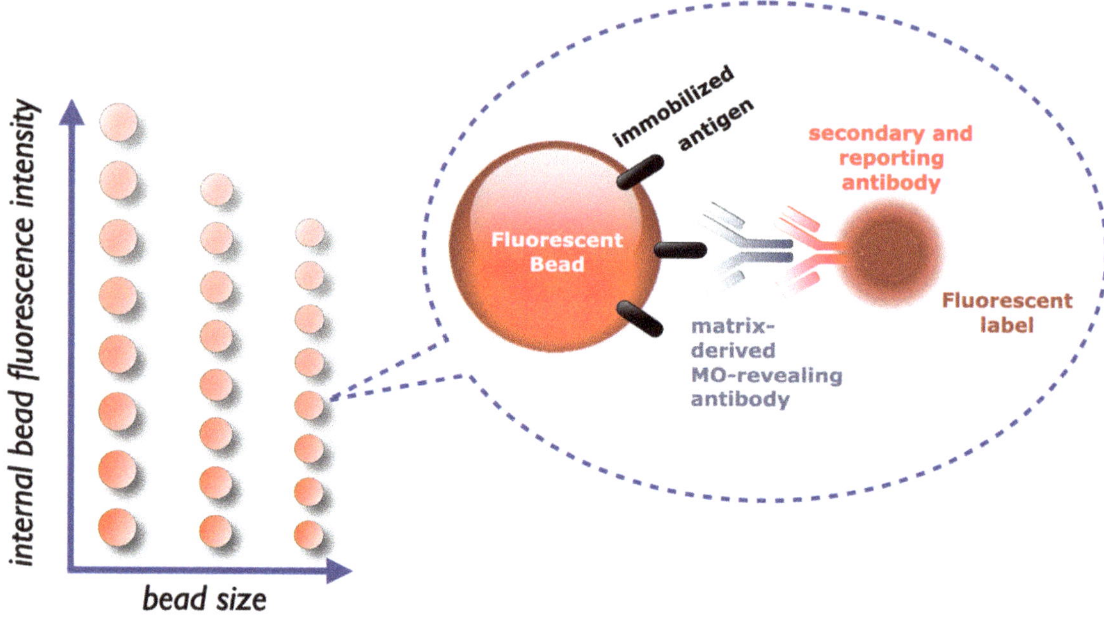

Figure 5. Illustration of an example of a bead or microsphere array used in BBA assays. In the left panel beads are discriminatory in two dimensions: a gradient of an internal color (Y-axis) and diameter of the microsphere (X-axis). Each bead is thus identifiable and can be coated distinctively (right panel). In this case, one of the beads picked from the series in the left panel, is carrying an antigen (black-colored rod shape) to which an immunoglobulin from the analytical matrix (grey-colored Y-shaped structure) is specifically bound. To enable detection of the sample-derived and captured antibody in a flow-cytometer, a secondary, labeled binder is allowed to attach (dark-red-colored Y-shaped structure with reddish orb). The assay is usually developed in Microtiter trays. In the instrument, bead-size, bead fluorescence and label fluorescence of each bead in the suspension are measured and reported. Note that many variations on this test principle are deployed.

The multiplexing feature has made BBA assays *en vogue* in many life-science disciplines, including food safety and veterinary diagnostics [148–150]. The technique is for example suitable to distinguish naturally infected animals from those immunized with modified vaccines in the DIVA approach, such as for foot-and-mouth disease and Rift Valley Fever virus [150].

Something to notice, however, the resolving power of the instrument may be insufficient to distinguish the spectral character of closely related beads, either by the interference of sample contaminants, poor calibration, and critical performance of the instrument. In addition, low-quality beads may have overlapping characteristics not distinguishable by a properly functioning instrument. One of these failures may result in "blurring" and "tunneling" of signals from the correct channel to the next measuring channel of another bead. Consequently, an analyte is identified and quantified incorrectly as another analyte (authors' experience).

Multiplexing offers a possibility to simultaneously detect: (i) complete, intact MOs with respect to multiple virulence factors, such as for serotyping, and/or (ii) multiple, different MOs or their freed moieties/excreted elements in a single sample in one analysis run. An example is the simultaneous analysis of staphylococcal enterotoxins SEA, SEB, and SHE in (cottage) cheese, meat broth, minced meat, milk, and omelet [151]. The performance of this BBA method was comparable to a frequently used commercial ELISA and the test was able to declare the products compliant or non-compliant with Commission Regulation 2073/2005 [3,151]. An overview of direct antigen BBA (multiplex) assays to screen a range of animal diseases and clinical biochemical parameters in various animals, including horses, rodents, ruminants, swine, and mycotoxins in feed is provided in [150].

Microspheres suitable for BBA assays come also as paramagnetic variants. The synergy with IMS increases sensitivity and specificity. An example of this approach is a direct antigen multiplex assay for the determination of pathogens and disease biomarkers of *Bacillus anthracis, Francisella tularensis, Yesinia pestis* causing anthrax, tularemia, and plague, respectively [152]. Another example is the same-day ($\leq$7 h) detection of *Escherichia coli* O157:H7 in spinach at a detection limit of 0.1 CFU/g using magnetic microspheres [153].

Not only beads are modified but the flow cytometer is altered as well. It may contain a magnetic field to immobilize magnetic microspheres in a layer for optical stability so that a CCD camera instead of lasers can be used and more beads at once can be analyzed [145,148]. An example of a direct antigen multiplexing assay using a CCD camera is the typing and classification of O26, O55, O78, O118, O124, O127, O128, O142, O145, and O157 antigens on pathogenic *Escherichia coli* [154]. *Mutatis mutandis*, bead arrays are also used in a planar format in so-called "-omics" technologies enabling thousands of tests simultaneously while each of these tests analyzes a range of molecules in parallel [149].

Indirect, instead of direct, antigen suspension-array assays have shown their merits for screening anti-*Salmonella* antibodies in *porcine* sera [155,156]. A BBA assay for anti-*Salmonella* antibodies in egg yolk to determine the *Salmonella* contamination status of a layer flock showed a good relationship with the infection status of the probed flock [157]. Here, the collection of eggs makes personnel taking blood samples redundant and contributes to better animal welfare. In fact, the collection of eggs for surveillance programs can be executed at egg-packing stations [157].

The legislative and mandatory method for finding the parasitic worm *Trichinella spiralis* is artificial digestion and microscopic examination of pooled pork samples [158]. The method is antiquated and in proficiency tests, the false-negative rate can vary between 11% and 100%, in particular when the parasitic load is low and when the parasites are in their pre-stadium of encapsulation [159,160]. In an animal experiment, an indirect antigen BBA showed 68% diagnostic sensitivity and 100% diagnostic specificity [161]. Despite the better performance, acceptance of indirect antigen assays is complicated because of the seroconversion window, discussed here above (Section 2.3.2.2), which may indeed declare meat from an individual animal falsely negative. Nevertheless, anti-*Trichinella spiralis* antibody screening is a high-quality aid for monitoring exposure to the worm and epidemiological studies [63].

Interestingly, the *Trichinella* BBA method was combined with finding antibodies against *Toxoplasma gondii* in a multiplex assay. This combination is not without reason. The *Toxoplasma gondii* parasite is an indicator of the level of hygiene in the pre-harvesting phase, as it reflects the contact of pigs with their extramural environment [159,160]. The parasites also share routes of infections, although *Toxoplasma gondii* more omnipotent than *Trichinella*, which makes *Toxoplasma* screening only more favorable. Moreover, *Toxoplasma gondii* is ranked as zoonotic MO with one of the highest disease burdens in The Netherlands [162] and USA [163], while it is very persistent and prevalent in the environment as well. Screening *Toxoplasma gondii* may not only help to intervene *Trichinella spiralis* but *Toxoplasma gondii* itself as well.

So far, many multiplex BBAs for pathogen analysis are not IA-based but are built on a combination of microspheres and analysis of bacterial genetic material [145]. A commercially available NASB suspension array multiplex kit for human stool detects *Campylobacter* sp., *Clostridium difficile* (toxin A/B), *Escherichia coli* O157, enterotoxigenic *Escherichia coli* (ETEC) LT/ST, Shiga-like toxin-producing *Escherichia coli* (STEC) stx1/stx2, *Salmonella* sp., *Shigella* sp., *Vibrio cholerae, Yersinia enterocolitica*, human adenovirus serotypes 40 and 41, norovirus GI and GII, rotavirus A, *Giardia, Cryptosporidium* and *Entamoeba histolytica* [145].

2.6.2.2. Immunosensors

Biosensors exist since 1962 [164] and hitherto include immunosensors, genosensors, aptasensors, phagosensors, etc. Here, the type of sensor is referring to the ligand attached to the sensor surface, namely immunoglobulins, oligo- or polynucleotides, aptamers, and bacteriophages, respectively. The word "sensor", i.e., *sentire* (Latin), is referring to something that "perceives" changes in its environment. The instruments are coming into vogue and have evolved at a tremendous pace. In line with this evolution, instruments biosensing (pathogenic) bacteria have attracted increasing attention reflected by almost 11,000 scientific publications since 1990 (12 publications in that year) of which half in the last six years (Figure 6). Roughly a quarter of the publications deal with the analysis of bacteria directly and indirectly in food or animals producing food. It is fairly impossible to give a comprehensive overview of all developments and applications. This paper does not desire to discuss limit-of-detections (using mostly spiked samples), chemical, microbiological and technical details, it aims to place them in the context of current and future practice of producing safe food. If the reader is interested in such details, he/she is kindly referred to reviews attempting to give such exclusive overviews as in recent [165–180].

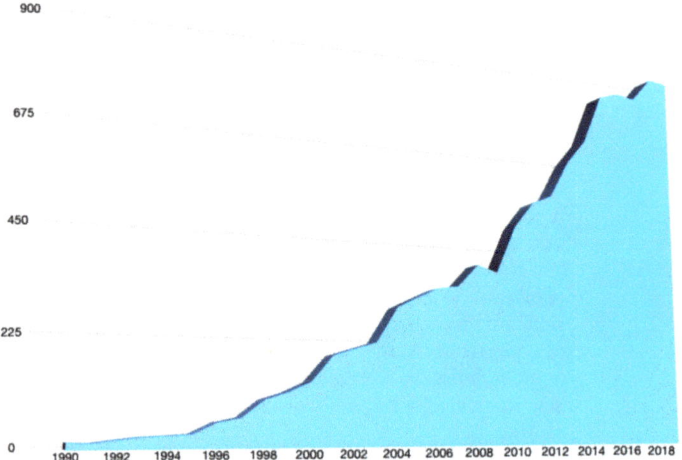

Figure 6. Number of publications per annum returned by PubMed when using search terms "biosensor" and "bacteria" [181].

There are multiple rationalizations for the spectacular growth of biosensor methods in food microbiology. Biosensors promise fast, ambulant, real-time measurements by a portable device. Many of the systems can split a flow over multiple channels on a single sensor or can combine multiple sensor surfaces in a single flow giving the means to combine the analysis of more than one analyte, indicator, parts, or whole cells. Multi-analyte testing not only involves the determination of e.g., *Salmonella* and *Listeria* in a single analysis run but the determination of species, subspecies, subsubspecies, and even strain as well. Such analytical aims might be accomplished in combination with predictive rationalized mixtures of multiple individual antisera with known binding profiles [66].

Biosensors contain an interface, the sensor, which includes a transducer generating an electronic, quantifiable, and recordable signal. An electronic signal is generated directly or indirectly when binding reactions at the sensor surface change its acoustic, electrochemical, optical, piezoelectric, or thermometric behavior. The reaction is that of the analyte with biologically derived or biomimetic material, such as enzymes, antibodies, cell receptors, lectins, and other (engineered) polypeptides, aptamers, (lipo)polysaccharides, or poly/oligonucleotides. The binders are immobilized on the sensing surface causing the following reaction [182]:

$$L + A \longleftrightarrow LA + B_{label} \longleftrightarrow LAB_{label} + \text{effectuator} \rightarrow \text{SENSOR DETECTABLE EFFECT}$$

Is the reaction in biosensors relying on the use of labels, where the effectuator is a substrate, photon or (conductive) particle etc. providing a measurable response in combination with the label.

or:

$$L + A \longleftrightarrow LA \rightarrow \text{SENSOR DETECTABLE EFFECT}$$

in the case of label-free biosensing.

In these formulae L is a ligand, A is analyte and B_{label} is the labeled binder, such as a labeled anti-analyte antibody.

Many of the machines produce a real-time signal, thus an observable signal while the reaction at the sensor surface is taking place. In fact, this distinguishes many immunosensors from most other IA techniques which are black-box, end-point measurements. Quite some immunosensors function without labeling a molecule to obtain a signal, which is common in all non-biosensor IA methods. Label-free measurements are preferred, as (non-reproducible) labeling chemistry potentially degrading biomolecules can be skipped. Furthermore, labels are notorious for steric hindrance and attachments to active sites destroying the function of the modified molecule. Surface plasmon resonance (SPR) biosensors are an example of such label-free systems and offer the possibility of multiplexing by two-dimensional arrays or multiple channels flowing and probing simultaneously [183]. Besides label-free, SPR measurements are real-time, and a skilled operator predicts the type of binding, for example, binding of IgG or IgM to an immobilized antigen, by judging the binding dynamics while the reaction occurs and the sensorgram is recorded (authors' observation). Another advantage of real-time measurements is that a run can be stopped untimely if no reaction occurs. This increases sample throughput in particular when the negative sample rate is high also because regeneration of the surface can be omitted.

When aiming for direct immunosensor assays, it should be realized that the size of a rod-shaped bacterial cell is typically 1 μm × 3–5 μm. In some devices, the channels are not much wider than these measures and bacteria can clog up the fluidic parts. In addition, the bacterial size can limit mass transport in a flow, i.e., diffusion of the whole bacteria to the sensor interface, giving a lower response than expected. To overcome this limitation, some methods use boiled bacteria to degrade the cells and enhance sensor response [184,185].

A bulky cell or a large cell part captured to sensing surface hinders sterically the capturing of another cell (part) [186]. In addition, when captured, shear forces generated by the laminar flow in the instrument can be stronger than the energy of the antigen-antibody bond further reducing sensitivity by loss of complexes [96,186]. Other considerations when developing and using direct antigen immunosensor assays are the length of the linker molecule that attaches the anti-target antibody to the sensor surface. Variation of the binding chemistry has a significant influence on selectivity and sensitivity for bacterial cells [166].

Unfortunately, in SPR biosensors, the size and mass of a captured bacteria do not contribute completely to a final response. SPR sensing takes namely place within approximately 700 nm distance from the sensor surface while bacteria are much larger [185,186]. The sensing electromagnetic field decays exponentially until this distance. In addition, immobilization chemistries cut another 10 to 100 nm of a specific signal near the surface layer, i.e., the most sensitive part. A solution for these signal limitations has been developed through so-called (grating-coupled) long-range surface plasmons enhancing the sensitivity for *Escherichia coli* 4 orders of magnitude [187]. Despite all challenges, 53 *Salmonella enterica* serovars were detected in avian feces and breast meat at 22 CFU/g or 1.7 10^3 CFU per test portion, while 30 non-*Salmonella* species did not give interference [184]. Other examples of successful direct bacterial SPR detection are that of *Campylobacter jejuni* sp. Jejuni, *Es-*

cherichia coli O157:H7, *Listeria monocytogenes, Staphylococcus aureus, Vibrio cholerae* O1, and *Yersinia enterocolitica* [185].

An indirect antigen SPR immunosensor method was developed to verify the *Salmonella* vaccination status of chickens [186,188]. Using this method involving LPS (O) and flagellar (H) antigen ligands, a linear response in chicken serum was acquired for antibodies against *Salmonella enteritidis* and *S. typhimurium*. Immunoglobulins against *Salmonella enterica* serovar Enteritidis were gauged in egg yolk using an SPR immunosensor to assess the infection status of non-vaccinated layer hens [189]. The method showed favorable sensitivity and specificity performance compared to ELISA methods. Similarly, an indirect SPR antigen assay based on immobilized LPS isolated from *Salmonella enterica* Typhimurium was used to probe *porcine* sera [190].

Like SPR, ellipsometric, surface-enhanced Raman scattering (SERS) and many other optical biosensors [175], impedimetric biosensors can be used label-free. Impedimetric biosensors respond to changes in effective resistance as a result of interfacial events. Sensitivities are as low as 10 CFU/mL of *Bacillus cereus, E. coli* O157 and K12, *Listeria monocytogenes, Salmonella, Shigella dysenteriae*, etc. in various food products, including chicken and milk (see for an overview [173]).

Amperometric and voltametric biosensors (Figure 4) are operated with non-labeled or labeled molecules using different voltametric regimes and reporting limits of detections of 10 CFU/mL and higher for various pathogenic bacteria in different food products, including dairy, eggs, fish, poultry and, remarkably, licorice [173,191]. An overview of developments and applications of electrochemical immunosensors to detect directly *Salmonella enterica* can be found in [192].

Most electrochemical biosensors rely on a label that is often horseradish peroxidase (HRP) [176]. The enzyme catalyzes the decomposition of various substrates while producing detectable electrons. Although some of these systems meet impressive analytical sensitivities, they have relatively long development times and/or are have a limited range of applicability. In an intriguing voltametric biosensor application, the ligand, *viz.* immobilized, specific peptides, is the recognition site and substrate for *Staphylococcus aureus-* and *Listeria monocytogenes*-specific proteases which cause an electrochemical response upon proteolytic cleavage of the ligand [193].

An attractive development in the sensor field is the accessibility of printing technologies enabling (in house) reproduction of electronics at low cost [194–196]. Furthermore, the availability, usability of, and possibility to integrate (magnetic) nanomaterials, carbon nanotubes, or nano rods in biosensors in, in particular, electrochemical biosensors is noteworthy [168,171,197]. Nanomaterials offer large surface-to-volume ratios and thereby enhanced interactions and reactions and thus selectivity, sensitivity, and speed. Not only interaction is improved, but nanomaterials also progress the function of the transducer in some sensor types [166]. Examples are the detection of *Escherichia coli, Listeria* spp., *Salmonella* spp., *Streptococcus* spp. and *Vibrio parahaemolyticus* in various food matrices [166,172].

Of the non-immunoglobulin binders in combination with biosensors, aptamers and molecular imprinted polymers (MIPs) are attracting the most attention to capture pathogens, including bacteria and viruses [198–202]. The alternative binders are relatively expensive and it can take effort and time to develop and produce sufficient amounts of the pure binder. The xenobiotic molecules are generally more stable than antibodies towards oxidation and heat. Although not degraded by proteases, aptamers are susceptible to nucleases. A potential beneficial characteristic in electrochemical biosensors is the combined negative charge formed by the phosphate constituents in the oligonucleotide backbone of aptamers [166].

Very interesting and exciting is that quite some studies explored the mounting of a smartphone onto biosensors. In this way, the mobile phone is a power source, detection facilitator (LED flashlight), detector (camera), signal converter (computer), and signal recorder (compute and memory storage) part in a sophisticated, portable biosensor system [203–205].

2.6.2.3. Microfluidic Devices

Further evolved immunosensors arose from combing and integrating different mechanical and biological techniques for sample preparation, processing, and analysis in a single miniaturized liquid flowing device which is generally known as lab-on-a-chip (LoaC). Solvent application is through micropumps or capillary force and uses hydrophobic stops. In these microfluidic devices, the analyte is immune-concentrated and detected using measuring principles as described above for biosensors (Section 2.6.2.2). In general, these compact platforms offer improved limit-of-detection, reduce human errors, enhance the accuracy of measurements, and are less affected by food matrix components [206]. However, the practice is wilful due to complex food matrices and the small sample volume to meet required selectivity and diagnostic sensitivity. A solution to increase surface area is the integration of anti-pathogen ABs-conjugated (magnetic) nanoparticles, which were used in combination with a LoaC to detect *Escherichia coli* and *Salmonella typhimurium* [207].

Concurrently, attempts are made to give these small systems also multiplexing abilities, which is a challenge by the requirement of the growing number of channels and valves with the number of analytes in usually two-dimensional systems [208]. Indirect antigen assay cassettes have been engineered for detection for anti-HIV and anti-*Treponema pallidum* (cause of syphilis) antibodies in 1 µL whole blood giving results in 20 min. The "mChip" was tested on hundreds of humans in situ in the field in Rwanda and performed equally well as the reference tests in a laboratory [209]. The assay is not suitable for high throughput.

Alternatively, lab-on-a-disc (LoaD) uses centrifugal forces to "pump" solvents and sample over the device. Although most LoaD applications involve LAMP for pathogen detection, IA assays in combination with SPR detection and smartphones have been described as well [210]. For screening purposes, LoaDs appear an interesting alternative for LoaC, as it offers high-throughput possibilities with more flexibility in assay design and multiplexing [208]. In the context of this review, there is a LoaD feature superior to for example LoaC systems: LoaD is virtually indifferent to sample properties like viscosity and surface tension. In the authors' hands, these properties, which tend to vary considerably per biosample, were often an issue, even after substantial dilution, in developing and validating advanced IA assays.

Generally, the microfluidic device technology is not matured for full routine applications (yet). There are a few unresolved issues which include problems with reproducibility, specificity, and stability and which have to be tackled before microfluidic systems can be considered to have passed their infancy.

2.6.3. Concluding Remarks on IA Assays

Whatever the binder and whatever the test configuration, in all cases the ligand should be immobilized correctly and reproducible to the interface, which can be a sensor surface or solid phase of carriers as in for example BBAs, ELISAs, LFDs. This means, that the ligand is preferably attached covalently or near covalency as in a streptavidin-biotin linkage, and its antigenic determinant or binding complementary regions are orientated towards the liquid in which the analyte is present. The attached ligand should not be inactivated by the immobilization reaction and the ligand should not autoconglomerate/denature on the sensor surface.

Another issue to note is the label used to acquire a detectable and recordable signal. Enzymic conversion of a substrate into a detectable signal follows Michaelis–Menten kinetics. The range of linearity and the window between the minimum and maximum signal is limited. Action limits, the limit of detection, and thresholds levels are often based on optical density in standardized ELISA protocols. Extended ranges are offered by other labels such as fluorophores.

Furthermore, AB-AG binding reaction dynamics are different in ELISA, BBA, LoaC, or immunosensor. Consequently, results are usually not intercomparable without intensive mathematical data massage. To obtain congruency of method outcomes, a significant number of results generated by the alternative and by the reference method have to be regressed to acquire a reliable formula for superimposition. This is a challenging and complex task, as both reference and alternative methods show, besides different reaction and signal dynamics, variance of results. This issue is also addressed by certification and accreditation bodies that request to meet their norms based on traditional, reference methods.

Concerning the difficulty of interpretation of results, it is remarkable how many failures are reported for the apparent fail-safe and easy-to-use LFD. Subjective visual interpretation of the test lines, in particular when they are vague, causes quite some ambiguities [211]. Vagueness can be a result of a contaminated sample, such as blood in nasal fluids in the case of SARS-CoV-2 LFDs. Other failures include inaccurate use of a dropper and thus varying amounts of applied sample. Users applying several loads of sample to the same LFD device deteriorating the carrier are also observed. These issues are not to discourage, but to make the reader aware of the unexpected when developing, evaluating, or executing an assay.

2.7. Available Test Principles Other Than Immunoaffinity

Despite the focus on immunoaffinity techniques, a few alternative test principles are discussed shortly as they are closely related or because they can be combined (also referred to as 'hyphenated') with IA techniques.

2.7.1. Bacteriophages

Bacteriophages have several intrinsic and desirable characteristics which make them very suitable for the detection and identification of bacteria [212,213]. Their natural specificity and ease of production and use make them very attractive for food bacteriology and promise some solutions for current analytical challenges. In particular, endolysins, i.e., peptidoglycan hydrolases recognizing specifically the bacterial peptidoglycan layer, are of much interest for the design of innovative affinity assays detecting bacteria [214]. Nevertheless, although much progress has been made, only a few bacteriophage-based methods have turned into (clinical) diagnostic devices so far [213]. A major drawback is namely the genetic drift and shift of the target bacteria and of the bacteriophage itself, which, if one thinks about it, is logical or otherwise the bacteria and/or the bacteriophage had not survived and existed anymore. Furthermore, most bacteriophages are very specific and not suitable to screen the huge diversity in bacterial *genera*, species, subspecies, and strains.

2.7.2. Nucleic Acid

Genetic (isothermal) amplification and detection technologies have developed at an impressive pace in food microbiology [215]. They comprise second, also called "next", and third-generation sequencing technologies. In general, NASB techniques will outperform the analytical sensitivity and specificity characteristics of IA assays. Speciation for example is better by PCR than by IA. When tracing viral pathogens in the food-chain in the case that (serum) antibodies are not available anymore for an indirect antigen test, NASB approaches are factually only suitable.

Despite better analytical sensitivity, the diagnostic sensitivity of PCR can be unsatisfactory. A single tube nested-PCR, of which positive results were confirmed by qPCR, found only 18 *Toxoplasma gondii*-positive goats and sheep, while the indirect antigen ELISA found 79 of the 278 sampled animals [81]. Sensitivity of *Toxoplasma gondii* PCR was criticized earlier [83,216,217]. In these studies, indirect antigen ELISA of serum from naturally and experimentally infected swine clearly outperformed direct, semi-nested PCR and qPCR methods [83]. While rigorous sample cleanup increases PCR sensitivity [218], the procedure is unattainable for testing on large scale [159].

An important disadvantage of NASB assays which can hinder inspection protocols is that they generally do not make a difference between viable and non-viable MOs, in most cases bacteria and parasites, without prior appropriate sample treatment. Degradation of two genes of defunct *Yersinia enterocolitica* comparable to 1 log unit reduction took between 0.5 h in chicken meat and 120.5 h on rinsed pork [219]. In other words, a NASB test without proper sample cleanup or treatment does not discriminate well which products represent a real health threat upon a positive signal.

Decontamination of fresh meat will leave traces of the nucleic acid signature of the killed pathogen(s) possibly recognized unintentionally by a NASB method. In addition, NASB tests, and actually other alternative measuring principles as well, are not suitable to test products against legislative norms expressed in CFUs. The norm for *Campylobacter* in broilers meat, 1000 CFU/g (Table 3), cannot be verified with a NASB method, although droplet digital PCR (ddPCR) seems to offer a solution for this problem (see next paragraph).

Another problem is the inhibiting sensitivity of the enzymes towards components, such as bile salts, EDTA, acriflavine, $MgCl_2$, high protein, and fat concentrations, in complex analytical matrices, like feces or food. A solution for this problem comes from (droplet) digital PCR techniques producing and analyzing ideally a single bacterial cell per droplet [220]. The platform is likely able to quantify foodborne bacterial pathogens without sample treatment [221], although it has not been demonstrated how the approach can test the obligatory 10–25 g sample for the presence of unwanted pathogens. In addition, the technology has not convincingly demonstrated yet whether it can discriminate viable from non-viable pathogens either.

The high specificity of molecular methods can be a disadvantage as well. Recently, the analytical risk of a high accuracy became apparent by the SARS-CoV-2 UK variant discovered beginning of December 2020. This variant was poorly picked up by the S-gene target RT-qPCR assay, which was used commonly in routine laboratories [222]. The probe used in this assay attached weakly to the new variant target.

2.7.3. Physicochemical Approaches

Physicochemical instrumental techniques to analyze MOs have been used for a long time, but not routinely. For example, gas chromatography can profile fatty acids of food-borne bacteria [223]. This instrument combined with mass spectrometry (GC-MS) can trace sensitively and specifically microbial markers in/from food, including metabolites and volatile organic compounds (VOCs) [224,225].

In clinical microbiology and increasingly in food microbiology, matrix-assisted laser-desorption ionization-time of flight (MALDI-TOF) mass spectrometry (MS) is becoming a routine methodology to detect and identify pathogens (and spoilage MOs). Alternatively, electrospray ionization (ESI) MS has also found its way into microbiological investigations [226]. These techniques bring characteristic (fragments of) biomolecules of the MO into the gas phase after they are ionized. Separation of the ions on basis of their charge and molecular mass delivers a unique mass spectrum revealing the molecular makeover of an MO in qualitative and quantitative terms. The resulting mass spectra of microbiological samples are very complex and reference databases, data massage, and evaluation software for a workable interpretation are imperative. Nevertheless, these techniques are increasingly able to distinguish morphologically and even genetically similar (sub)species.

Besides spectrometry, variations of Raman and Fourier-transform infra-red spectroscopies, sometimes also hyphenated with other techniques discussed here, are recurrent technologies which fingerprint whole bacterial cells [52]. For example, microfluidics-integrated SERS using gold-surfaced nanoparticles coated with MABs against *Listeria monocytogenes* also capturing *L. innocua*, detected the pathogen in less than 2 min at 10^4–10^6 CFU/mL [227].

In general, reported advanced (hyphenated-)spectroscopic techniques are apparently still in an experimental phase. They need purified pathogens at relatively high concentrations. Field applications, i.e., running routine analyses in an agricultural, veterinary, or food safety laboratory, and diagnostic validations of any spectroscopic technique in food are difficult to find. Despite what the developers often claim and promise, namely fast, powerful and reliable alternatives for traditional procedures, spectroscopic methods need lengthy (pre-) enrichment steps and have not shown convincing diagnostic sensitivities or specificities (yet).

3. Reliability of Results

Analytical methods blooming from the "food microbiology" field in the 20th century were formally and internationally classified in ISO 07.100.30. This list slowly extended with, at the moment of their inclusion, alternative methods. Anno 2021, this list contains 119 protocols for validation, classic culturing, EIA, HPLC (of histamine as a result of bacterial activity), and qPCR methods for various foodborne pathogens.

It was not without effort to convince the authorities to acknowledge quality assurance and validation protocols for innovative alternative methods. Nowadays, various programs for validation of alternative methods, including LFDs, BBAs, immunosensors, involving detection or determination of MOs in the food-chain have been initiated worldwide. Most of these programs are based on ISO 16140. Examples are programs offered by AFNOR (Paris, France), MicroVal (Delft, The Netherlands), NordVal International (Nordnes, Norway), AOAC-RI (Rockville, MD, USA), and UKAS (Staines-upon-Thames, UK). These and other national or regional accreditation bodies, such as the American Association of Veterinary Laboratory Diagnosticians [228], Collaborating Veterinary Laboratories [229], and European Association of Veterinary Laboratory Diagnosticians [230], will help maintain the validity of diagnostic results from e.g., (novel) IA procedures.

3.1. Method Validation

Before any validation or verification process starts, the user should answer the question of whether the test will deliver the requested results in the given circumstances, such as geographical region [231]. The latter refers to a situation not as in Gertrude Stein's 1913 poem "Sacred Emily": "A rose is a rose is a rose", but *bovine* samples from Northern European or from Southern African cattle will give other test results [231]. A similar phenomenon was observed for samples from Dutch and North-German pigs in a BBA assay (dr. Bergwerff, unpublished).

Evaluation of an assay to determine its fitness (performance characteristics) for particular use in the diagnosis, monitoring, surveillance or trade, is essential to ascertain the integrity and reliability of test results. After all, public health and socio-economic impact are often too high to allow avoidable false results. Furthermore, test results should be transparent, understandable, acceptable, and legitimate for any third party, including for a court of law when results or safety of products are disputed. In the case assays are used to test compliance with legislative norms, a validation study is compulsory and not non-binding. Many governmental and non-governmental, national, and international organizations and institutions have devised obligatory validation and verification programs according to recognized international quality standards (ISO). These programs should maintain administrative, management, and technical performances for various diagnostic tests at a high level of accuracy and uniformity. An excellent overview of method validation of assays for the determination of immune status after vaccination and of infections in animals is given in [231]. A lucid description of issues involving calibration and standardization of immunoassays is found in [232].

The variables that affect an assay's performance can be grouped into three categories:

- Sample: Host-MO interactions affecting the composition of the sample and its analyte concentration. Surprisingly, this can thus be different from one geographical region to another.
- Assay system: Physical, chemical, biological, managerial, acclimation, housing, and technician-related factors affecting the capacity of the assay to detect a specific analyte. Here, sources of errors are not necessarily random and independent. They are for example laboratory effects, method bias, matrix variation effects, random and systemic errors of measurement, run effects, or bias.
- Other sources of errors that affect the capacity of a test result to predict accurately the contamination status of animals, food products, plants, or populations relative to the analyte in question.

With bioagents, such as MOs, the "true" content in complex biomatrices, such as feces, serum, or food, can never be determined exactly. Outcomes from an alternative method are therefore commonly compared with those of the gold standard or reference method. To pass validation, this comparison has to demonstrate equal performance in terms of relative specificity (SP), relative sensitivity (SE), and relative accuracy (AC). Here, lies a hidden potential disappointment. Alternative methods with e.g., increased sensitivity or increased specificity compared to their respective reference methods produce more "false positives" or "false negatives", respectively.

Commonly, a reference method is recognized, respected, and results undisputed. The reference method is the outcome of a culmination of reaching a consensus in the scientific field of the best available assay at the time of evaluation under given, reasonable circumstances. This is a long and complex process to converge opinions and to reach a common agreement of understanding on a reference standard between the food industry, (veterinary) public health, and food-safety regulators, in which Codex Alimentarius, OIE, and other NGO's play an important intermediary role. The finally agreed benchmark is therefore per definition dated. Accordingly, the gold standard is not a perfect and most accurate method [233]. On the contrary, reference methods are frequently challenged. A well-known example is the several orders of magnitude of discrepancy between the viable bacterial plate count, the gold standard, and the microscopic count of bacteria [234]. Another example is the gold standard for subtyping foodborne bacteria for over two decades, namely pulsed-field gel electrophoresis (PFGE), while other well-validated superior methods are available [235]. Consequently, improved performance by an innovative assay is frustrated by comparison with the benchmark as it will give *a priori* worse relative SE, SP, and AC.

Compared to a classic, reference ELISA using colorimetry of an enzyme-generated chromophore, an alternative method using fluorophores provides a signal order of magnitude more linear. Consequently, responses outside the linear range of a chromophoric ELISA do not congruent and inside this range, it is probably with another steepness. This is not acceptable in validation programs and not for accreditation bodies either (dr. Bergwerff, unpublished).

Another diagnostic validation hurdle is the multitude of information that a (real-time) multiplex assay produces, while reference methods are typically traditional, single endpoint measurements. For example, indirect antigen *Salmonella enterica* ELISAs produce a result per sample which is a resultant of immunoglobulins reacting to (a mixture of) immobilized O-antigens in a single well. The O-antigen mixture represents serogroups B, C1 and in some versions supplemented with serogroups D and E [156]. Multiplex assays can provide information for each serogroup independently. In a BBA, each serogroup is represented, for example, by a single microsphere. However, the reaction dynamics and kinetics on a coated surface of a microsphere compared to those in a static well, even when agitated, are significantly different. The sum of fluorescent responses of each microsphere does not coincide with the resultant colorimetric response in a well using an identical sample (dr. Bergwerff, unpublished).

Despite good repeatability and reproducibility performances of an assay, special attention should be given to so-called assay drift within or between runs, and within and between days/weeks [236]. This comes often unnoticed in microbiological assays as, unlike most chemical assays, internal standards in food microbiology are difficult to define and/or are not stable (under assay conditions and/or by e.g., freeze-thawing cycles). For this and other reasons, when a test passed a validation program and used to determine contaminations in animals and/or food products, its validity has to be verified ongoingly in a proficiency test or "ring test", although this test also assesses the quality of laboratory and operational factors [229].

3.2. Predictive Values

A result of an indirect antigen test is not an endpoint indicator of an infection or contamination of an individual animal or a food product as explained above. An endpoint is acquired by repeated analyses (monitoring), or evaluation of a set of results reflecting a herd/flock or batch (eggs). However, evaluation of the contamination level of batches or herds/flocks based on tests applied to single products or individual animals is complex. An aspect that is regularly out of view of inspectors and analysts is the prevalence (persistence) and incidence (spreading) of some pathogens, in particular when these are very low. Prevalence depends often on a geographical region. *Trichinella* spp. in swine raised under (un)controlled conditions is in the EU virtually 0% (0–0.0003% in the period 2014–2018) with few human cases [18]. All positive cases involved pigs which were raised under non-controlled housing conditions, in particular in the Member States Croatia (58 per 9.5 10^5), France (5 per 6.0 10^5), Italy (8 per 1.6 10^5), Poland (39 per 22.7 10^6), Romania (134 per 2.5 10^5), Spain (5 per 27.7 10^6) [18]. *Salmonella* prevalence in swine herds in Scandinavian countries is near 0% as well [237]. A low prevalence requires other assay characteristics. After all, the outcome of all (obligatory) tests performed on pathogens with a negligible occurrence in a population is prevalently "negative". In that case, assay sensitivity is of less importance, whereas, reciprocally, assay specificity is gaining importance, i.e., rate of false positives. A false-positive result can have namely dramatic effects, such as the condemning of herds, closure of farms (also those near the false-positive herd), and even the closing of borders for the international trade of animals and their products. In other words, the negative and positive predictive value of an assay is of utmost importance given the prevalence of a pathogenic MO.

There is a relationship between sensitivity and specificity through the threshold of the assay so that the chance of false-positive results can be reduced. A consequence of such reduction is that the false-negative rate increases concordantly. However, in contrast to human testing, animals are grouped in herds and are sampled and tested in clusters. Assessment of a cluster will normalize the results and blot out the impact of false negatives and false positives. On basis of prevalence and desired confidence level, statistics will help an inspector to determine sample size and how many samples are 'allowed' to be positive in a cluster before it has to be considered positive. The interested reader is referred to e.g., [110,111] for further information.

4. Example: *Salmonella* Detection in the Pork-Production Chain

To illustrate the methodological challenges of pathogen screening to warrant food safety, the monitoring of *Salmonella* in pork as part of intervention and eradication programs is discussed here. If hygiene is assured further down-stream, intervention is most successful in the primary sector of the pork production chain. The intervention relies greatly on testing of the pathogen, and timely and adequate acting upon a positive finding. However, when choosing the wrong strategy, the most excellent analytical method will deliver a wrong conclusion on a carrier status as explained below.

Ideally, bacteriological examination of feces from multiple animals will deliver an accurate infection status of a pig herd at the moment of sampling. Positive isolation and confirmation of *Salmonella* will leave little doubt to its presence in the sampled animal. However, a negative bacteriological test result should be interpreted prudently. After oral exposure, *Salmonella* migrates rapidly to the caecum and, consequently, it can become undetected by bacteriological investigations of feces. Furthermore, competitive saprophytic microorganisms, such as coliforms, dominate *Salmonella* which usually becomes injured and therefore problematic to culture (*cf.* Section 2.3.3.1). Consequently, fecal *Salmonella enterica* shedding is usually intermittent, at low numbers, and observed predominantly in the first half of the fattening period [238–240]. A delicate consequence is that when non-shedders are slaughtered non-hygienically, carcasses can be a source of contamination in the slaughter line and further down the food-chain.

Direct antigen detection of *Salmonella* is thus difficult in subclinically infected herds, which are much more frequently encountered than herds with clinical signs [240]. The diagnostic sensitivity of microbiological culture methods of fecal samples, which are considered reference methods, is relatively low in a practical setting, namely 50–60% compared to that of the fecal samples collected at the abattoir [241]. As a result, the negative predictive value of direct antigen screening of herds is poor and the correlation of bacteriological results with the risk status of the herd is weak [242,243].

On the other hand, anti-*Salmonella* antibodies persist beyond the time of infection and indirect antigen testing reveals a historical exposure per definition. Serological data do thus not correlate closely to the apparent bacteriological burden at the time of sampling [244]. While fecal *Salmonella* (intermittent) shedding is observed in the first half, seroconversion occurs generally during the last third of the fattening, i.e., the pre-harvest period [238,239]. However, a *Salmonella* infection during transport to an abattoir or in the lairage does not result in a timely seropositive reaction, as there is an insufficient interval to induce a detectable immune response before slaughter. Maternal anti-*Salmonella* antibodies, for example, persist more than eight weeks in the off-spring and cannot be distinguished from the pig's own antibodies in that period. This fact is of importance as most piglets are weaned just three weeks after birth. They arrive in new pens/groups and an introduction of a new *Salmonella* infection cannot be distinguished on serological data until eight weeks of age [238].

Despite these contraindications and the apparent asynchronous bacterial infection and serological status [244], the prevalence of seropositive swine on herd level at slaughter correlates significantly with *Salmonella* in rectal contents and mesenterial lymph nodes [245,246]. Actually, indirect antigen assays are considered to have a higher diagnostic sensitivity than direct antigen tests [240,247], in particular as these tests can reveal latent carriers or intermittent shedders. The *antemortem* serological status of the herd gives thus a good estimate of the risk of *Salmonella enterica* spp. in pig products [245,248]. Despite a conversion window of days for each animal, the varying individual infection onsets on a herd level, varying individual seroconversion intervals, and varying immune response intensities will give a rather reliable overall indirect antigen testing outcome. Pig farms have to acquire a *Salmonella* status classification based on direct and on indirect antigen monitoring regularly (three-monthly in some countries). Classification of herds is in accordance with a certain percentage of positive animals and determines whether farms are required to increase prevention activities and reduce *Salmonella* prevalence [249,250]. This successful system was developed and introduced by Denmark in the 1990s [77,251].

This example of screening does not advertise exclusive indirect antigen testing. Feed, water, and environment are integrally part of monitoring systems to maintain biosecurity and they are commonly investigated using direct antigen tests, including NASB tests.

5. General Discussion and Conclusions

5.1. Responsive and Smart Monitoring and Control

Foodborne infections and food spoilage are persistent global problems that keep on giving analytical microbiology a pivotal role in prevention, intervention, and outbreak control. The population of vulnerable people grow vastly, and eating habits, agricultural and manufacturing practices, and climate change. *Ergo*, there is a growing "opportunity" for unfamiliar pathogens to emerge and for familiar pathogens to re-emerge. Irrefutably, a firm need to assess and prevent infections and contaminations in the food-chain will hold on.

Methods, design of food inspection, and intervention regimes stem from the end of the 19th and the beginning of the 20th century. After a long standstill since the enforcement of the first general food laws, food inspection and control of food hazards was scrutinized and modernization advocated [252,253]. As we are continuously striving "to die young in life as late as possible" (after late Prof. Bob Kroes, Utrecht University, Utrecht, The Netherlands [254]), it became clear that new ways of thinking about how to deliver improved monitoring to secure food-safety are required. The one-health concept (OH), although the term was already coined in 1964, emerged from this. It advertises to secure food safety starting at the farm into the chain by warranting pathogen-free feed, water, environment, and healthy animals [255]. It is in fact an inter-professional area linking the human-animal-environmental health triad on which was elaborated by Prof. Frans van Knapen and co-workers (Utrecht University, Utrecht, The Netherlands) since the end of the last century [256]. As a result, the EU is the first supranational government worldwide implying the "farm-to-fork" approach and to require visual-only inspection for all swine herds slaughtered meeting certain epidemiological and animal rearing conditions [257]. The system obliges producers to provide so-called food-chain information (FCI), including data revealing herd's health status and zoonotic risks. The status, i.e., herd risk level, is based for a great part on the outcomes of indirect antigen (serological) monitoring, including that of *Mycobacterium avium*, *Toxoplasma gondii*, and *Salmonella*. Apparently, as in inspection and control systems in non-food sectors, big data analysis gets a foothold in the food-chain as well. In fact, the first signs of involvement of blockchain technology fueled by supermarkets are visible.

In the scope of this review, this reorientation on keeping food free from contamination means that indirect antigen methods will survive despite upcoming other measuring principles, such as genotyping methods. The poor diagnostic sensitivity of current NASB methods, as highlighted above, makes them in several cases unfit for routine livestock screening in novel responsive, as opposed to inflexible bureaucratic, inspection systems. On the other hand, environmental monitoring, which is essential in an OH approach, is best served with the availability of multi-analyte direct antigen detection systems with a wide range of applicability, which also includes NASB methodology. It should be able to deliver data before the product is transported/delivered to the next link in the food-chain, which means not rapid *per se*.

Besides having mitigation strategies in place at crucial points in the chain, modernization also requires risk-based sampling, monitoring, and effective, targeted surveillance. After all, *Salmonella*, for example, is more of a problem in West-Europe than in North-Europe, while *Taenia saginata* and *Trichinella spiralis* have a higher prevalence in East-Europe than in the rest of the continent.

5.2. Prediction, Indicators, and Prevention of Sherlock's Holmes Statistics

One of the strategies is controlling biohazards using predictive microbiology to quantify microbial ecology (of foods) of bacteria [258]. The use of loggers of, for example, temperature, retrospective analysis of data supports the prediction of the contamination status of food [259]. This approach may lead to, net, less sampling, and less laboratory MO-testing, but testing at higher quality and more to-the-point instead.

Actually, there is also a statistical reason for reconsideration of our conventional monitoring systems. An ever-increasing higher security level demanded by the consumer is obstructed by what is called "Sherlock's statistics". A low prevalence level and an increasing confidence level require more and more samples to deliver the requested reliability. For example, at a 20% prevalence in a population or batch, only 11, 14, or 21 samples are needed to reach a confidence level of 90%, 95%, and 99%, respectively. However, given a prevalence of 0.1%, 2302, 2995, and 4603 samples, respectively, are necessary to meet the same confidence levels [260]. Actually, in *Trichinella spiralis* monitoring programs each and every pig carcass has to be investigated by a so-called artificial digestion method by law to meet the required highest confidence level [261] at extremely low prevalence in most countries [262].

To make *Trichinella spiralis* monitoring even more complicated, the sensitivity of the digestion method can be, depending on the burden, depending on the stage of infection in the animal and whether samples are pooled, as low as 40% [159,160]. An immune response-based method may deliver more reliable infection status data not of an individual carcass but a group [63,159]. In the case of a low prevalence of a pathogen with a serious health threat, such as *Trichinella spiralis*, the solution can be monitoring of antibodies in a herd against a pathogen occurring in the environment with a higher prevalence as well. Like *Trichinella* spp., *Toxoplasma gondii* is an indicator of contact with the environment. Infection routes are shared at which those for *Toxoplasma* are wider and more extensive than those for *Trichinella*. In other words, *Toxoplasma* infections are an excellent indicator of farm hygiene, contact with the environment, and good practices (GAP, GHP). Finding a *Toxoplasma*-positive result indicates increased risk for several other public health-threatening zoonoses, including *Trichinella* [159]. Humoral responses to these bioagents are dynamically independent and co-infections do not influence each other [263] so that screening anti-*Toxoplasma* antibodies can support and improve pork monitoring and control programs e.g., *Trichinella* spp. [159]. It is an intriguing strategy that should be elaborated on for all farm animals and all relevant MOs if pathogens-free has to be warranted at a low prevalence and at a high confidence level.

5.3. The Weak Link

A chain is no stronger than its weakest link. Whatever effort upstream in the food-chain, threats are clear and present downstream in mass kitchens and domestic situations. Factually, the relative contribution of contaminations in the third and quaternary sectors (Figure 2) are already causing the majority of foodborne infections with norovirus as the champion of all causative bioagents. The relative contribution in this last part of the food-chain will increase as an intervention in the (pre-)harvesting and processing phases become even more effective. Changing eating habits, poor personal and kitchen hygiene, pets, pests, etc. contribute substantially to all foodborne infections. *Yersinia enterocolitica* and, in particular, *Listeria monocytogenes* seem to have adapted themselves to modern food production and preparation. They grow at refrigerator temperatures and are persistent unwanted guests in processing facilities and households. Is there a more prominent role for food producers, authorities, and/or dietitians to educate the consumer [42,43]? Can kitchen equipment be engineered more smartly so that the chance of contamination is reduced and in such a way that cleaning every corner and edge is easier?

Currently, spoilage/freshness indicators are increasingly applied to food packages (see [264] for an overview) in which IA-based methodology is also used. However, only a few fail-safe, rapid methods, as it were modern food-tasters like at the courts of kings and emperors, are available to laypersons to check the safety of food and adequate hygiene in institutional kitchens or at home. Here, assay developers find a new, growing market for cheap, stable IA methods. LFD technology or alike platforms seem to have a very good starting position for this market.

5.4. Fool's Gold?

The need to analyze multiple MOs simultaneously in a virtually indefinite number of different analytical complex matrices is outweighing the requirement for more sensitivity and shorter time-to-results. An animal or food product, let alone a herd or a batch of products, is not a potential carrier of a single species, (sub)subspecies, but can be infected/contaminated with quite some bacteria, bacterial toxins, parasites, and viruses (*cf.* Figure 1). It says that there is a preference, at least in field laboratories, for protocols enabling multi-analyte and high-throughput screening more than for methods providing more speed and sensitivity.

Sensu lato, magnificent speed-of-analysis, and impressive analytical sensitivities are demonstrated, and almost all developers claim that their innovation will replace laborious, cost-inefficient, time-consuming traditional methods. A majority of these developments show indeed excellent qualitative or quantitative analytical advancements, but they fail to report transparently the evaluations of diagnostic sensitivity and diagnostic specificity. Sophisticated, sensitive NASB techniques will namely most likely give false-negative feces results as e.g., *Salmonella* hides intracellularly in lymph nodes in swine of three months and older [240] (*cf.* Section 4). Moreover, many improvements are specialistic as well, they are described for only one or a small set of MOs in one or a few analytical matrices. These investigated matrices are often spiked and not naturally infected for the development/validation/performance studies. Many microbiologists can explain to these developers that there is a world of difference between spiked samples and samples from natural infections.

Developments giving means to simple multiplexing analysis have considerable market potential. They have an unexpected advantage which can be "fool's gold" or a "driver" (*cf.* Table 5). Up to 22% of acute hospitalized gastroenteritis cases are caused by two or more pathogenic MOs [265,266]. Food products may be contaminated with more than one pathogen. In fact, more than one *Salmonella enterica* serovar in poultry and swine, and curli- and non-curli-producing *Escherichia coli* O157 strains in beef were observed regularly in the author's laboratory (dr. Bergwerff, unpublished results).

Table 5. Relevance versus differentiation criteria of (innovative, alternative) analytical food microbiology methods for routine laboratories. The various method parameters that make up the overall profile are separated into four categories that reflect their relative attractiveness to users and differentiation from assays that are already in use. The completion of the quadrants is empiric and the weight of criteria may vary between readers.

High	Antes [a]	Drivers [b]
Relevance	• Equivalent accuracy to reference methods. • Relevant matrices (farm, feed, abattoir, food, and beverage). • Required validation(s), such as AFNOR, UKAS. • Reference labs for new users to contact. • Robust system.Information handling—LIMS connectivity • Security of access and data (CFR 21 pt. 11 etc.). • Stoichiometric results compared to the reference method. • CE Mark	• Lower overall costs than other tests. • Multiplexing relevant combinations of Mos. • Time-to-result in a working shift. • Fits factory sampling and laboratory routines. • Closed system: sample in-result out • No enrichment or otherwise universal enrichment broth. • Can detect, identify and confirm within the same test. • Smaller footprint than reference tests. • Flexible number of tests per day. • Possibility to enumerate. • Single platform for direct and indirect antigen testing.
	Neutrals [c]	Fool's Gold [d]
Low	• Low waste. • Fashionable design. • Product source.	• Fast time-to-result in cases where time is not of the essence (e.g., farm surveillance). • High sensitivity in cases where the normal.
	Low	Differentiation	*High*

[a] Antes, important but provided by all methods; [b] Drivers, important and which highly differentiate from those of reference methods; [c] Neutrals, irrelevant; [d] Fool's Gold, distinctive, but which do not convince users to switch.

An ultimate solution seems a holistic approach using a chip coated with probes for more than 12,000 archaea, bacteria, fungi, protozoa, and viruses suggested for veterinary use as well [267]. The analytical sensitivity of this technology is approximately 100–1000 genome copies but it does not deliver verifiable quantitative data. Of interest here is how the specificity of the multiplexing chip is secured. An increasing number of probes will also swell the chance of (weak) a-specific binding and thus a good chance that one or more of the 12,000 probes give a false-positive result. It will affect the overall false-positive rate. But even true positive responses do not imply safety relevancy per se without quantitative data of the detected MOs. In some cases, for example, PRRS in swine, the impact of a positive sample can be dramatic, *viz.* immediate closure and isolation of the farm from which the sample was collected. Danish field laboratories for example demand therefore a diagnostic specificity for PRRS near 100%.

Routine laboratories need reliable universal methodologies to acquire all relevant answers with a single technology. The desire arises not only from a cost-efficiency point of view but also for other reasons, such as comparability of results, participation in proficiency programs with a method which is considered not to be unusual.

5.5. Validity and Comparability of Results

The higher specificity of genotyping methods is a clear advantage over IA methods. Although IGs and AGs react with a relatively high degree of specificity in IA assays, other MOs can share (similar) antigenic structures with the target organism giving false-positive outcomes. Therefore, direct and indirect antigen tests are regarded as presumptive. A positive result in a direct IA antigen test has to be confirmed through e.g., conventional culture procedures. A positive outcome from a single indirect antigen test is considered without much value. Reliability of serology is obtained by repeated sampling (monitoring) or analysis of a series of random samples to assess a group (herd or flock). Having noted this, the specificity of NASB assays can be too strict as demonstrated by their failure to pick up SARS-CoV-2 mutants [222].

Analytical food microbiology seems to be based increasingly on the determination of a genetic code [268] of in particular bacteria and viruses. Despite the high analytical sensitivity of these techniques, their diagnostic sensitivity has to be scrutinized carefully. After all, a test has to show e.g., the absence of a disease-causative agent in a 25 g sample, or in 1 to 4 g tissue in the case of some parasites and has to make a difference between viable and non-viable pathogens. Isolation of a few bacterial cells from a 25 g sample to deliver a portion of only a few microliters or less is still practicably impossible on a routine basis. Even IMS does not deliver sufficient recoveries in complex food matrices. To mitigate these analytical hurdles, sample preparation, such as inoculation of a broth and culturing of the bacterium or virus, is needed and this will delay inherently the time-to-result. This sample preparation is often suitable for NASB and IA methods so that a final decision on an analytical approach is determined on criteria listed in Tables 4 and 5.

A difficulty with IA-based and NASB techniques is that they do not uniformly express results. The optical density of an ELISA result is not comparable with the dynamics of a fluorescent signal in a BBA assay or with a cycle threshold (CT) number in a qPCR method. *Salmonella* surveillance programs rely commonly on indirect antigen ELISA analyses and use thresholds expressed in OD values. It prevents the introduction of for example BBA methodology showing other, from the authors' point of view much better, signal dynamics. This incomparability is also apparent from the difficulty to verify a result produced by an alternative method against legislative norms expressed in, for example, colony forming units per mass or volume unity (Table 6). The norm for *Campylobacter* in broiler meat, 1000 CFU/g (Table 2), cannot easily be verified with a PCR-based method, or with any other non-traditional method for that matter.

Table 6. Comparison of characteristics of direct and indirect antigen methods (*cf.* Table 4). The table only lists those items which differ significantly.

Traditional (Direct Antigen)	Alternative (Indirect Antigen)
Laboratory-bound	Point-of-care possible
Intensive material use	Relative expensive material
TTR long	TTR usually short
Reproducibility moderate	Reproducibility mostly better
Almost no instruments	Needs more devices
Difficult to automate	Possibilities to automate

Despite all advancements, the ideal method (Table 4) does not exist. For a new alternative method to get implemented it has to have a profile that provides maximum end-user satisfaction in combination with the greatest possible differentiation from technologies already in use in his laboratory while results are comparable and stoichiometric (Table 5).

5.6. Weighing the Investment in New Methodology

In a field laboratory, the decision-maker has little attention to the scientific incentives for the acquisition of new technology. He/she assumes that the scientific performance has been approved by her/his laboratory co-workers. In her/his final decision, she/he is interested almost exclusively in the continuity of the laboratory and in overall costs. When asked, reduction in TTR is not as important as cost reduction in many routine laboratories (dr. Bergwerff, unpublished). When a big test producer with its headquarters in France, started offering a 24 h instead of a 48 h *Salmonella* test, only 10–15% of the users converted to the new test. This was because the majority of *Salmonella* tests are for screening/monitoring purposes [268] in which case a difference of a day does not outweigh the higher costs of a faster method.

Of all pathogen tests used worldwide, almost 50% gauge *Salmonella* in processed food and meat (each approximately 40% of the total market). The majority of these tests are traditional plating or petrifilm^{TM} tests in Europe and North America while genotyping assays are more popular in Asia [268]. Although the balance of types of tests will change over time, it is probably disappointing for the novel assay developer. Routine laboratories are reluctant to switch to new technologies as tolerance for failures -read economic impact on the organization- is very limited. After all, the margins of the added value of food are very small and the cost reduction of an alternative testing technology must be very convincing. The price, not costs, for a *Salmonella* test, varies between less than €1 (indirect antigen) to approximately €4 (direct antigen) in field laboratories Europe-wide (dr. Bergwerff, unpublished). The price must not only deliver profit e.g., new investments but should also cover all costs for the laboratory, including the costs of purchasing the test, overhead, (training of) personnel, sample handling-time, sample collection, storage, and preparation, laboratory infrastructure (footprint), depreciation of equipment, operation failure (down-time), false-result rates, (equipment) maintenance, etc.

Concerning (frequency of) down-time, in several instruments, parts, such as sensor chips, have to be replaced or regenerated for each analysis cycle. Without losing critical analysis capacity, regeneration of sensor surface was up to 300 cycles in a *Salmonella* indirect SPR immunosensor test [188]. A sensor can bleed or the regeneration conditions degrade and inactivate the ligand. Actually, many publications do not report the stability of the prepared measuring interface/instrument. Replacing sensor chips or other hardware, also in the case of single-use, is lost time and not attractive for a routine laboratory commonly receiving hundreds to thousands of samples daily. The reader notices here a subtle difference between TTR and throughput: A slow high-throughput method can be more efficient than the fastest (single analyte) test.

Down-time can also be caused by unforeseen factors. Meat drip is the analytical matrix of choice in national monitoring programs of some countries and is collected by butchers or inspectors (not laboratory personnel). As muscle has more value than fat tissue, it is tempting to collect samples with high-fat content. These samples give not only measurement interference, but also clog tubing, needles, and pipette tips in liquid handlers and other instruments.

5.7. Bioprepology

The fat issue in meat drip is a bridge to an important but often neglected subject: sampling, sample handling, and sample preparation. Although Prof. Chris Elliott (Queen's University Belfast, Belfast, UK) not exclusively emphasizes the subject, he states that "food analysis is all about sample treatment" and coined therefore the term: "bioprepology". The term expresses not only the gravity of a crucial step in any food analysis procedure, and in fact, in any life science method, it also legitimizes the existence of a separate, specialistic scientific discipline. The holy grail in food analysis is namely a universal and minimal sample pre-treatment for a great variety of (complex) matrices, while the resulting test portion still reflects integrally the animal, plant, or food product from which the sample was taken. In many cases, especially direct testing of MOs, sample treatment is a bottleneck for fast, inexpensive, and easy-to-use (IA) assays.

When sampling a contaminated carcass, one should realize that a swab may give a false-negative result. First of all, MOs are not homogeneously distributed over a carcass, and swabbing a too small surface on a single, improper location may give a wrongful negative result. Therefore, surface areas for swabbing are regulated. Secondly, after culling and by stressing the bacterium, *Campylobacter* retracts from fecal spillage into the skin by adhering and invading epithelial cells [269]. But even with a proper sample, *Campylobacter* may remain unnoticed when it has converted into a non-culturable coccoid state (VBNC) [269].

5.8. Conclusions and Messages

- A great part of the *ante* and *postmortem* monitoring comprises indirect antigen assays gauging specific antibodies in serum, meat juice, and oral fluids.
- When (intracellularly) hidden or low body burden, IA assays outperform other analytical techniques, including NASB methods.
- IA assays offer a chief advantage over NASB assays: they can detect acellular biomolecules, including toxins, uncovering a (past) infection.
- The largest part of analyses worldwide involves the *ante*- and *post-mortem* monitoring of MOs in the (pre-)harvesting phase of the food-chain. Almost 50% of all tests involve measuring *Salmonella*. IMS plays an important role.
- Whatever analytical sensitivity, analytical specificity, and other test characteristics, the applied assay should fit the purpose while it is clear when and where it is used in the food-chain.
- Novel methods should be presented with data from field samples, not from spiked or polished reference samples exclusively.
- Integrated chain control and One Health principles in combination with risk-based sampling are imperative to combat effectively current and (re-)emerging pathogens while increasing the safety level of food.
- Successful intervention on the guidance of (environmental) monitoring will also protect (families of) farmers and food-workers as a good health and safety practice.
- The need for more speed and sensitivity is modest and not prominent in field laboratories, albeit results within a working shift are highly desired.
- Mutual comparison of results produced by a gamut of alternative analysis systems and comparison with reference methods is an unsolved challenge.
- In the case of group assessment, routine laboratories prefer high diagnostic specificity, multiplexing, and high throughput, but convincing low all-inclusive costs even more.
- All steps between the decision to sample and conversion of a sample into a test portion need continuous and careful attention or the analysis result becomes less reliable or even worthless. Sampling, sample treatment, and sample processing are of cardinal importance.
- In spite of the numerous innovative techniques that evolved over the last decades, only a few have been authorized for screening, monitoring, and control programs.
- The food analysis field is conservative for several understandable reasons, not only because of financial risks. Routine laboratories are bound to accreditation and providing results as if generated by reference methods.

6. Future Perspectives

The outer limits of food science including food microbiology and in particular food microbiological screening techniques are not clear. Whether direct and indirect antigen assays will keep claiming an important role in securing food safety as they did for decades, is uncertain. Nevertheless, there is a mindful trend for methods offering ease-to-use, in situ results, full risk assessment (multiplex), and high sample throughput in a true culture-independent way. What is the heuristic direction towards the best technology in the analysis of foodborne MOs? A glance at the end-user requirements may help:

(1) Test performances that are compliant with local, national, and international (ISO) quality standards.
(2) Enabling the use of (relatively) easily available, preferably non-invasive, samples, such as egg, feces, hair/feathers, saliva/mucus, urine.

(3) Able to analyze simultaneously different types of analytes, such as cells, oligo-/polynucleotides, oligo-/polypeptides, organic metabolites, toxins, in a single run.

(4) Easy-to-use or automated platform demanding minimal user involvement.

(5) Giving accurate results instantaneously (i.e., within 10 min in a PoC situation or otherwise in a work shift).

(6) At low costs, while,

(7) Using a robust, reliable portable multiplex point-of-care testing device (xPOCT) and reagents that have a long shelf-life at ambient temperatures, and which are easily disposed of after use.

New technologies, in particular those based on nucleic acid sequences, seem to give a peek of the future: detection and identification of even the fastidious and/or non-cultivable bioagents in a rapid, cheap, and reliable way. The advent of for example (droplet) digital PCR techniques [270], which are less affected by enzyme inhibitors in complex analytical matrices such as food or feces [220], may supersede IA techniques as soon as they are also able to deliver good diagnostic specificity and diagnostic sensitivity. The technique also promises a possibility to quantify bacteria reliably so that it may replace the counting of colonies in plating methods.

Considering the last half-century, the design and synthesis of antigens, of (derivatives of) immunoglobulins, and chromophores, and fluorophores has progressed slowly compared to NASB technology. The promise of immunosensors in food microbiology has not been fulfilled (yet) despite the investments and research efforts put in the development of this type of instrument. In the end, the decision to change testing systems is determined by legislative restrictions, (international) trade agreements, and costs of sample collection to result in interpretation and reporting. The food analysis market is a very conservative world. In the practice of an ordinary (commercial) food-safety laboratory, applied methods evolved unhurried.

On-site determination of different analytes from a single specimen in a single run (xPOCT), is gaining increasing attention in clinical diagnostics [208]. This development will certainly find its way to surveillance systems in the austere environments of the food-chain like many other advancements in human medical settings did before. After all, intervention at the farm is the best approach to prevent contamination down the production line.

An interesting development is the detection of volatile and non-volatile biomarkers for pathogens that ferment materials in or on animals, feed, food products, or plants. Screening of such indicators may support and even improve surveillance programs. E-noses or dedicated miniaturized MS devices, for example, can sample air and select suspect farms and processing plants for further confirmatory analyses. Such devices can sample livestock houses continuously and on multiple locations and activate alarms and automatic sampling through internet-of-things (IoT). It will relieve laboratory testing while making it more effective when alarms have to be confirmed. The implementation of machine-learning is perhaps to come. It may help to integrate seamlessly methods in various contexts and applications, which are producing uninterrupted unfathomable high volumes of raw analysis data and gathering other information in the food-chain.

Ultimately, we strive for utopian sci-fi methods: technology like the Star-Trek tricorder device [271] which gives non-invasively, instantaneous, and easy-to-interpret results and which can be operated by anyone not only by a skilled person like Doctor Leonard McCoy!

Author Contributions: A.A.B. writing, figure design and drawing, original draft, review, and editing; S.B.D. writing and review. All authors have read and agreed to the published version of the manuscript.

Funding: Preparation: Writing, editing, and completion of this manuscript and delivery of a full publication did not receive funding, and the work was performed without any commercial interest or commercial benefit whatsoever.

Acknowledgments: Philippe Delahaut and Riccardo Marega are offered thanks for giving the authors the unique opportunity to deliver this review. Hugh Ballantine Dykes is acknowledged for introducing the authors in the 'relevance versus differentiation matrix of methods' as presented in Table 5.

Conflicts of Interest: All authors declare that they have no conflicting or dual interests.

Abbreviations

AB	antibody
AC	(relative) accuracy
AFNOR	Association Française de Normalisation
AG	antigen
AgBP	antigen-binding protein
AOAC-RI	Association of Official Agricultural Chemists Research Institute
ATP	adenosine triphosphate
BBA	bead-based assay
BSE	*bovine* spongiform encephalopathy
CE	Conformité Européenne (standard mark)
CFU	colony-forming units
CPA	cross-priming amplification
CPS	capsular polysaccharides
CSF	classic swine fever
CT	cycle threshold (as in qPCR)
DALY	disability-adjusted life year
DES	diethylstilbestrol
DIVA	differentiating vaccinated from infected animals
EF	extracellular factor (of *Streptococcus suis*)
EFSA	European Food Safety Authority
EIA	enzyme immunoassay
ELFA	enzyme-linked immunofluorescent assay
ELISA	enzyme-linked immunosorbent assay
ESI	electrospray ionization
FCI	food-chain information
FSO	food-safety objective
GC	gas-chromatography
GAP	good agricultural practice
GHP	good hygiene practice
GMP	good manufacturing practice
GVP	good veterinary practice
HACCP	hazard analysis critical control points
HAV	hepatitis A virus
HEV	hepatitis E virus
HRP	horseradish peroxidase
IA	immunoaffinity
ICT	immunochromatographic test
IG	immunoglobulin
IgA	immunoglobulin class A
IgG	immunoglobulin class G
IgM	immunoglobulin class M
IMS	immunomagnetic separation
iPCR	immuno-PCR
iqPCR	real-time immunoquantative PCR
IR	infra-red
ISO	International Organization for Standardization
LAMP	loop-mediated isothermal amplification
LFIA	lateral flow immunoassay
LIMS	laboratory information management system
LFD	lateral flow device

LoaC	lab-on-a-chip
LoaD	lab-on-a-disc
LOD	limit-of-detection
LPS	lipopolysaccharides
MAB	monoclonal antibody
MALDI-TOF	matrix-assisted laser-desorption ionization time-of-flight
MIP	molecular imprinted polymer
MO	micro-organism
MS	mass spectrometry
NASB	nucleic acid sequence-based
NASBA	nucleic acid sequence-based amplification
NGO	non-governmental organization
NMR	nuclear magnetic resonance
NPV	negative predictive value
OD	optical density
OH	one health
OIE	Office International des Epizooties (World Organization for Animal Health)
PAB	polyclonal antibody
PFGE	pulsed-field gel electrophoresis
PoC	point-of-care
ppb	parts per billion
PRRS	*porcine* reproduction and respiratory syndrome
PRV	pseudorabies virus
qPCR	quantitative (real-time) polymerase chain reaction
RCA	rolling circle amplification
RPA	recombinase polymerase amplification
RT	reverse transcription
RT-PCR	reverse transcription polymerase chain reaction
SDA	strand displacement amplification
SE	(relative) sensitivity
SERS	surface-enhanced Raman-spectroscopy
SP	(relative) specificity
SPR	surface plasmon resonance
STEC	Shiga toxin-producing *Escherichia coli*
TTR	time-to-result
VBNC	viable but non-culturable
UKAS	United Kingdom Accreditation Service
VOC	volatile organic compound
WHO	World Health Organization
xPOCT	multiplexed point-of-care testing

References

1. Martyushev, L.M.; Seleznev, V.D. Maximum entropy production principle in physics, chemistry and biology. *Phys. Rep.* **2006**, *426*, 1–45. [CrossRef]
2. Silva, C.; Annamalai, K. Entropy Generation and Human Aging: Lifespan Entropy and Effect of Physical Activity Level. *Entropy* **2008**, *10*, 100–123. [CrossRef]
3. Commission of The European Communities. Commission Regulation (EC) No 2073/2005 of 15 November 2005 on microbiological criteria for foodstuffs (consolidated version including Commission Regulation amendments and corrections). *Off. J. Eur. Union* **2005**, *L338*, 1–26.
4. Janssen, R.; Krogfelt, K.A.; Cawthraw, S.A.; Van Pelt, W.; Wagenaar, J.A.; Owen, R.J. Host-pathogen interactions in *Campylobacter* infections: The host perspective. *Clin. Microbiol. Rev.* **2008**, *21*, 505–518. [CrossRef] [PubMed]
5. Mishu Allos, B.; Iovine, N.M.; Blaser, M.J. *Campylobacter jejuni* and Related Species. In *Mandell, Douglas, and Bennett's Principles and Practice of Infectious Diseases*, 8th ed.; Bennett, J.E., Dolin, R., Blaser, M.J., Eds.; W.B. Saunders: Philadelphia, PA, USA, 2015; Chapter 2018, Volume 2, pp. 2485–2493.e4. [CrossRef]
6. Mangen, M.J.J.; Havelaar, A.H.; de Wit, G.A. Campylobacteriosis and Sequelae in the Netherlands—Estimating the Disease Burden and the Cost-of-Illness, National Institute for Public Health and the Environment, Centre for Prevention and Health Services Research (RIVM), RIVM Report 250911004/2004. Available online: https://www.rivm.nl/bibliotheek/rapporten/250911004.pdf (accessed on 16 December 2020).

7.   Overgaauw, P.A.M. Parasite risks from raw meat-based diets for companion animals. *Companion Anim.* **2020**, *25*, 261–267. [CrossRef]
8.   Carter, M.E.; Quinn, P.J. *Salmonella* infections in dogs and cats. In *Salmonella in Domestic Animals*; Wray, C., Wray, A., Eds.; CABI Publishing: Wallingford, UK, 2000; pp. 231–244.
9.   Nemser, S.M.; Doran, T.; Grabenstein, M.; McConnell, T.; McGrath, T.; Pamboukian, R.; Smith, A.C.; Achen, M.; Danzeisen, G.; Kim, S.; et al. Investigation of *Listeria, Salmonella,* and toxigenic *Escherichia coli* in various pet foods. *Foodborne Pathog. Dis.* **2014**, *11*, 706–709. [CrossRef]
10.   Peeples, E.H., Jr. Meanwhile, Humans Eat Pet Food. *The New York Times*, 16 December 1975, p. 39. Available online: https://www.nytimes.com/1975/12/16/archives/meanwhile-humans-eat-pet-food.html (accessed on 17 December 2020).
11.   Greig, J.D.; Todd, E.C.D.; Bartleson, C.A.; Michaels, B.S. Outbreaks where food workers have been implicated in the spread of foodborne disease. Part 1. Description of the problem, methods, and agents involved. *J. Food Prot.* **2007**, *70*, 1752–1761. [CrossRef]
12.   Battelli, G. Zoonoses as occupational diseases. *Vet. Ital.* **2008**, *44*, 601–609. [PubMed]
13.   Rabozzi, G.; Bonizzi, L.; Crespi, E.; Somaruga, C.; Sokooti, M.; Tabibi, R.; Vellere, F.; Brambilla, G.; Colosio, C. Emerging zoonoses: The "One Health approach". *Saf. Health Work* **2012**, *3*, 77–83. [CrossRef]
14.   Food Safety. Available online: https://www.who.int/news-room/fact-sheets/detail/food-safety (accessed on 2 December 2020).
15.   Lanata, C.F.; Fischer-Walker, C.L.; Olascoaga, A.C.; Torres, C.X.; Aryee, M.J.; Black, R.E. Global causes of diarrheal disease mortality in children <5 years of age: A systematic review. *PLoS ONE* **2013**, *8*, e72788. [CrossRef]
16.   European Food Safety Authority and European Centre for Disease Prevention and Control. The European Union summary report on trends and sources of zoonoses, zoonotic agents and food-borne outbreaks in 2017. *EFSA J.* **2018**, *16*, 1–262. [CrossRef]
17.   Kirk, M.D.; Pires, S.M.; Black, R.E.; Caipo, M.; Crump, J.A.; Devleesschauwer, B.; Döpfer, D.; Fazil, A.; Fischer-Walker, C.L.; Hald, T.; et al. World Health Organization Estimates of the Global and Regional Disease Burden of 22 Foodborne Bacterial, Protozoal, and Viral Diseases, 2010: A Data Synthesis. *PLoS Med.* **2015**, *12*, e1001921. [CrossRef]
18.   European Food Safety Authority and European Centre for Disease Prevention and Control. The European Union One Health 2018 Zoonoses Report. *EFSA J.* **2019**, *17*, 1–276. [CrossRef]
19.   European Food Safety Authority and European Centre for Disease Prevention and Control. The European Union One Health 2019 Zoonoses Report. *EFSA J.* **2021**, *19*, 6406. [CrossRef]
20.   Foley, S.L.; Lynne, A.M.; Nayak, R. *Salmonella* challenges: Prevalence in swine and poultry and potential pathogenicity of such isolates. *J. Anim. Sci.* **2008**, *86*, E149–E162. [CrossRef]
21.   Healy, B.; Cooney, S.; O'Brien, S.; Iversen, C.; Whyte, P.; Nally, J.; Callanan, J.J.; Fanning, S. *Cronobacter* (*Enterobacter sakazakii*): An opportunistic foodborne pathogen. *Foodborne Pathog. Dis.* **2010**, *7*, 339–350. [CrossRef]
22.   Fusco, V.; Abriouel, H.; Benomar, N.; Kabisch, J.; Chieffi, D.; Cho, G.-S.; Franz, C.M.A.P. Opportunistic Food-Borne Pathogens. In *Food Safety and Preservation*; Grumezescu, A.M., Holban, A.M., Eds.; Academic Press: London, UK, 2018; pp. 269–306. [CrossRef]
23.   Lipman, L.J.A.; Van Knapen, F. Voedselinfecties en intoxicaties. In *Inleiding tot de Levensmiddelenhygiëne—Achtergrond en Feiten*, 2nd ed.; Lipman, L.J.A., Ruiter, A., Eds.; Reed Business: Amsterdam, The Netherlands, 2011; pp. 89–114. (In Dutch)
24.   Roberts, T.; Murrell, K.D.; Marks, S. Economic losses caused by foodborne parasitic diseases. *J. Parasitol.* **1994**, *10*, 419–423. [CrossRef]
25.   Chengat Prakashbabu, B.; Marshall, L.R.; Crotta, M.; Gilbert, W.; Johnson, J.C.; Alban, L.; Guitian, J. Risk-based inspection as a cost-effective strategy to reduce human exposure to cysticerci of *Taenia saginata* in low-prevalence settings. *Parasites Vectors* **2018**, *11*, 257. [CrossRef]
26.   Laranjo-González, M.; Devleesschauwer, B.; Gabriël, S.; Dorny, P.; Allepuz, A. Epidemiology, impact and control of bovine cysticercosis in Europe: A systematic review. *Parasites Vectors* **2016**, *9*, 81. [CrossRef]
27.   Jansen, F.; Dorny, P.; Berkvens, D.; Van Hul, A.; Van den Broeck, N.; Makay, C.; Praet, N.; Gabriël, S. Assessment of the repeatability and border-plate effects of the B158/B60 enzyme-linked-immunosorbent assay for the detection of circulating antigens (Ag-ELISA) of *Taenia saginata*. *Vet. Parasitol.* **2016**, *227*, 69–72. [CrossRef]
28.   Regional Office for Europe of the World Health Organization. *The Burden of Foodborne Diseases in the WHO European Region*; Kruse, H., Ed.; Copenhagen, Denmark, 2017. Available online: https://www.euro.who.int/__data/assets/pdf_file/0005/402989/50607-WHO-Food-Safety-publicationV4_Web.pdf (accessed on 22 January 2021).
29.   Teunis, P.F.M.; Le Guyader, F.S.; Liu, P.; Ollivier, J.; Moe, C.L. Noroviruses are highly infectious but there is strong variation in host susceptibility and virus pathogenicity. *Epidemics* **2020**, *32*, 100401. [CrossRef] [PubMed]
30.   Potasman, I.; Paz, A.; Odeh, M. Infectious outbreaks associated with bivalve shellfish consumption: A worldwide perspective. *Clin. Infect. Dis.* **2002**, *35*, 921–928. [CrossRef]
31.   European Food Safety Authority Panel on Biological Hazards; Ricci, A.; Allende, A.; Bolton, D.; Chemaly, M.; Davies, R.; Fernandez Escamez, P.S.; Herman, L.; Koutsoumanis, K.; Lindqvist, R.; et al. Scientific Opinion on the public health risks associated with hepatitis E virus (HEV) as a food-borne pathogen. *EFSA J.* **2017**, *15*, 4886. [CrossRef]
32.   Gofflot, S.; El Moualij, B.; Zorzi, D.; Melen, L.; Roels, S.; Quatpers, D.; Grassi, J.; Vanopdenbosch, E.; Heinen, E.; Zorzi, W. Immuno-Quantitative Polymerase Chain Reaction for Detection and Quantitation of Prion Protein. *J. Immunoassay Immunochem.* **2004**, *25*, 241–258. [CrossRef] [PubMed]

33. Eskola, M.; Kos, G.; Elliott, C.T.; Hajšlová, J.; Mayar, S.; Krska, R. Worldwide contamination of food-crops with mycotoxins: Validity of the widely cited 'FAO estimate' of 25%, Critical Reviews in Food Science and Nutrition. *Crit. Rev. Food Sci. Nutr.* **2020**, *60*, 2773–2789. [CrossRef] [PubMed]

34. Noor, R. Insight to Foodborne Diseases: Proposed Models for Infections and Intoxications. *Biomed. Biotechnol. Res. J.* **2019**, *3*, 135–139. [CrossRef]

35. International Society for Infectious Diseases, ProMed. Available online: https://promedmail.org (accessed on 28 January 2021).

36. Ferreira, V.; Wiedmann, M.; Teixeira, P.; Stasiewicz, M.J. *Listeria monocytogenes* Persistence in Food-Associated Environments: Epidemiology, Strain Characteristics, and Implications for Public Health. *J. Food Prot.* **2014**, *77*, 150–170. [CrossRef] [PubMed]

37. Azizoglu, R.O.; Osborne, J.; Wilson, S.; Kathariou, S. Role of Growth Temperature in Freeze-Thaw Tolerance of *Listeria* spp. *Appl. Environ. Microbiol.* **2009**, *75*, 5315–5320. [CrossRef] [PubMed]

38. Thévenot, D.; Dernburg, A.; Vernozy-Rozand, C. An updated review of *Listeria monocytogenes* in the pork meat industry and its products. *J. Appl. Microbiol.* **2006**, *101*, 7–17. [CrossRef]

39. Osimani, A.; Aquilanti, L.; Clementi, F. Salmonellosis associated with mass catering: A survey of European Union cases over a 15-year period. *Epidemiol. Infect.* **2016**, *144*, 3000–3012. [CrossRef]

40. Hugas, M.; Beloeil, P.A. Controlling *Salmonella* along the in the European Union—Progress over the last ten years. *Euro Surveill.* **2014**, *19*, 20804. [CrossRef]

41. Beumer, R.R.; Kusumaningrum, H. Kitchen hygiene in daily life. *Int. Biodeterior. Biodegrad.* **2003**, *51*, 299–302. [CrossRef]

42. Anderson, J.B.; Shuster, T.A.; Hansen, K.E.; Levy, A.S.; Volk, A. A camera's view of consumer food-handling behaviors. *J. Am. Diet. Assoc.* **2004**, *104*, 186–191. [CrossRef] [PubMed]

43. Burke, T.; Young, I.; Papadopoulos, A. Assessing food safety knowledge and preferred information sources among 19–29 year olds. *Food Control* **2016**, *69*, 83–89. [CrossRef]

44. Koch, S.; Lohmann, M.; Geppert, J.; Stamminger, R.; Epp, A.; Böl, G.-F. Kitchen Hygiene in the Spotlight: How Cooking Shows Influence Viewers' Hygiene Practices. *Risk Anal.* **2020**, *41*, 131–140. [CrossRef]

45. Satin, M. *Death in the Pot: The Impact of Food Poisoning on History*, 2nd ed.; Prometheus Books: New York, NY, USA, 2009; p. 258.

46. Dixon, B. *Power Unseen: How Microbes Rule the World*; W.H. Freeman & Company Limited: Oxford, UK, 1994; p. 237.

47. Satin, M. History Foodborne Disease—Part I—Ancient History. In *Encyclopedia of Food Safety*, 1st ed.; Motarjemi, Y., Moy, G., Todd, E., Eds.; Volume 1 History, Science and Methods; Academic Press: London, UK, 2014; p. 2304.

48. Foster, E.M. Historical overview of key issues in food safety. *Emerg. Infect. Dis.* **1997**, *3*, 481–482. [CrossRef]

49. Stephany, R.W. Hormonal growth promoting agents in food producing animals. In *Handbook of Experimental Pharmacology*; Thieme, D., Hemmersbach, P., Eds.; Doping in Sports; Springer: Berlin/Heidelberg, Germany, 2010; Volume 195, pp. 355–367. [CrossRef]

50. Gribble, G.W. Food chemistry and chemophobia. *Food Sec.* **2013**, *5*, 177–187. [CrossRef]

51. Hameed, S.; Xie, L.; Ying, Y. Conventional and emerging detection techniques for pathogenic bacteria in food science: A review. *Trends Food Sci. Technol.* **2018**, *81*, 61–73. [CrossRef]

52. Saravanan, A.; Senthil Kumar, P.; Hemavathy, R.V.; Jeevanantham, S.; Kamalesh, R.; Sneha, S.; Yaashikaa, P.R. Methods of detection of food-borne pathogens: A review. *Environ. Chem. Lett.* **2021**, *19*, 189–207. [CrossRef]

53. Visansirikul, S.; Kolodziej, S.A.; Demchenko, A.V. *Staphylococcus aureus* capsular polysaccharides: A structural and synthetic perspective. *Org. Biomol. Chem.* **2020**, *18*, 783. [CrossRef]

54. Vinogradov, E.; Conlan, J.W.; Perry, M.B. Serological cross-reaction between the lipopolysaccharide O-polysaccharide antigens of *Escherichia coli* O157:H7 and strains of *Citrobacter freundii* and *Citrobacter sedlakii*. *FEMS Microbiol. Lett.* **1998**, *190*, 157–161. [CrossRef] [PubMed]

55. Gal-Mor, O.; Boyle, E.C.; Grassl, G.A. Same species, different diseases: How and why typhoidal and non-typhoidal *Salmonella enterica* serovars differ. *Front. Microbiol.* **2014**, *5*, 391. [CrossRef] [PubMed]

56. Paoli, G.C.; Kleina, L.G.; Brewster, J.D. Development of *Listeria monocytogenes*–specific immunomagnetic beads using a single-chain antibody fragment. *Foodborne Pathog. Dis.* **2007**, *4*, 74–83. [CrossRef] [PubMed]

57. Gaastra, W. Micro-organismen. In *Inleiding tot de Levensmiddelenhygiëne—Achtergrond en Feiten*, 2nd ed.; Lipman, L.J.A., Ruiter, A., Eds.; Reed Business: Amsterdam, The Netherlands, 2011; pp. 31–70. (In Dutch)

58. Bergwerff, A.A.; Van Dam, G.J.; Rotmans, J.P.; Deelder, A.M.; Kamerling, J.P.; Vliegenthart, J.F. The immunologically reactive part of immunopurified circulating anodic antigen from *Schistosoma mansoni* is a threonine-linked polysaccharide consisting of -> 6)-(beta-D-GlcpA-(1 -> 3))-beta-D-GalpNAc-(1 -> repeating units. *J. Biol. Chem.* **1994**, *269*, 31510–31517. [CrossRef]

59. Sun, G.G.; Wang, Z.Q.; Liu, C.Y.; Jiang, P.; Liu, P.-D.; Wen, H.; Qi, X.; Wang, L.; Cui, J. Early serodiagnosis of trichinellosis by ELISA using excretory–secretory antigens of *Trichinella spiralis* adult worms. *Parasites Vectors* **2015**, *8*, 484. [CrossRef]

60. Morales-Yanez, F.J.; Sariego, I.; Vincke, C.; Hassanzadeh-Ghassabeh, G.; Polman, K.; Muyldermans, S. An innovative approach in the detection of *Toxocara canis* excretory/secretory antigens using specific nanobodies. *Int. J. Parasitol.* **2019**, *49*, 635–645. [CrossRef] [PubMed]

61. Roitt, I.; Brostoff, J.; Male, D. *Immunology*, 6th ed.; Mosby: London, UK, 2001; p. 480.

62. Janeway, C.A., Jr.; Travers, P.; Walport, M.; Shlomchik, M.J. *Immunobiology: The Immunesystem in Health and Disease*, 6th ed.; Elsevier Ltd.: Oxford, UK, 2005; p. 823.

63. Bruschi, F.; Gómez-Morales, M.A.; Hill, D.E. International Commission on Trichinellosis: Recommendations on the use of serological tests for the detection of Trichinella infection in animals and humans. *Food Waterborne Parasitol.* **2019**, *14*, e00032. [CrossRef]
64. Leavy, O. The birth of monoclonal antibodies. *Nat. Immunol.* **2016**, *17*, S13. [CrossRef]
65. Ehrlich, P.H.; Moyle, W.R. Cooperative immunoassays: Ultrasensitive assays with mixed monoclonal antibodies. *Science* **1983**, *221*, 279–281. [CrossRef]
66. Marega, R.; Desroche, N.; Huet, A.-C.; Paulus, M.; Suarez Pantaleon, C.; Larose, D.; Arbault, P.; Delahaut, P.; Gillard, N. A general strategy to control antibody specificity against targets showing molecular and biological similarity: *Salmonella* case study. *Sci. Rep.* **2020**, *10*, 18439. [CrossRef] [PubMed]
67. Petrenko, V.A.; Vodyanoy, V.J. Phage display for detection of biological threat agents. *J. Microbiol. Methods* **2003**, *53*, 253–262. [CrossRef]
68. Nguyen, X.H.; Trinh, T.-L.; Vu, T.-B.-H.; Le, Q.-H.; To, K.-A. Isolation of phage-display library-derived scFv antibody specific to *Listeria monocytogenes* by a novel immobilized method. *J. Appl. Microbiol.* **2018**, *124*, 591–597. [CrossRef] [PubMed]
69. Ritchie, R.F. The foundations of immunochemistry. In *The Immunoassay Handbook—Theory and Applications of Ligand Binding, ELISA and Related Techniques*, 4th ed.; Wild, D., Ed.; Elsevier: Amsterdam, The Netherlands, 2013; Chapter 4.1, pp. 339–356. [CrossRef]
70. Swildens, B. Detection and Transmission of Extracellular Factor Producing *Streptococcus suis* Serotype 2 Strains in Pigs. Ph.D. Thesis, Utrecht University, Utrecht, The Netherlands, 2009.
71. Brown, E.; Lawson, S.; Welbon, C.; Gnanandarajah, J.; Li, J.; Murtaugh, M.P.; Nelson, E.A.; Molina, R.M.; Zimmerman, J.J.; Rowland, R.R.R.; et al. Antibody response to porcine reproductive and respiratory syndrome virus (PRRSV) nonstructural proteins and implications for diagnostic detection and differentiation of PRRSV types I and II. *Clin. Vaccine Immunol.* **2009**, *16*, 628–635. [CrossRef]
72. World Organization for Animal Health. *Campylobacter jejuni* and *Campylobacter coli*. In *OIE Manual of Diagnostic Tests and Vaccines for Terrestrial Animals (Mammals, Birds and Bees)*, 6th ed.; OIE Biological Standards Commission, Ed.; Office International Des Epizooties: Paris, France, 2008; Volume 2, Chapter 2.9.3, pp. 1185–1191.
73. AbuOun, M.; Manning, G.; Cawthraw, S.A.; Ridley, A.; Ahmed, I.H.; Wassenaar, T.M.; Newell, D.G. Cytolethal distending toxin (CDT)-negative *Campylobacter jejuni* strains and anti-CDT neutralizing antibodies are induced during human infection but not during colonization in chickens. *Infect. Immun.* **2005**, *73*, 3053–3062. [CrossRef]
74. Berndt, A.; Wilhelm, A.; Jugert, C.; Pieper, J.; Sachse, K.; Methner, U. Chicken cecum immune response to *Salmonella enterica* serovars of different levels of invasiveness. *Infect. Immun.* **2007**, *75*, 5993–6007. [CrossRef] [PubMed]
75. Berk, P.A.; Van der Heijden, H.M.J.F.; Mooijman, K.A. On Behalf of the European CommissionComparability of Different ELISAs on the Detection of *Salmonella* spp. Antibodies in Meat Juice and Serum, RIVM Report 330604007. 2008. Available online: https://www.rivm.nl/bibliotheek/rapporten/330604007.pdf (accessed on 14 December 2020).
76. Vico, J.P.; Mainar-Jaime, R.C. The use of meat juice or blood serum for the diagnosis of *Salmonella* infection in pigs and its possible implications on *Salmonella* control programs. *J. Vet. Diagn. Investig.* **2011**, *23*, 528–531. [CrossRef] [PubMed]
77. Mousing, J.; Thode Jensen, P.; Halgaard, C.; Bager, F.; Feld, N.; Nielsen, B.; Nielsen, J.P.; Beth-Nielsen, S. Nation-wide *Salmonella enterica* surveillance and control in Danish slaughter swine herds. *Prev. Vet. Med.* **1997**, *29*, 247–261. [CrossRef]
78. Meemken, D.; Tangemann, A.H.; Meermeier, D.; Gundlach, S.; Mischok, D.; Greiner, M.; Klein, G.; Blaha, T. Establishment of serological herd profiles for zoonoses and production diseases in pigs by "meat juice multi-serology". *Prev. Vet. Med.* **2014**, *113*, 589–598. [CrossRef]
79. Dzierzon, J.; Oswaldi, V.; Merle, R.; Langkabel, N.; Meemken, D. High Predictive Power of Meat Juice Serology on the Presence of Hepatitis E Virus in Slaughter Pigs. *Foodborne Pathog. Dis.* **2020**, *17*, 687–692. [CrossRef]
80. Berger-Schoch, A.E.; Bernet, D.; Doherr, M.G.; Gottstein, B.; Frey, C.F. *Toxoplasma gondii* in Switzerland: A serosurvey based on meat juice analysis of slaughtered pigs, wild boar, sheep and cattle. *Zoonoses Public Health* **2011**, *58*, 472–478. [CrossRef] [PubMed]
81. Gazzonis, A.L.; Zanzani, S.A.; Villa, L.; Manfredi, M.T. *Toxoplasma gondii* infection in meat-producing small ruminants: Meat juice serology and genotyping. *Parasitol. Int.* **2020**, *76*, 102060. [CrossRef]
82. Wallander, C.; Frössling, J.; Vågsholm, I.; Burrells, A.; Lundén, A. "Meat juice" is not a homogeneous serological matrix. *Foodborne Pathog. Dis.* **2015**, *12*, 280–288. [CrossRef]
83. Hill, D.E.; Chirukandoth, S.; Dubey, J.P.; Lunney, J.K.; Gamble, H.R. Comparison of detection methods for *Toxoplasma gondii* in naturally and experimentally infected swine. *Vet. Parasitol.* **2006**, *141*, 9–17. [CrossRef]
84. Atkinson, B.M.; Bearson, B.L.; Loving, C.L.; Zimmerman, J.J.; Kich, J.D.; Bearson, S.M.D. Detection of *Salmonella*-specific antibody in swine oral fluids. *Porc. Health Manag.* **2019**, *5*, 29. [CrossRef]
85. Prickett, J.R.; Zimmerman, J.J. The development of oral fluid-based diagnostics and applications in veterinary medicine. *Anim. Health Res. Rev.* **2010**, *11*, 207–216. [CrossRef] [PubMed]
86. Bjustrom-Kraft, J.; Christopher-Hennings, J.; Daly, R.; Main, R.; Torrison, J.; Thurn, M.; Zimmerman, J. The use of oral fluid diagnostics in swine medicine. *J. Swine Health Prod.* **2018**, *26*, 262–269.
87. Fosgate, G.T. Practical sample size calculations for surveillance and diagnostic investigations. *J. Vet. Diagn. Investig.* **2009**, *21*, 3–14. [CrossRef] [PubMed]
88. Freuling, C.M.; Müller, T.F.; Mettenleiter, T.C. Vaccines against pseudorabies virus (PrV). *Vet. Microbiol.* **2017**, *206*, 3–9. [CrossRef] [PubMed]

89. Leifer, I.; Everett, H.; Hoffmann, B.; Sosan, O.; Crooke, H.; Beer, M.; Blome, S. Escape of classical swine fever C-strain vaccine virus from detection by C-strain specific real-time RT-PCR caused by a point mutation in the primer-binding site. *J. Virol. Methods* **2010**, *166*, 98–100. [CrossRef]

90. Silva, N.F.D.; Neves, M.M.P.S.; Magalhães, J.M.C.S.; Freire, C.; Delerue-Matos, C. Electrochemical immunosensor towards invasion-associated protein p60: An alternative strategy for *Listeria monocytogenes* screening in food. *Talanta* **2020**, 120976. [CrossRef] [PubMed]

91. Balasubramanian, S.; Amamcharla, J.; Shin, J.E. Possible application of electronic nose systems for meat safety: An overview. In *Electronic Noses and Tongues in Food Science*, 1st ed.; Rodriguez, M.M.L., Ed.; Part I; Academic Press: London, UK, 2016; Chapter 7, pp. 59–71.

92. Gorski, L. Selective enrichment media bias the types of *Salmonella enterica* strains isolated from mixed strain cultures and complex enrichment broths. *PLoS ONE* **2012**, *7*, e34722. [CrossRef]

93. Löfström, C.; Krause, M.; Josefsen, M.H.; Hansen, F.; Hoorfar, J. Validation of a same-day real-time PCR method for screening of meat and carcass swabs for *Salmonella*. *BMC Microbiol.* **2009**, *9*, 85. [CrossRef]

94. Delibato, E.; Rodriguez-Lazaro, D.; Gianfranceschi, M.; De Cesare, A.; Comin, D.; Gattuso, A.; Hernandez, M.; Sonnessa, M.; Pasquali, F.; Sreter-Lancz, Z.; et al. European validation of Real-Time PCR method for detection of *Salmonella* spp. in pork meat. *Int. J. Food Microbiol.* **2014**, *184*, 134–138. [CrossRef]

95. Srisa-Art, M.; Boehle, K.E.; Geiss, B.J.; Henry, C.S. Highly sensitive detection of *Salmonella typhimurium* using a colorimetric paper-based analytical device coupled with immunomagnetic separation. *Anal. Chem.* **2018**, *90*, 1035–1043. [CrossRef]

96. Brewster, J.D. Isolation and concentration of Salmonellae with an immunoaffinity column. *J. Microbiol. Methods* **2003**, *55*, 287–293. [CrossRef]

97. Lee, K.-M.; Runyon, M.; Herrman, T.J.; Phillips, R.; Hsieh, J. Review of *Salmonella* detection and identification methods: Aspects of rapid emergency response and food safety. *Food Control* **2015**, *47*, 264–276. [CrossRef]

98. Saengthongpinit, C.; Bergwerff, A.A.; Van Knapen, F.; Amavisit, P.; Sirinarumitr, T.; Sakpuaram, T. Comparison of immunomagnetic separation and multiplex PCR assay for detection of *Campylobacter jejuni* and *Campylobacter coli* in chicken meat. *Kasetsart J.* **2007**, *41*, 696–704.

99. Ozalp, V.C.; Bayramoglu, G.; Erdem, Z.; Arica, M.Y. Pathogen detection in complex samples by quartz crystal microbalance sensor coupled to aptamer functionalized core–shell type magnetic separation. *Anal. Chim. Acta* **2015**, *853*, 533–540. [CrossRef] [PubMed]

100. Huo, X.; Chen, Q.; Wang, L.; Cai, G.; Qi, W.; Xia, Z.; Wen, W.; Lin, J. Continuous-flow separation and efficient concentration of foodborne bacteria from large volume using nickel nanowire bridge in microfluidic chip. *Micromachines* **2019**, *10*, 644. [CrossRef]

101. Chen, C.-C.; Tu, Y.-Y.; Chang, H.-M. Thermal stability of bovine milk immunoglobulin g (igg) and the effect of added thermal protectants on the stability. *J. Food Sci.* **2000**, *65*, 188–193. [CrossRef]

102. Bergwerff, A.A. New strategies for detecting *Salmonella* and sulphonamides. *World Poult.* **2006**, *22*, 2–3.

103. Videnska, P.; Sisak, F.; Havlickova, H.; Faldynova, M.; Rychlik, I. Influence of *Salmonella enterica* serovar Enteritidis infection on the composition of chicken cecal microbiota. *BMC Vet. Res.* **2013**, *9*, 140. [CrossRef] [PubMed]

104. Blondel, C.J.; Yang, H.-J.; Castro, B.; Chiang, S.; Toro, C.S.; Zaldívar, M.; Contreras, I.; Andrews-Polymenis, H.L.; Santiviago, C.A. Contribution of the Type VI Secretion System Encoded in SPI-19 to Chicken Colonization by *Salmonella enterica* Serotypes Gallinarum and Enteritidis. *PLoS ONE* **2010**, *5*, e11724. [CrossRef] [PubMed]

105. Kothary, M.H.; Babu, U.S. Infective dose of foodborne pathogens in volunteers: A review. *J. Food Saf.* **2001**, *21*, 49–73. [CrossRef]

106. Smith, J.L. Infectious dose and an aging population: Susceptibility of the aged to foodborne pathogens. In *Foodborne Pathogens, Food Microbiology and Food Safety*; Gurtler, J., Doyle, M., Kornacki, J., Eds.; Springer: Cham, Switzerland, 2017; pp. 451–468. [CrossRef]

107. Todd, E.C.D.; Greig, J.D.; Bartleson, C.A.; Michaels, B.S. Outbreaks where food workers have been implicated in the spread of foodborne disease. Part 4. Infective doses and pathogen carriage. *J. Food Prot.* **2008**, *71*, 2339–2373. [CrossRef] [PubMed]

108. Mead, G.; Lammerding, A.M.; Cox, N.; Doyle, M.P.; Humbert, F.; Kulikovskiy, A.; Panin, A.; Pinheiro Do Nascimento, V.; Wierup, M. The *Salmonella* on Raw Poultry Writing Committee. Scientific and technical factors affecting the setting of *Salmonella* criteria for raw poultry: A global perspective. *J. Food Prot.* **2010**, *73*, 1566–1590. [CrossRef]

109. Eggenkamp, A.E.; Bergwerff, A.A. Microbiologie in het laboratorium. In *Inleiding tot de Levensmiddelenhygiëne—Achtergrond en Feiten*, 2nd ed.; Lipman, L.J.A., Ruiter, A., Eds.; Reed Business: Amsterdam, The Netherlands, 2011; pp. 144–165. (In Dutch)

110. Reist, M.; Jemmi, T.; Stärk, K.D.C. Policy-driven development of cost-effective, risk-based surveillance strategies. *Prev. Vet. Med.* **2012**, *105*, 176–184. [CrossRef]

111. Martin, S.W.; Shoukri, M.; Thorburn, M.A. Evaluating the health status of herds based on tests applied to individuals. *Prev. Vet. Med.* **1992**, *14*, 33–43. [CrossRef]

112. Sahin, O.; Kassem, I.I.; Shen, Z.; Lin, J.; Rajashekara, G.; Zhang, Q. *Campylobacter* in poultry: Ecology and potential interventions. *Avian Dis.* **2015**, *59*, 185–200. [CrossRef]

113. Hugas, M.; Tsigarida, E. Pros and cons of carcass decontamination: The role of the European Food Safety Authority. *Meat Sci.* **2008**, *78*, 43–52. [CrossRef]

114. European Food Safety Authority Panel on Biological Hazards; Ricci, A.; Allende, A.; Bolton, D.; Chemaly, M.; Davies, R.; Fernández Escámez, P.S.; Girones, R.; Herman, L.; Koutsoumanis, K.; et al. Scientific opinion on the guidance on the requirements for the development of microbiological criteria. *EFSA J.* **2017**, *15*, 5052. [CrossRef]

115. Barlow, S.M.; Boobis, A.R.; Bridges, J.; Cockburn, A.; Dekant, W.; Hepburn, P.; Houben, G.F.; König, J.; Nauta, M.J.; Schuermans, J.; et al. The role of hazard- and risk-based approaches in ensuring food safety. *Trends Food Sci. Technol.* **2015**, *46*, 176–188. [CrossRef]

116. Bahrndorff, S.; Rangstrup-Christensen, L.; Nordentoft, S.; Hald, B. Foodborne Disease Prevention and Broiler Chickens with Reduced Campylobacter Infection. *Emerg. Infect. Dis.* **2013**, *19*, 425–430. [CrossRef] [PubMed]

117. Finley, R.; Reid-Smith, R.; Weese, J.S. Human health implications of *Salmonella*-contaminated natural pet treats and raw pet food. *Clin. Infect. Dis.* **2006**, *42*, 686–691. [CrossRef] [PubMed]

118. Notermans, S.; Wernars, K. Immunological methods for detection of foodborne pathogens and their toxins. *Int. J. Food Microbiol.* **1991**, *12*, 91–102. [CrossRef]

119. Candlish, A.A.G. Immunological methods in food microbiology. *Food Microbiol.* **1991**, *8*, 1–14. [CrossRef]

120. Wild, D. (Ed.) *The Immunoassay Handbook*, 4th ed.; Elsevier: Amsterdam, The Netherlands, 2013. [CrossRef]

121. Köhler, W. Zentralblatt fur Bakteriologie—100 years ago Agglutination and filament formation of *Proteus* bacteria and maternal-fetal transfer of agglutinins. *Int. J. Med. Microbiol.* **2001**, *290*, 643–646. [CrossRef]

122. Gracias, K.S.; McKillip, J.L. A review of conventional detection and enumeration methods for pathogenic bacteria in food. *Can. J. Microbiol.* **2004**, *50*, 883–890. [CrossRef]

123. Schmidt, B.; Sankaran, S.; Stegemann, L.; Strassert, C.A.; Jonkheijm, P.; Voskuhl, J. Agglutination of bacteria using polyvalent nanoparticles of aggregation-induced emissive thiophthalonitrile dyes. *J. Mater. Chem. B* **2016**, *4*, 4732–4738. [CrossRef]

124. Denayer, S.; Delbrassinne, L.; Nia, Y.; Botteldoorn, N. Food-borne outbreak investigation and molecular typing: High diversity of *Staphylococcus aureus* strains and importance of toxin detection. *Toxins* **2017**, *9*, 407. [CrossRef]

125. Abel, E.S. Preliminary studies on the detection of *Bacillus cereus* and its toxins: Comparing conventional and immunological assays with a direct polymerase chain reaction method. *Curr. J. Appl. Sci. Technol.* **2019**, *36*, 1–9. [CrossRef]

126. Fletcher, P.; Logan, N.A. Improved cytotoxicity assay for *Bacillus cereus* diarrhoeal enterotoxin. *Lett. Appl. Microbiol.* **1999**, *28*, 394–400. [CrossRef] [PubMed]

127. Chart, H.; Willshaw, G.A.; Cheasty, T. Evaluation of a reversed passive latex agglutination test for the detection of verocytotoxin (VT) expressed by strains of VT-producing *Escherichia coli*. *Lett. Appl. Microbiol.* **2001**, *32*, 370–374. [CrossRef] [PubMed]

128. O'Farrell, B. Evolution in lateral flow-based immunoassay systems. In *Lateral Flow Immunoassay*; Wong, R., Tse, H., Eds.; Springer Science & Business Media: New York, NY, USA, 2008; pp. 1–34.

129. Bird, C.B.; Hoerner, R.J.; Restaino, L. Reveal 8–Hour test system for detection of *Escherichia coli* O157:H7 in raw ground beef, raw beef cubes, and lettuce rinse: Collaborative study. *J. AOAC Int.* **2001**, *84*, 719–736. [CrossRef]

130. Bird, C.B.; Hoerner, R.J.; Restaino, L. Comparison of the Reveal 20–hour method and the BAM culture method for the detection of *Escherichia coli* O157:H7 in selected foods and environmental swabs: Collaborative study. *J. AOAC Int.* **2001**, *84*, 737–751. [CrossRef]

131. Feldsine, P.T.; Falbo-Nelson, M.T.; Brunelle, S.L.; Forgey, R.L. Visual Immunoprecipitate Assay (VIP) for Detection of Enterohemorrhagic *Escherichia coli* (EHEC) 0157:H7 in Selected Foods: Collaborative Study. *J. AOAC Int.* **1997**, *80*, 517–529. [CrossRef]

132. Feldsine, P.T.; Mui, L.A.; Forgey, R.L.; Kerr, D.E.A. Equivalence of Visual Immunoprecipitate Assay (VIP®) for *Salmonella* for the Detection of Motile and Nonmotile *Salmonella* in All Foods to AOAC Culture Method: Collaborative Study. *J. AOAC Int.* **2000**, *83*, 888–902. [CrossRef]

133. Feldsine, P.T.; Kerr, D.E.; Shen, G.; Lienau, A. Comparative Validation Study to Demonstrate the Equivalence of a Minor Modification to AOAC Method 997.03 Visual Immunoprecipitate (VIP®) for *Listeria* to the Reference Culture Method. *J. AOAC Int.* **2009**, *92*, 1421–1425. [CrossRef]

134. Li, J.; Macdonald, J. Multiplexed lateral flow biosensors: Technological advances for radically improving point-of-care diagnoses. *Biosens. Bioelectron.* **2016**, *83*, 177–192. [CrossRef]

135. Chang, L.; Li, J.; Wang, L. Immuno-PCR: An ultrasensitive immunoassay for biomolecular detection. *Anal. Chim. Acta* **2016**, *910*, 12–24. [CrossRef]

136. Sadat Tabatabaei, M.; Islam, R.; Ahmed, M. Applications of gold nanoparticles in ELISA, PCR, and immuno-PCR assays: A review. *Anal. Chim. Acta* **2021**, *1143*, 250–266. [CrossRef]

137. Rajkovic, A.; El Moualij, B.; Fikri, Y.; Dierick, K.; Zorzi, W.; Heinen, E.; Uner, A.; Uyttendaele, M. Detection of *Clostridium botulinum* neurotoxins A and B in milk by ELISA and immuno-PCR at higher sensitivity than mouse bio-assay. *Food Anal. Methods* **2012**, *5*, 319–326. [CrossRef]

138. Rajkovic, A.; El Moualij, B.; Uyttendaele, M.; Brolet, P.; Zorzi, W.; Heinen, E.; Foubert, E.; Debevere, J. Immunoquantitative Real-Time PCR for Detection and Quantification of *Staphylococcus aureus* Enterotoxin B in Foods. *Appl. Environ. Microbiol.* **2006**, *72*, 6593–6599. [CrossRef] [PubMed]

139. Fischer, A.; Von Eiff, C.; Kuczius, T.; Omoe, K.; Peters, G.; Becker, K. A quantitative real-time immuno-PCR approach for detection of staphylococcal enterotoxins. *J. Mol. Med.* **2007**, *85*, 461–469. [CrossRef] [PubMed]

140. Grangar, V.; Curiale, M.S.; D'onorio, A.; Schultz, A.; Johnson, R.L.; Atrache, V. VIDAS® enzyme-linked Immunofluorescent assay for detection of *Listeria* in Foods: Collaborative study. *J. AOAC Int.* **2000**, *83*, 903–918. [CrossRef]

141. Karuppannan, A.K.; De Castro, A.M.M.G.; Opriessnig, T. Recent advances in veterinary diagnostic virology. In *Advanced Techniques in Diagnostic Microbiology*, 3rd ed.; Tang, Y.W., Stratton, C., Eds.; Applications; Springer: Cham, Switzerland, 2018; Volume 2, pp. 317–344. [CrossRef]

142. Olopoenia, L.A.; King, A.L. Widal agglutination test—100 years later: Still plagued by controversy. *Postgrad Med. J.* **2000**, *76*, 80–84. [CrossRef]

143. Kennedy, D.; Wilkinson, M.G. Application of flow cytometry to the detection of pathogenic bacteria. *Curr. Issues Mol. Biol.* **2017**, *23*, 21–38. [CrossRef]

144. Robinson, J.P. Overview of flow cytometry and microbiology. *Curr. Protoc. Cytom.* **2018**, *84*, e37. [CrossRef]

145. Reslova, N.; Michna, V.; Kasny, M.; Mikel, P.; Kralik, P. xMAP technology: Applications in detection of pathogens. *Front. Microbiol.* **2017**, *8*, 55. [CrossRef]

146. Vignali, D.A.A. Multiplexed particle-based flow cytometric assays. *J. Immunol. Methods* **2000**, *243*, 243–255. [CrossRef]

147. Kellar, K.L.; Iannone, M.A. Multiplexed microsphere-based flow cytometric assays. *Exp. Hematol.* **2002**, *30*, 1227–1237. [CrossRef]

148. Graham, H.; Chandler, D.J.; Dunbar, S.A. The genesis and evolution of bead-based multiplexing. *Methods* **2019**, *158*, 2–11. [CrossRef]

149. Foroutan Parsa, S.; Vafajoo, A.; Rostami, A.; Salarian, R.; Rabiee, M.; Rabiee, N.; Rabiee, G.; Tahriri, M.; Yadegari, A.; Vashaee, D.; et al. Early diagnosis of disease using microbead array technology: A review. *Anal. Chim. Acta* **2018**, *1032*, 1–17. [CrossRef] [PubMed]

150. Christopher-Hennings, J.; Araujo, K.P.C.; Souza, C.J.H.; Fang, Y.; Lawson, S.; Nelson, E.A.; Clement, T.; Dunn, M.; Lunney, J.K. Opportunities for bead-based multiplex assays in veterinary diagnostic laboratories. *J. Vet. Diagn. Investig.* **2013**, *25*, 671–691. [CrossRef]

151. Shepelyakovskaya, A.; Rudenko, N.; Karatovskaya, A.; Shchannikova, M.; Shulcheva, I.; Fursova, K.; Zamyatina, A.; Boziev, K.; Oleinikov, V.; Brovko, F. Development of a bead-based multiplex assay for the simultaneous quantification of three staphylococcal enterotoxins in food by flow cytometry. *Food Anal. Methods* **2020**, *13*, 1202–1210. [CrossRef]

152. Mechaly, A.; Vitner, E.; Levy, H.; Weiss, S.; Bar-David, E.; Gur, D.; Koren, M.; Cohen, H.; Cohen, O.; Mamroud, E.; et al. Simultaneous immunodetection of anthrax, plague, and tularemia from blood cultures by use of multiplexed suspension arrays. *J. Clin. Microbiol.* **2018**, *56*, e01479-17. [CrossRef] [PubMed]

153. Leach, K.M.; Stroot, J.M.; Lim, D.V. Same-day detection of *Escherichia coli* O157:H7 from spinach by using electrochemiluminescent and cytometric bead array biosensors. *Appl. Environ. Microbiol.* **2010**, *76*, 8044–8052. [CrossRef]

154. Liebsch, C.; Rödiger, S.; Böhm, A.; Nitschke, J.; Weinreich, J.; Fruth, A.; Roggenbuck, D.; Lehmann, W.; Schedler, U.; Juretzek, T.; et al. Solid-phase microbead array for multiplex O-serotyping of *Escherichia coli*. *Microchim. Acta* **2017**, *184*, 1405–1415. [CrossRef]

155. Bergwerff, A.A.; Bokken, G.C.A.M.; Gortemaker, B.G.M. Immobilisation of Antigenic Carbohydrates to Support Detection of Pathogenic Microorganisms. Patent Application U.S. 20090081638 A1, 26 March 2009. Available online: https://patents.google.com/patent/US20090081638A1/en#patentCitations (accessed on 18 January 2021).

156. Van der Wal, F.J.; Achterberg, R.P.; Maassen, C.B.M. A bead-based suspension array for the detection of *Salmonella* antibodies in pig sera. *BMC Vet. Res.* **2018**, *14*, 226. [CrossRef]

157. Thomas, M.E.; Klinkenberg, D.; Bergwerff, A.A.; Van Eerden, E.; Stegeman, J.A.; Bouma, A. Evaluation of suspension array analysis for detection of egg yolk antibodies against *Salmonella* Enteritidis. *Prev. Vet. Med.* **2010**, *95*, 137–143. [CrossRef]

158. European Commission of the European Union. Regulation (EU) 2015/1375 of 10 August 2015 laying down specific rules on official controls for *Trichinella* in meat. *Off. J. Eur. Union* **2015**, *L212*, 7.

159. Bokken, G.C.A.M. Concurrent Monitoring of Trichinella and Toxoplasma Infections in Pigs from Controlled Housing Systems. Ph.D. Thesis, Utrecht University, Utrecht, The Netherlands, 2017. Available online: https://dspace.library.uu.nl/handle/1874/351669 (accessed on 15 January 2021).

160. Jiang, P.; Wang, Z.-Q.; Cui, J.; Zhang, X. Comparison of artificial digestion and Baermann's methods for detection of *Trichinella spiralis* pre-encapsulated larvae in muscles with low-level infections. *Foodborne Pathog. Dis.* **2012**, *9*, 27–31. [CrossRef] [PubMed]

161. Bokken, G.C.A.M.; Bergwerff, A.A.; Van Knapen, F. A novel bead-based assay to detect specific antibody responses against *Toxoplasma gondi* and *Trichinella spiralis* simultaneously in sera of experimentally infected swine. *BMC Vet. Res.* **2012**, *8*, 36. [CrossRef]

162. Havelaar, A.H.; Van Rosse, F.; Bucura, C.; Toetenel, M.A.; Haagsma, J.A.; Kurowicka, D.; Heesterbeek, J.A.P.; Speybroeck, N.; Langelaar, M.F.M.; Van der Giessen, J.W.B.; et al. Prioritizing emerging zoonoses in the Netherlands. *PLoS ONE* **2010**, *5*, e13965. [CrossRef]

163. Batz, M.; Hoffmann, S.; Morris, J., Jr. Ranking the Risks: The 10 Pathogen-Food Combinations with the Greatest Burden on Public Health, Emerging Pathogens Institute at University of Florida. 2011. Available online: http://hdl.handle.net/10244/1022 (accessed on 15 January 2021).

164. Guilbault, G.; Kramer, D.; Cannon, P., Jr. Electrical determination of organophosphorous compounds. *Anal. Chem.* **1962**, *34*, 1437–1439. [CrossRef]

165. Duffy, G.F.; Moore, E.J. Electrochemical immunosensors for food analysis: A review of recent developments. *Anal. Lett.* **2017**, *50*, 1–32. [CrossRef]

166. Evtugyn, G.A.; Shamagsumova, R.V.; Hianik, T. Biosensors for detection mycotoxins and pathogenic bacteria in food. In *Nanobiosensors*; Grumezescu, A.M., Ed.; Academic Press: London, UK, 2017; pp. 35–92. [CrossRef]

167. Cinti, S.; Volpe, G.; Piermarini, S.; Delibato, E.; Palleschi, G. Electrochemical biosensors for rapid detection of foodborne *Salmonella*: A critical overview. *Sensors* **2017**, *17*, 1910. [CrossRef]

168. Bhardwaj, N.; Bhardwaj, S.K.; Nayak, M.K.; Mehta, J.; Kim, K.-H.; Deep, A. Fluorescent nanobiosensors for the targeted detection of foodborne bacteria. *Trends Anal. Chem.* **2017**, *97*, 120e135. [CrossRef]

169. Das, B.; Balasubramanian, P.; Jayabalan, R.; Lekshmi, N.; Thomas, S. Strategies behind biosensors for food and waterborne pathogens. In *Quorum Sensing and Its Biotechnological Applications*; Kalia, V., Ed.; Springer: Singapore, 2018; pp. 107–141. [CrossRef]

170. Ali, A.A.; Altemimi, A.B.; Alhelfi, N.; Ibrahim, S.A. Application of biosensors for detection of pathogenic food bacteria: A review. *Biosensors* **2020**, *10*, 58. [CrossRef]

171. Yunus, G.; Kuddus, M. Electrochemical biosensor for food borne pathogens: An overview. *Carpath. J. Food Sci. Technol.* **2020**, *12*, 5–16. [CrossRef]

172. Silva, N.F.D.; Magalhães, J.M.C.S.; Freire, C.; Delerue-Matos, C. Electrochemical biosensors for *Salmonella*: State of the art and challenges in food safety assessment. *Biosens. Bioelectron.* **2018**, *99*, 667–682. [CrossRef]

173. Riu, J.; Giussani, B. Electrochemical biosensors for the detection of pathogenic bacteria in food. *Trends Anal. Chem.* **2020**, *126*, 115863. [CrossRef]

174. Liébana, A.; Brandão, D.; Alegret, S.; Pividori, M.I. Electrochemical immunosensors, genosensors and phagosensors for *Salmonella* detection. *Anal. Methods* **2014**, *6*, 8858–8873. [CrossRef]

175. Khansili, N.; Rattu, G.; Krishna, P.M. Label-free optical biosensors for food and biological sensor applications. *Sens. Actuators B Chem.* **2018**, *265*, 35–49. [CrossRef]

176. Zourob, M.; Elwary, S.; Turner, A. (Eds.) *Principles of Bacterial Detection: Biosensors, Recognition Receptors and Microsystems*; Springer: New York, NY, USA, 2008. [CrossRef]

177. Gehring, A.G.; Tu, S.-I. High-throughput biosensors for multiplexed food-borne pathogen detection. *Annu. Rev. Anal. Chem.* **2011**, *4*, 151–172. [CrossRef] [PubMed]

178. Bozal-Palabiyik, B.; Gumustas, A.; Ozkan, S.A.; Uslu, B. Biosensor-based methods for the determination of foodborne pathogens. In *Handbook of Food Bioengineering—Foodborne Diseases*; Holban, A.M., Grumezescu, A.M., Eds.; Academic Press: London, UK, 2018; Chapter 12, pp. 379–420. [CrossRef]

179. Kumar Mishra, G.; Barfidokht, A.; Tehrani, F.; Kumar Mishra, R. Food Safety Analysis Using Electrochemical Biosensors. *Foods* **2018**, *7*, 141. [CrossRef]

180. Poltronieri, P.; Mezzolla, V.; Primiceri, E.; Maruccio, G. Biosensors for the detection of food pathogens. *Foods* **2014**, *3*, 511–526. [CrossRef] [PubMed]

181. National Center for Biotechnology Information (PubMed). Available online: https://pubmed.ncbi.nlm.nih.gov (accessed on 20 November 2020).

182. Bergwerff, A.A. Moderne microbiologische technieken. In *Inleiding tot de—Achtergrond en Feiten*, 2nd ed.; Lipman, L.J.A., Ruiter, A., Eds.; Reed Business: Amsterdam, The Netherlands, 2011; pp. 167–178. (In Dutch)

183. Situ, C.; Mooney, M.H.; Elliott, C.T.; Buijs, J. Advances in surface plasmon resonance biosensor technology towards high-throughput, food-safety analysis. *Trends Anal. Chem.* **2010**, *29*, 1305–1315. [CrossRef]

184. Bokken, G.C.A.M.; Corbee, R.J.; Van Knapen, F.; Bergwerff, A.A. Immunochemical detection of *Salmonella* group B, D and E using an optical surface plasmon resonance biosensor. *FEMS Microbiol. Lett.* **2003**, *222*, 75–82. [CrossRef]

185. Taylor, A.D.; Ladd, J.; Homola, J.; Jiang, S. Surface plasmon resonance (SPR) sensors for the detection of bacterial pathogens. In *Principles of Bacterial Detection: Biosensors, Recognition Receptors and Microsystems*; Zourob, M., Elwary, S., Turner, A., Eds.; Springer: New York, NY, USA, 2008; Part II, Chapter 5, pp. 83–108. [CrossRef]

186. Bergwerff, A.A.; Van Knapen, F. Surface plasmon resonance biosensors for detection of pathogenic microorganisms: Strategies to secure food and environmental safety. *J. AOAC Int.* **2006**, *89*, 826–831. [CrossRef]

187. Wang, Y.; Knoll, W.; Dostalek, J. Bacterial pathogen surface plasmon resonance biosensor advanced by long range surface plasmons and magnetic nanoparticle assays. *Anal. Chem.* **2012**, *84*, 8345–8350. [CrossRef] [PubMed]

188. Jongerius-Gortemaker, B.G.M.; Goverde, R.L.J.; Van Knapen, F.; Bergwerff, A.A. Surface plasmon resonance (BIACORE) detection of serum antibodies against *Salmonella enteritidis* and *Salmonella typhimurium*. *J. Immunol. Meth.* **2002**, *266*, 33–44. [CrossRef]

189. Thomas, A.; Bouma, A.; Van Eerden, E.; Landman, W.J.M.; Van Knapen, F.; Stegeman, A.; Bergwerff, A.A. Detection of egg yolk antibodies reflecting *Salmonella enteritidis* infections using a surface plasmon resonance biosensor. *J. Immunol. Methods* **2006**, *315*, 68–74. [CrossRef] [PubMed]

190. Datta Mazumdar, S.; Barlen, B.; Kramer, T.; Keusgen, M. A rapid serological assay for prediction of *Salmonella* infection status in slaughter pigs using surface plasmon resonance. *J. Microbiol. Methods* **2008**, *75*, 545–550. [CrossRef] [PubMed]

191. Palchetti, I.; Mascini, M. Amperometric biosensors for pathogenic bacteria detection. In *Principles of Bacterial Detection: Biosensors, Recognition Receptors and Microsystems*; Zourob, M., Elwary, S., Turner, A., Eds.; Springer: New York, NY, USA, 2008; Part II, Chapter 13, pp. 299–312. [CrossRef]

192. Melo, A.M.A.; Alexandre, D.L.; Furtado, R.F.; Borges, M.F.; Figueiredo, E.A.T.; Biswas, A.; Cheng, H.N.; Alves, C.R. Electrochemical immunosensors for *Salmonella* detection in food. *Appl. Microbiol. Biotechnol.* **2016**, *100*, 5301–5312. [CrossRef]

193. Eissa, S.; Zourob, M. Ultrasensitive peptide-based multiplexed electrochemical biosensor for the simultaneous detection of *Listeria monocytogenes* and *Staphylococcus aureus*. *Microchim. Acta* **2020**, *187*, 486. [CrossRef] [PubMed]

194. Wu, W. Inorganic nanomaterials for printed electronics: A review. *Nanoscale* **2017**, *9*, 7342–7372. [CrossRef]

195. Viswanathan, S.; Ranib, C.; Ho, J.A. Electrochemical immunosensor for multiplexed detection of food-borne pathogens using nanocrystal bioconjugates and MWCNT screen-printed electrode. *Talanta* **2012**, *94*, 315–319. [CrossRef]

196. Mohamed, H.M. Screen-printed disposable electrodes: Pharmaceutical applications and recent developments. *Trends Anal. Chem.* **2016**, *82*, 1–11. [CrossRef]

197. Freitas, M.; Viswanathan, S.; Nouws, H.P.A.; Oliveira, M.B.P.P.; Delerue-Matos, C. Iron oxide/gold core/shell nanomagnetic probes and CdS biolabels for amplified electrochemical immunosensing of *Salmonella typhimurium*. *Biosens. Bioelectron.* **2014**, *51*, 195–200. [CrossRef]

198. Birnbaumer, G.M.; Lieberzeit, P.A.; Richter, L.; Schirhagl, R.; Milnera, M.; Dickert, F.L.; Bailey, A.; Ertl, P. Detection of viruses with molecularly imprinted polymers integrated on a microfluidic biochip using contact-less dielectric microsensors. *Lab. Chip* **2009**, *9*, 3549–3556. [CrossRef] [PubMed]

199. BelBruno, J.J. Molecularly imprinted polymers. *Chem. Rev.* **2019**, *119*, 94–119. [CrossRef] [PubMed]

200. Cui, F.; Zhou, Z.; Zhou, H.S. Molecularly imprinted polymers and surface imprinted polymers based electrochemical biosensor for infectious diseases. *Sensors* **2020**, *20*, 996. [CrossRef]

201. Mukama, O.; Sinumvayo, J.P.; Shamoon, M.; Shoaib, M.; Mushimiyimana, H.; Safdar, W.; Bemena, L.; Rwibasira, P.; Mugisha, S.; Wang, Z. An update on aptamer-based multiplex system approaches for the detection of common foodborne pathogens. *Food Anal. Methods* **2017**, *10*, 2549–2565. [CrossRef]

202. Majdinasab, M.; Hayat, A.; Marty, J.L. Aptamer-based assays and aptasensors for detection of pathogenic bacteria in food samples. *Trends Anal. Chem.* **2018**, *107*, 60–77. [CrossRef]

203. Zhu, H.; Sikora, U.; Ozcan, A. Quantum dot enabled detection of *Escherichia coli* using a cell-phone. *Analyst* **2012**, *137*, 2541–2544. [CrossRef]

204. Zeinhom, M.M.A.; Wang, Y.; Sheng, L.; Du, D.; Li, L.; Zhu, M.-J.; Lin, Y. Smart phone based immunosensor coupled with nanoflower signal amplification for rapid detection of *Salmonella* Enteritidis in milk, cheese and water. *Sens. Actuators B Chem.* **2018**, *261*, 75–82. [CrossRef]

205. de Dieu Habimana, J.; Ji, J.; Sun, X. Minireview: Trends in optical-based biosensors for point-of-care bacterial pathogen detection for food safety and clinical diagnostics. *Anal. Lett.* **2018**, *51*, 2933–2966. [CrossRef]

206. Kant, K.; Shahbazi, M.-A.; Dave, V.P.; Ngo, T.A.; Chidambara, V.A.; Than, L.Q.; Bang, D.D.; Wolff, A. Microfluidic devices for sample preparation and rapid detection of foodborne pathogens. *Biotechnol. Adv.* **2018**, *36*, 1003–1024. [CrossRef]

207. Agrawal, S.; Morarka, A.; Bodas, D.; Paknikar, K.M. Multiplexed detection of waterborne pathogens in circular microfluidics. *Appl. Biochem. Biotechnol.* **2012**, *167*, 1668–1677. [CrossRef]

208. Dincer, C.; Bruch, R.; Kling, A.; Dittrich, P.S.; Urban, G.A. Multiplexed point-of-care testing—xPOCT. *Trends Biotechnol.* **2017**, *35*, 728–742. [CrossRef] [PubMed]

209. Chin, C.; Laksanasopin, T.; Cheung, Y.; Steinmiller, D.; Linder, V.; Parsa, H.; Wang, J.; Moore, H.; Rouse, R.; Umviligihozo, G.; et al. Microfluidics-based diagnostics of infectious diseases in the developing world. *Nat. Med.* **2011**, *17*, 1015–1019. [CrossRef] [PubMed]

210. Miyazaki, C.M.; Kinahan, D.J.; Mishra, R.; Mangwanya, F.; Kilcawley, N.; Ferreira, M.; Ducrée, J. Label-free, spatially multiplexed SPR detection of immunoassays on a highly integrated centrifugal Lab-on-a-Disc platform. *Biosens. Bioelectron.* **2018**, *119*, 86–93. [CrossRef]

211. Posthuma-Trumpie, G.A.; Korf, J.; Van Amerongen, A. Lateral flow (immuno)assay: Its strengths, weaknesses, opportunities and threats. A literature survey. *Anal. Bioanal. Chem.* **2009**, *393*, 569–582. [CrossRef]

212. Ahovan, Z.A.; Hashemi, A.; De Plano, L.M.; Gholipourmalekabadi, M.; Seifalian, A. Bacteriophage based biosensors: Trends, outcomes and challenges. *Nanomaterials* **2020**, *10*, 501. [CrossRef]

213. Schofield, D.; Sharp, N.J.; Westwater, C. Phage-based platforms for the clinical detection of human bacterial pathogens. *Bacteriophage* **2012**, *2*, 105–121. [CrossRef] [PubMed]

214. Schmelcher, M.; Loessner, M.J. Bacteriophage endolysins: Applications for food safety. *Curr. Opin. Biotechnol.* **2016**, *37*, 76–87. [CrossRef] [PubMed]

215. Zhong, J.; Zhao, X. Isothermal amplification technologies for the detection of foodborne pathogens. *Food Anal. Methods* **2018**, *11*, 1543–1560. [CrossRef]

216. Garcia, J.L.; Gennari, S.M.; Machado, R.M.; Navarro, I.T. *Toxoplasma gondii*: Detection by mouse bioassay, histopathology, and polymerase chain reaction in tissues from experimentally infected pigs. *Exp. Parasitol.* **2006**, *113*, 267–271. [CrossRef] [PubMed]

217. Bezerra, R.A.; Carvalho, F.S.; Guimarães, L.A.; Rocha, D.S.; Silva, F.L.; Wenceslau, A.A.; Albuquerque, G.R. Comparison of methods for detection of *Toxoplasma gondii* in tissues of naturally exposed pigs. *Parasitol. Res.* **2012**, *110*, 509–514. [CrossRef]

218. Opsteegh, M.; Langelaar, M.; Sprong, H.; Den Hartog, L.; De Craeye, S.; Bokken, G.; Ajzenberg, D.; Kijlstra, A.; Van der Giessen, J. Direct detection and genotyping of *Toxoplasma gondii* in meat samples using magnetic capture and PCR. *Int. J. Food Microbiol.* **2010**, *139*, 193–201. [CrossRef]

219. Wolffs, P.; Norling, B.; Rådström, P. Risk assessment of false-positive quantitative real-time PCR results in food, due to detection of DNA originating from dead cells. *J. Microbiol. Methods* **2005**, *60*, 315–323. [CrossRef] [PubMed]

220. Salipante, S.J.; Jerome, K.R. Digital PCR—An emerging technology with broad applications in microbiology. *Clin. Chem.* **2020**, *66*, 117–123. [CrossRef] [PubMed]

221. Jang, M.; Jeong, S.W.; Bae, N.H.; Song, Y.; Lee, T.J.; Lee, M.K.; Lee, S.J.; Lee, K.G. Droplet-based digital PCR system for detection of single-cell level of foodborne pathogens. *BioChip J.* **2017**, *11*, 329–337. [CrossRef]

222. European Centre for Disease Prevention and Control. *Rapid Increase of a SARS-CoV-2 Variant with Multiple Spike Protein Mutations Observed in the United Kingdom*; ECDC: Stockholm, Sweden, 20 December 2020.

223. Whittaker, P.; Fry, F.S.; Curtis, S.K.; Al-Khaldi, S.F.; Mossoba, M.M.; Yurawecz, M.P.; Dunkel, V.C. Use of fatty acid profiles to identify food-borne bacterial pathogens and aerobic endospore-forming bacilli. *J. Agric. Food Chem.* **2005**, *53*, 3735–3742. [CrossRef]

224. Pinu, F.R. Early detection of food pathogens and food spoilage microorganisms: Application of metabolomics. *Trends Food Sci. Technol.* **2016**, *54*, 213–215. [CrossRef]

225. Wang, Y.; Liu, S.; Pu, Q.; Li, Y.; Wang, X.; Jiang, Y.; Yang, D.; Yang, Y.; Yang, J.; Sun, C. Rapid identification of *Staphylococcus aureus*, *Vibrio parahaemolyticus* and *Shigella sonnei* in foods by solid phase microextraction coupled with gas chromatography–mass spectrometry. *Food Chem.* **2018**, *262*, 7–13. [CrossRef] [PubMed]

226. Harris, G.A.; Galhena, A.S.; Fernández, F.M. Ambient Sampling/Ionization Mass Spectrometry: Applications and Current Trends. *Anal. Chem.* **2011**, *83*, 4508–4538. [CrossRef]

227. Rodríguez-Lorenzo, L.; Garrido-Maestu, A.; Bhunia, A.K.; Espiña, B.; Prado, M.; Dieǵuez, L.; Abalde-Cela, S. Gold nanostars for the detection of foodborne pathogens via surface-enhanced Raman scattering combined with microfluidics. *ACS Appl. Nano Mater.* **2019**, *2*, 6081–6086. [CrossRef]

228. American Association of Veterinary Laboratory Diagnosticians. Available online: http://www.aavld.org (accessed on 28 January 2021).

229. Collaborating Veterinary Laboratories. Available online: http://www.covetlab.org (accessed on 28 January 2021).

230. European Association of Veterinary Laboratory Diagnosticians. Available online: http://www.eavld.org (accessed on 28 January 2021).

231. World Organization for Animal Health. Principles of validation of diagnostic assays for infectious diseases. In *OIE Manual of Diagnostic Tests and Vaccines for Terrestrial Animals (Mammals, Birds and Bees)*, 6th ed.; OIE Biological Standards Commission, Ed.; Office International Des Epizooties: Paris, France, 2008; Volume 1, Chapter 1.1.4, pp. 34 35.

232. Wild, D.; Sheehan, C. Standardization and calibration. In *The Immunoassay Handbook*, 4th ed.; Wild, D., Ed.; Elsevier: Amsterdam, The Netherlands, 2013; pp. 315–322. [CrossRef]

233. Cardoso, J.R.; Pereira, L.M.; Iversen, M.D.; Ramos, A.L. What is gold standard and what is ground truth? *Dent. Press J. Orthod.* **2014**, *19*, 27–30. [CrossRef]

234. Cundell, T. The limitations of the colony-forming unit in microbiology. *Eur. Pharm. Rev.* **2015**, *20*, 11–13.

235. Tang, S.; Orsi, R.H.; Luo, H.; Ge, C.; Zhang, G.; Baker, R.C.; Stevenson, A.; Wiedmann, M. Assessment and comparison of molecular subtyping and characterization methods for *Salmonella*. *Front. Microbiol.* **2019**, *10*, 1591. [CrossRef]

236. Azadeh, M.; Sondag, P.; Wang, Y.; Raines, M.; Sailstad, J. Quality controls in ligand binding assays: Recommendations and best practices for preparation, qualification, maintenance of lot to lot consistency, and prevention of assay drift. *AAPS J.* **2019**, *21*, 89. [CrossRef]

237. Bonardi, S. *Salmonella* in the pork production chain and its impact on human health in the European Union. *Epidemiol. Infect.* **2017**, *145*, 1513–1526. [CrossRef] [PubMed]

238. Belœil, P.A.; Chauvin, C.; Proux, K.; Rose, N.; Queguiner, S.; Eveno, E.; Houdayer, C.; Rose, V.; Fravalo, P.; Madec, F. Longitudinal serological responses to *Salmonella enterica* of growing pigs in a subclinically infected herd. *Prev. Vet. Med.* **2003**, *60*, 207–226. [CrossRef]

239. Berends, B.R.; Urlings, H.A.P.; Snijders, J.M.A.; Van Knapen, F. Identification and quantification of risk factors in animal management and transport regarding *Salmonella* spp. in pigs. *Int. J. Food Microbiol.* **1996**, *30*, 37–53. [CrossRef]

240. Lo, F.; Wong, D.M.A.; Hald, T. (Eds.) *Salmonella in Pork (SALINPORK): Pre-Harvest and Harvest Control Options Based on Epidemiologic, Diagnostic and Economic Research*; Contract No: FAIR1 CT95-0400. Copenhagen, Denmark, 2002. Available online: https://cordis.europa.eu/project/id/FAIR950400 (accessed on 28 January 2021).

241. Baggesen, D.L.; Wegener, H.C.; Bager, F.; Stege, H.; Christensen, J. Herd prevalence of *Salmonella enterica* infections in Danish slaughter pigs determined by microbiological testing. *Prev. Vet. Med.* **1996**, *26*, 201–213. [CrossRef]

242. Nollet, N.; Maes, D.; Duchateau, L.; Hautekiet, V.; Houf, K.; Van Hoof, J.; De Zutter, L.; De Kruif, A.; Geers, R. Discrepancies between the isolation of *Salmonella* from mesenteric lymph nodes and the results of serological screening in slaughter pigs. *Vet. Res.* **2005**, *36*, 545–555. [CrossRef]

243. Farzan, A.; Friendship, R.; Dewey, C. Evaluation of enzyme-linked immunosorbent assay (ELISA) tests and culture for determining *Salmonella* status of a pig herd. *Epidemiol. Infect.* **2007**, *135*, 238–244. [CrossRef] [PubMed]

244. Wilhelm, E.; Hilbert, F.; Paulsen, P.; Smulders, F.J.M.; Rossmanith, W. *Salmonella* diagnosis in pig production: Methodological problems in monitoring the prevalence in pigs and pork. *J. Food Prot.* **2007**, *70*, 1246–1248. [CrossRef] [PubMed]

245. European Food Safety Authority Task Force on Zoonoses. Report of the Task Force on Zoonoses Data Collection on the analysis of the baseline survey on the prevalence of *Salmonella* in slaughter pigs, in the EU, 2006–2007. Part A: Salmonella prevalence estimates. *EFSA J.* **2008**, *135*, 1–111.

246. Swanenburg, M.; Berends, B.R.; Urlings, H.A.; Snijders, J.M.; Van Knapen, F. Epidemiological investigations into the sources of *Salmonella* contamination of pork. *Berl. Munch. Tierarztl. Wochenschr.* **2001**, *114*, 356–359.

247. Mainar-Jaime, R.C.; Casanova-Higes, A.; Andrés-Barranco, S.; Vico, J.P. Looking for new approaches for the use of serology in the context of control programmes against pig salmonellosis. *Zoonoses Public Health* **2018**, *65*, e222–e228. [CrossRef] [PubMed]

248. Felin, E.; Hälli, O.; Heinonen, M.; Jukola, E.; Fredriksson-Ahomaa, M. Assessment of the feasibility of serological monitoring and on-farm information about health status for the future meat inspection of fattening pigs. *Prev. Vet. Med.* **2019**, *162*, 76–82. [CrossRef]

249. Snary, E.L.; Munday, D.K.; Arnold, M.E.; Cook, A.J.C. Zoonoses action plan *Salmonella* monitoring programme: An investigation of the sampling protocol. *J. Food Prot.* **2010**, *73*, 488–494. [CrossRef]

250. Hanssen, E.J.M.; Swanenburg, M.; Maassen, C.B.M. The Dutch *Salmonella* Monitoring Programme for Pigs and Some Recommendations for Control Plans in the Future. In Proceedings of the 7th International Symposium on the Epidemiology & Control of Foodborne Pathogens in Pork, Verona, Italy, 9–11 May 2007; pp. 169–172. [CrossRef]

251. Wegener, H.C.; Hald, T.; Wong, L.F.; Madsen, M.; Korsgaard, H.; Bager, F.; Gerner-Smidt, P.; Mølbak, K. *Salmonella* control programs in Denmark. *Emerg. Infect. Dis.* **2003**, *9*, 774–780. [CrossRef] [PubMed]

252. Berends, B.R.; Snijders, J.M.; Van Logtestijn, J.G. Efficacy of current EC meat inspection procedures and some proposed revisions with respect to microbiological safety: A critical review. *Vet. Rec.* **1993**, *133*, 411–415. [CrossRef]

253. Snijders, J.M.A.; Van Knapen, F. Prevention of human diseases by an integrated quality control system. *Livest. Prod. Sci.* **2002**, *76*, 203–206. [CrossRef]

254. Robert, M.K. Memorium. *Crit. Rev. Toxicol.* **2007**, *37*, 627. [CrossRef]

255. Kahn, L.H. Perspective: The one-health way. *Nature* **2017**, *543*, S47. [CrossRef]

256. Van Knapen, F. Veterinary public health: Past, present, and future. *Vet. Q.* **2000**, *22*, 61–62. [CrossRef] [PubMed]

257. Riess, L.E.; Hoelzer, K. Implementation of visual-only swine inspection in the European Union: Challenges, opportunities, and lessons learned. *J. Food Prot.* **2020**, *83*, 1918–1928. [CrossRef]

258. Messens, W.; Hempen, M.; Koutsoumanis, K. Use of predictive modelling in recent work of the Panel on Biological Hazards of the European Food Safety Authority. *Microb. Risk Anal.* **2018**, *10*, 37–43. [CrossRef]

259. McMeekin, T.; JBowman, J.; McQuestin, O.; Mellefont, L.; Ross, T.; Tamplin, M. The future of predictive microbiology: Strategic research, innovative applications and great expectations. *Int. J. Food Microbiol.* **2008**, *128*, 2–9. [CrossRef] [PubMed]

260. Bergwerff, A.A. Rapid assays for detection of residus of veterinary drugs. In *Rapid Methods for Biological and Chemical Contaminants in Food and Feed*; Van Amerongen, A., Barug, D., Lauwaars, M., Eds.; Wageningen Academic Publishers: Wageningen, The Netherlands, 2005; pp. 259–292.

261. Commission of The European Communities. Commission Regulation (EC) No 2075/2005 of 5 December 2005 laying down specific rules on official controls for *Trichinella* in meat (consolidated version including Commission Regulation amendments and corrections). *Off. J. Eur. Union* **2005**, *L338*, 60–82.

262. Pozio, E. World distribution of *Trichinella* spp. infections in animals and humans. *Vet. Parasitol.* **2007**, *149*, 3–21. [CrossRef]

263. Bokken, G.C.A.M.; Van Eerden, E.; Opsteegh, M.; Augustijn, M.; Graat, E.A.M.; Franssen, F.F.J.; Görlich, K.; Buschtöns, S.; Tenter, A.M.; Van der Giessen, J.W.B.; et al. Specific serum antibody responses following a *Toxoplasma gondii* and *Trichinella spiralis* co-infection in swine. *Vet. Parasitol.* **2012**, *184*, 126–132. [CrossRef]

264. Yousefi, H.; Su, H.-M.; Imani, S.M.; Alkhaldi, K.; Filipe, C.D.M.; Didar, T.F. Intelligent food packaging: A review of smart sensing technologies for monitoring food quality. *ACS Sens.* **2019**, *4*, 808–821. [CrossRef]

265. Friesema, I.H.M.; De Boer, R.F.; Duizer, E.; Kortbeek, L.M.; Notermans, D.W.; Smeulders, A.; Bogerman, J.; Pronk, M.J.H.; Uil, J.J.; BrInkman, K.; et al. Aetiology of acute gastroenteritis in adults requiring hospitalization in The Netherlands. *Epidemiol. Infect.* **2012**, *140*, 1780–1786. [CrossRef]

266. Jansen, A.; Stark, K.; Kunkel, J.; Schreier, E.; Ignatius, R.; Liesenfeld, O.; Werber, D.; Göbel, U.B.; Zeitz, M.; Schneider, T. Aetiology of community-acquired, acute gastroenteritis in hospitalised adults: A prospective cohort study. *BMC Infect. Dis.* **2008**, *8*, 143. [CrossRef]

267. Thissen, J.B.; Be, N.A.; McLoughlin, K.; Gardner, S.; Rack, P.G.; Shapero, M.H.; Rowland, R.R.R.; Slezak, T.; Jaing, C.J. Axiom Microbiome Array, the next generation microarray for high-throughput pathogen and microbiome analysis. *PLoS ONE* **2019**, *14*, e0212045. [CrossRef]

268. Strategic Consulting. *Microbiology Testing in the Global Food Industry*, 8th ed.; 2013. Available online: https://www.strategic-consult.com/product/food-micro-eighth-edition-microbiology-testing-global-food-industry/ (accessed on 23 January 2021).

269. Bolton, D.J. *Campylobacter* virulence and survival factors. *Food Microbiol.* **2015**, *48*, 99–108. [CrossRef] [PubMed]

270. An, X.; Zuo, P.; Ye, B.-C. A single cell droplet microfluidic system for quantitative determination of food-borne pathogens. *Talanta* **2020**, *209*, 120571. [CrossRef] [PubMed]

271. Moschou, D. How close are we to a real Star Trek-style medical tricorder? *Conversation*. 16 June 2017. Available online: https://theconversation.com/how-close-are-we-to-a-real-star-trek-style-medical-tricorder-77977 (accessed on 11 December 2020).

Review

Suitability of High-Resolution Mass Spectrometry for Routine Analysis of Small Molecules in Food, Feed and Water for Safety and Authenticity Purposes: A Review

Maxime Gavage, Philippe Delahaut * and Nathalie Gillard

CER Groupe, Rue du Point du Jour 8, 6900 Marloie, Belgium; m.gavage@cergroupe.be (M.G.); n.gillard@cergroupe.be (N.G.)
* Correspondence: p.delahaut@cergroupe.be; Tel.: +32-(0)-84-31-00-90

Abstract: During the last decade, food, feed and environmental analysis using high-resolution mass spectrometry became increasingly popular. Recent accessibility and technological improvements of this system make it a potential tool for routine laboratory work. However, this kind of instrument is still often considered a research tool. The wide range of potential contaminants and residues that must be monitored, including pesticides, veterinary drugs and natural toxins, is steadily increasing. Thanks to full-scan analysis and the theoretically unlimited number of compounds that can be screened in a single analysis, high-resolution mass spectrometry is particularly well-suited for food, feed and water analysis. This review aims, through a series of relevant selected studies and developed methods dedicated to the different classes of contaminants and residues, to demonstrate that high-resolution mass spectrometry can reach detection levels in compliance with current legislation and is a versatile and appropriate tool for routine testing.

Keywords: routine testing; high-resolution mass spectrometry; food; feed; water; veterinary drug residues; natural toxins; pesticides; food authenticity

Citation: Gavage, M.; Delahaut, P.; Gillard, N. Suitability of High-Resolution Mass Spectrometry for Routine Analysis of Small Molecules in Food, Feed and Water for Safety and Authenticity Purposes: A Review. *Foods* **2021**, *10*, 601. https://doi.org/10.3390/foods10030601

Academic Editor: Isabel Sierra Alonso

Received: 12 February 2021
Accepted: 10 March 2021
Published: 12 March 2021

Publisher's Note: MDPI stays neutral with regard to jurisdictional claims in published maps and institutional affiliations.

1. Introduction

Food and feed analysis is essential to guaranty their quality, authenticity and safety. Analytical strategies have been developed for decades to evaluate food and feed composition and nutritional value and to detect the presence of undesirable or harmful compounds or foodborne pathogens. The analysis of chemical substances in food and feed is a challenging task, given the multitude of matrices encountered and the disparate properties of targeted contaminants [1]. Moreover, some of these substances must be detected and/or quantified at trace levels with sufficient accuracy and robustness.

In Europe, a high level of consumer protection is required by Article 152 of the Treaty establishing the European Community [2]. To reach this high level of health protection, a risk analysis procedure based on scientific evaluation and including factors such as the feasibility of control underpins Community legislation. The aim is to establish the optimal balance between the risks and benefits of substances that are used intentionally, focusing on the reduction of contaminants. The legislation separately considers different classes of chemical substances, including contaminants and residues. The legislation on contaminants is based on scientific advice and the principle that contaminant levels should be kept as low as can be reasonably achieved following good working practices. Maximum levels have been set for certain contaminants (e.g., mycotoxins, dioxins, heavy metals, nitrates and chloropropanols) in order to protect public health [3]. The legislation on residues of veterinary medicinal products used in food-producing animals and on residues of plant protection products (pesticides) provides for scientific evaluation before respective products are authorised. If necessary, maximum residue limits (MRLs) are established, and in some cases, the use of substances is prohibited [4,5]. This present review focus on the

European legislation. Complementary information on the regulation and safety assessment of food substances in various countries and jurisdictions can be found in the review of Magnuson and co-workers [6].

Water is one of the most important resources, and its preservation is a major challenge, regarding both the environment and humans. Synthetic pesticides used intensively in agriculture can enter surface waters, mainly due to runoff driven by precipitation or irrigation. Pharmaceuticals for both human and veterinary purposes are excreted, and the unaltered parent compounds or their metabolites and can be deposited in environmental waters as a consequence of incomplete elimination by wastewater-treatment plants. However, efforts are being undertaken to develop new systems to degrade pharmaceutical products in wastewater treatment plants [7]. This mixture of chemicals, pharmacologically active compounds and their transformation products are potentially harmful to aquatic life and humans when they enter drinking water. The justified concern over this hazard has led to the development of analytical methods for measuring freshwater contamination.

For years, mass spectrometry has been considered the most suitable analytical technique for the detection of multiple compounds in food, feed and water. Coupled to liquid chromatography (LC), high-performance LC and ultra-high performance LC (HPLC, UHPLC) or gas chromatographic (GC) separation with an ionization source such as electrospray (ESI), a large number of mass spectrometry-based methods were developed to comply with updated regulations. Most of the developed methods use triple-quadrupole instruments, and the best-performing of these are able to sensitively and accurately detect and quantify more than 1000 compounds in a single analysis [8–11]. These instruments are defined as low-resolution mass spectrometers, with a typical resolution of approximately 1 atomic mass unit for quadrupole analysers [12]. The combination of chromatographic separation and the use of multiple reaction monitoring (MRM) mode, working like a noise-reducing double filter, allow enhanced sensitivity and selectivity. However, the use of triple quadrupole instruments has proven to have some limitations, such as a limited number of compounds targeted during the analysis. The sensitivity of MRM methods strongly depends on the length of the chosen dwell-time. Therefore, the more transitions to be monitored, the shorter the resulting dwell-time and the poorer the obtained sensitivity [13]. This sensitivity issue of multiple-compounds methods can be balanced by the use of retention time-based MRM windows. In addition, the use of triple quadrupole instruments and MRM methods is limited to targeted analysis. To effectively apply this approach, the structure of the compound must be characterized before its detection. Methods development can be time-consuming, and standards must be acquired to optimize compound-specific instrumental conditions, including transition selections, ion-source voltages, and collision energies [14]. MRM methods are, therefore, unable to screen for unknown compounds.

In the last decade, high-resolution mass spectrometry (HRMS) has become more accessible, particularly with the development of Orbitrap MS-based instruments and the improvements to time-of-flight (ToF) MS systems. As for triple quadrupole instruments, high-resolution mass spectrometers can be coupled to chromatographic separation units. Orbitrap and ToF systems are versatile instruments with fast scan velocities, sufficient dynamic ranges and the possibility of tandem mass spectrometry (MS/MS) when used as components of hybrid instruments (combining a quadrupole analyser with Orbitrap (Q-Orbitrap) and ToF (Q-ToF)). HRMS instruments are usually described as less sensitive than triple quadrupoles [15]. However, thanks to recent technical improvements such as the introduction of new ion transition devices or advances in detection technology [1], several studies presented similar sensitivities achieved by the two types of instrument [16–18]. Moreover, a higher resolution provides an enhanced selectivity when a large number of analytes are determined simultaneously and, for the best-performing instruments with sufficient resolving power, the potential to discriminate analytes from isobaric co-eluting sample matrix compounds. The use of ion mobility-coupled chromatographic separation and HRMS is particularly powerful for this purpose [19]. Complementary technical information on HRMS instruments including separation techniques, ionization and acquisition modes can be found in the review from Yan and co-workers [20].

The major advantage of HRMS systems is the ability to record a theoretically unlimited number of compounds in full-scan mode with additional structural information, using hybrid instruments. Acquired data can be processed using target analysis, suspect screening and non-target screening. Moreover, the high volume of stored full-scan or MS/MS data can be retrospectively analysed without sample re-injection. Finally, untargeted HRMS analysis can be combined with multivariate chemometric tools to efficiently extract relevant information from very complex datasets. This statistical analysis helps in the exploration of specific biomarkers that can categorize/differentiate the analysed samples. The chemical profiling of samples could be focused on the m/z values, which vary significantly from one sample category to another [21].

Thanks to these strengths and the ever-increasing number of new analytes that must be monitored, HRMS analyses are increasingly accepted for multi-residue analysis [15]. This type of instrument is, however, still rarely used for routine analysis by control laboratories. For instance, in 2019, only 4 of the 26 participants in 19 PU (study code) proficiency testing for the screening of antibiotic residues in pork muscle used HRMS. Additionally, in the 2019 Fapas® food chemistry proficiency test for the detection of avermectins and anthelmintics in bovine liver, only 3 out of the 32 participants used HRMS [22]. The proportion of laboratories using HRMS is higher for proficiency tests for pesticides in fruits and vegetables. For instance, 33% of the laboratories participating in EUPT-FV-SM11 [23], pesticide residues in red cabbage homogenate (with 67 participants in 2019), and EUPT-FV-SM10 [24], pesticide residues in green bean homogenate (with 69 participants in 2018), used HRMS.

To demonstrate the applicability of HRMS in routine analyses, several relevant examples were collected over the last decade. The selected studies are presented according to the type of targeted compounds, pesticides, veterinary drug residues and toxins. Given the high number of compounds that can be detected in single analysis, a section is dedicated to multi-class analysis. The analysis of water constitutes a significant portion of the HRMS-related literature. The selected studies were, however, limited to drinking water to maintain a focus on the 'food and feed' topic, excluding the environmental aspects of water. Finally, quite apart regulated compounds detection, the potential of HRMS-based analysis combined with chemometrics tools for food authenticity control is addressed.

2. Analysis of Pesticides

Synthetic pesticides play a major role in food and feed production. Their use has helped to immensely increase agricultural productivity and resist the ever-increasing demographical and economical pressure. Pesticides are used to protect crops, including fruits, vegetables, cereals and fibre plants, against insects, plant pathogens, weed and fungi. It has been estimated that nearly one-third of all agricultural products are produced using pesticides, and without them, the loss of fruits, vegetables and cereals from pest injury could increase to 78%, 54% and 32%, respectively [25]. However, the use of pesticides is associated with negative external effects, e.g., pollution of waterways and non-target ecosystems, risks for human health and costs for monitoring of residues on food.

In 2003, the work of Klein and Alder [26] was considered a masterpiece in the field of pesticide analysis. The LC-MS/MS method they developed, based on a triple quadrupole instrument, was able to screen and quantify 100 pesticides and metabolites in various crops. Nowadays, it is estimated that almost 1000 different pesticides could potentially be used in agriculture [14]. The main challenge is the ability to economically analyse the presence of these chemicals, their metabolites, and degradation products when precise knowledge of pesticide application or misuse is lacking.

Given the large number of compounds to detect, HRMS is particularly well-suited for this purpose. Zhibin Wang and co-workers [27] developed a qualitative screening and identification strategy for 317 pesticides in fruits and vegetables using LC-Q-ToF. The strategy is based on two injections of each sample extract. In the first chromatographic run, the single full-scan MS mode was performed and the sample was screened for possible target compounds. Potential contaminants were confirmed in a second chromatographic run under targeted MS/MS conditions in which the resulting product ion spectra were used to search a homemade MS/MS library. In studies from Jian Wang and co-workers, identification and quantification of pesticides were performed using the same approach. A sequential combination of a full MS scan for quantification and a data-dependent MS/MS scan for confirmation was used for the analysis of 166 pesticides in fruits and vegetables [28]. The method was later extended to 451 pesticides residues and validated via an evaluation of overall recovery, intermediate precision, limits of detection (LOD) and quantification (LOQ) and measurement uncertainty [29]. For the 10 studied matrices, 94.5% of the pesticides in fruits and 90.7% of those in vegetables had recoveries of between 81% and 110%; 99.3% of the pesticides in fruits and 99.1% of those in vegetables had an intermediate precision $\leq 20\%$; and 97.8% of the pesticides in fruits and 96.4% of those in vegetables showed a measurement uncertainty $\leq 50\%$.

In another study from Gómez-Ramos and co-workers [30], the authors used LC-Q-Orbitrap MS for the analysis of pesticide residues in fruit and vegetable commodities. The system was used to detect, identify and quantify, in various extracts (tomato, pepper, orange and green tea), 139 pesticides, all of which were included in the European Union Monitoring Program. Here, detection, identification and quantification were achieved within the same analysis, combining full scan data acquisition for detection and quantification and MS/MS data for identification. Extracts spiked with a mixture of the analysed pesticides at 10, 50, 100 or 500 µg/kg were analysed using both LC-Q-Orbitrap MS and a triple quadrupole instrument. A comparison of the results showed that these two systems have similar capabilities for quantification, with the advantage of a better selectivity for HRMS as well as the possibility to perform retrospective analysis. In a recent study by Kiefer and co-workers [31], the authors performed a suspect screening for over 300 pesticides and over 1100 pesticide transformation products in 31 Swiss groundwater samples. This study aimed to comprehensively assess the impact of agricultural pesticide application on groundwater quality. Suspect screening was combined with HRMS analysis to overcome the lack of reference material for most of the transformation products. The acquired data were used to search the suspect list containing the monoisotopic masses of expected compounds. The suspect hits were then checked for plausibility, with criteria including background interference, retention time, isotope pattern, ionization potential and MS/MS fragmentation. The authors demonstrated the importance of considering transformation products in analysis, with the total concentration of pesticide transformation products exceeding the total concentration of the active substances in 30 samples. One of the findings of the study was that the concentration of 15 transformation products of 9 pesticides exceeded 100 ng/L in at least one sample, demonstrating the importance of such an analysis.

In previous studies, reverse-phase liquid chromatography was used to separate the analytes before HRMS analysis. However, highly polar pesticides have poor retention with this type of column and are co-eluted with unwanted co-extractive substances. To successfully analyse highly polar pesticides and avoid derivatization steps or single-residue analysis, Gasparini and co-workers [32] developed a method based on HRMS and ion chromatography as a separating technique. The method was validated for the quantification of 11 highly polar molecules (four pesticides and relative metabolites) in fruit, cereals and honey. Several proficiency tests, used to verify procedure performance, demonstrated that the method is fit for the purpose of routine analysis in an official laboratory.

Gas chromatography combined with mass spectrometry was traditionally used for pesticide analysis. However, the use of this separation technique requires that the analytes are volatile and thermally stable. To extend analysis applicability to a wider range of compounds, without the need of prior derivatization, strategies gradually changed to liquid chromatography with similar performance [33]. Nevertheless, the properties of several pesticides are not compatible with LC separation, and GC has been required and used in some recent publications. The use of GC-Q-ToF, combining full scan with MS/MS experiments and using accurate mass analysis, was explored by Besil and co-workers [34] for the automated determination of 70 pesticide residues in fruits and vegetables. In addition to satisfying validation results at low targeted contamination levels (1, 5 and 10 µg/kg), the authors pointed out the limited dynamic range of the method as a potential limitation for quantification. However, this problem can be overcome by selecting characteristic ion fragments with lower abundance or sample dilution, which necessitate a second analysis. Vargas-Pérez and co-workers [35] proposed an application combining targeted and non-targeted approaches, using GC-Q-Orbitrap, for the multi-residue analysis of multiple pesticides in fruits and vegetables. The targeted method was successfully validated for 191 pesticides. When applied to real unknown samples, targeted and untargeted methods generated the same results. In addition, data acquired with the untargeted method were compared with a library containing more than 200,000 spectra (containing multiple classes of compounds, such as metabolites, drugs, small peptides, lipids or glycans, in addition to pesticides). Results were based on a search index score indicating the match quality between the library hit and deconvolved experimental spectrum.

As demonstrated by the numerous studies presented in this section and summarized in Table 1, HRMS became a key element in the analysis of pesticides. In the near future, thanks to the combination of separation methods coupled to HRMS, databases and software tools, it is estimated that methods capable of screening up to 1000 pesticides and metabolites will be achievable [14].

Table 1. Selected studies on pesticides analysis by high-resolution mass spectrometry (HRMS).

Instrument and Scanning Technique	Matrix	Number of Analytes	Method Sensitivity	Reference
HPLC-ESI-Q-ToF with full-scan MS suspect screening in the first injection and target MS/MS confirmation in the second injection	Cucumber and orange	317	48.9% of the analytes detected and 17.3% confirmed at 1 µg/kg 83.9% of the analytes detected and 77.6% confirmed at 10 µg/kg 98.1% of the analytes detected and 83.9% confirmed at 50 µg/kg	[27]
UHPLC-ESI-Q-Orbitrap with full-scan MS for screening and quantification in the first injection and target MS/MS confirmation in the second injection	Apple, banana, grape, orange, strawberry, carrot, potato, tomato, cucumber, and lettuce.	166	87.3–92.7% of the analytes with LOD and LOQ $\leq$ 5 µg/kg Most of the analytes with LOQ $\leq$ 10 µg/kg	[28]
UHPLC-ESI-Q-Orbitrap with full-scan MS for screening and quantification in the first injection and target MS/MS confirmation in the second injection	Apple, banana, grape, orange, strawberry, carrot, potato, tomato, cucumber, and lettuce.	451	85% of the analytes with LOD and LOQ $\leq$ 5 µg/kg Most of the analytes with LOQ $\leq$ 10 µg/kg	[29]
UHPLC-ESI-Q-Orbitrap with simultaneous full-scan MS and single MS/MS scan	Tomato, pepper, orange and green tea	139	>90% of the analytes with LOD and LOQ $\leq$ 10 µg/kg	[30]

Table 1. *Cont.*

Instrument and Scanning Technique	Matrix	Number of Analytes	Method Sensitivity	Reference
HPLC-ESI-Q-Orbitrap with simultaneous full-scan MS and MS/MS scans	Swiss groundwater samples	519 target analytes and 1256 suspect analytes	78% of the target analytes with LOQ ≤ 10 ng/L	[31]
Ion Chromatography ESI-Q-Orbitrap with simultaneous full-scan MS and MS/MS scans	Grapes, wheat and honey	11	7 analytes with LOQ ≤ 10 μg/kg, 9 analytes with LOQ ≤ 50 μg/kg, 11 analytes with LOQ ≤ 100 μg/kg	[32]
GC-Negative Chemical Ionization- Q-ToF with simultaneous full-scan MS and MS/MS scans	Tomato	70	76% of the analytes with LOD ≤ 1 μg/kg 57% of the analytes with LOQ ≤ 1 μg/kg 99% of the analytes with LOD ≤ 5 μg/kg 93% of the analytes with LOQ ≤ 5 μg/kg	[34]
GC-Electron Ionization- Q-Orbitrap with simultaneous full-scan MS and MS/MS scans	Pear, banana, watermelon and strawberry	191	LOD at 1 μg/kg in pear, banana, watermelon and 5 μg/kg in strawberry. LOQ at 5 μg/kg in all 4 matrices	[35]

LOD, limits of detection LOQ, limits of quantification.

3. Analysis of Veterinary Drug Residues

The use of veterinary drugs or veterinary medicinal products (such as antibiotics, antiprotozoals, anthelmintics, anti-inflammatory, corticosteroids or hormones substitutes) has become essential to providing a sufficient amount of food for the growing world population. For the purpose of increasing productivity, drugs improve the rate of weight gain, improve feed efficiency or prevent and treat diseases in food-producing animals [36]. However, the use of veterinary drugs is associated with health hazards for the consumer of animal food products, including meat, fat, milk, egg, fish, seafood, honey and derived products. The presence of veterinary drug residues in food might induce various effects, such as allergic reactions, carcinogenic or teratogenic mechanisms or antimicrobial resistance [37].

On the basis of the scientific assessment of the safety of those substances and to protect public health, the presence of veterinary drug residues in foodstuffs of animal origin is regulated by the European Union (Commission Regulation EC No 2377/90) [38] with imposed maximum residue limits (MRLs). Based on the lowest acceptable daily intake, together with metabolism and residue depletion studies, the MRLs of the residues are determined for each tissue, expressed in micrograms per kilogram on a fresh-weight basis [39]. In most of the cases, the MRLs are related to the parent compounds, but they could also be based on single metabolites or a mixture of compounds. Some substances, such as chloramphenicol or nitrofurans, are totally prohibited and MRLs are not established.

Pharmacokinetic parameters of administered veterinary drugs are extensively studied to establish a withdrawal period. This period between last administration and slaughter ensures that food from treated animals will not exceed the MRL and can be eaten safely by humans. This withdrawal period is drug related with specific absorption and elimination rates and also depends on the route of administration and the dosage regimen [40]. However, besides illegal use, a series of causes, such as producer mistakes, disease state or concomitant use of different drugs, may lead to failure to comply with MRLs. Therefore, veterinary drug residue analysis is required to verify compliance with MRLs and detect the presence of prohibited substances. Today, there are approximately 200 veterinary drug residues that must be controlled in foodstuffs [41].

Most of the current methods used for the analysis of multiple veterinary drugs residues are based on liquid chromatography coupled to tandem mass spectrometry, using a triple quadrupole mass analyser programmed to acquire data for selected ion transitions corresponding to the analytes of interest. In recent years, HRMS-based methods were proposed with the advantage of potentially analysing an unlimited number of compounds using a full-scan acquisition mode. This acquisition mode is also well-suited to monitoring drug metabolites that could be more stable and/or toxic than parent drugs but are seldom commercially available as reference substances [42]. In this case, the optimization of analyte-specific MS/MS transitions with triple quadrupole instruments is more complicated.

Nearly a decade ago, Romero-González and co-workers [43] demonstrated the potential of HRMS for the determination of veterinary drugs in milk and its applicability for routine analysis. They compared three methods (running time <4 min) based on Orbitrap, quadrupole time of flight and triple quadrupole instruments for the screening of 29 veterinary drugs from different classes. Overall better results, in terms of cut-off values and uncertainty regions, were obtained using the Orbitrap-based screening method. Additionally, the Orbitrap method showed good quantitative results for all the studied analytes, and the limits of quantification were, except for one compound, lower than the MRLs established for the European Union (EU). Berendsen and co-workers [44] organised an inter-laboratory study including 21 laboratories to evaluate the use of different low- and high-resolution MS techniques and acquisition modes, with respect to the selectivity of 100 veterinary drugs in liver tissue, muscle and urine. For complex matrices, they concluded that only targeted MS/MS monitoring a single product ion in HRMS using a (with a maximum of 5 ppm mass deviation), yields comparable selectivity and false positive and negative rates as triple quadrupole monitoring two product ions. Authors highlighted the clear advantage that data acquired with HRMS instruments can be retrospectively analysed. Another method, based on an Orbitrap analyser, was proposed by León and co-workers [45] for the multi-residue screening of 87 banned or unregulated veterinary drugs in urine. The method was validated, and the detection capability (CCβ) established levels were equal to or lower than the recommended concentrations established by EU reference laboratories. The authors concluded that HRMS is a powerful and reliable tool for the identification of substances in multi-class multi-residue analysis and could be used on a routine basis in the official food safety laboratories.

In more recent studies and published methods, the number of screened drugs increased, as did the number of matrices considered in validation. Staub Spörri and co-workers [46] developed, for instance, a method based on a time-of-flight instrument for the screening of 200 veterinary drugs in honey. Boix and co-workers [47] developed another screening method based on quadrupole time-of-flight mass spectrometry and covering 116 human and veterinary drugs. The method was validated in five types of animal feed at 0.02 and 0.2 mg per kg. The method was successfully applied to real feed samples, and two hormones banned in Europe were detected in some samples. The authors also evaluated the applicability of their screening method to quantitative analysis. Quantification was performed using calibration standards in solvents and relative responses to isotope-labelled internal standards (ILIS) for matrix effects correction. They concluded that, due to the strong matrix effects resulting from the matrix complexity and little sample manipulation, the analyte-labelled ILIS was required to ensure an adequate quantification. Kaklamanos and co-workers [48] considered another approach for the quantification of 48 antimicrobial agents from a wide range of chemical groups/families. Using an Orbitrap instrument, the target analytes were quantified using the standard addition approach. The authors argued that this approach is safer and easier, given the high complexity of animal feed and the lack of blank representative material. The use of a one-point standard addition was shown to be suitable for the accurate determination of the analytes and reduction of the workload, which constitutes a main advantage for using the method in routine analysis. Alcántara-Durán and co-workers [49] developed a multi-residue method, based on an Orbitrap instrument, for the analysis of 87 veterinary drugs in honey, veal muscle, egg and milk. By optimizing the parameters of the method and applying a dilution factor of 100 to the sample, matrix effects were completely removed for all the compounds and matrices tested. Sensitivity decrease due to sample dilution was balanced by downscaling the size of liquid separations using nanoflow liquid chromatography. The considered veterinary drugs were all detected at a level below their corresponding MRLs. Nanoflow liquid chromatographic methods gradients are usually longer than those of (ultra) high-performance liquid chromatography. However, the use of this approach and the complete elimination of matrix effects makes the use of matrix-matched calibration or the standard addition method unnecessary, a valuable feature considering the potential savings resulting from its implementation in laboratories. Other recent studies [50–52] demonstrate the potential of HRMS for the detection and quantification, in routine analysis, of a wide range of veterinary drug residues in multiple matrices.

The studies presented in this section and summarized in Table 2 are based on the detection of residues of known drugs. However, new drugs and new methods of application are being developed to overcome the detection of fraudulent practices. In this context, Dervilly-Pinel and co-workers [53,54] developed and validated a method to screen for β-agonists in bovines, based on a pure metabolomics approach. The method combined three biomarkers and bioinformatics to formulate a discriminant function to predict β-agonist treatment in bovines in an inexpensive, accurate, feasible, high-throughput test. Despite efforts to identify these three biomarkers, two of them remained unresolved. The untargeted workflow was, therefore, accredited to implement the innovative screening tool, demonstrating the power of HRMS in the analysis of veterinary drugs.

Table 2. Selected studies on veterinary drug residues analysis by HRMS.

Instrument and Scanning Technique	Matrix	Number of Analytes	Method Sensitivity	Reference
UHPLC-ESI-Orbitrap with full-scan MS	Bovine urine	87	33.3% of the analytes with LOD $\leq$ 1 µg/L 98.9% of the analytes with LOD $\leq$ 10 µg/L	[45]
UHPLC-ESI-ToF with full-scan MS	Honey	200	75.5% of the analytes with LOD $\leq$ 20 µg/kg 89.3% of the analytes with LOD $\leq$ 50 µg/L	[46]
UHPLC-ESI-Q-ToF with full-scan MS suspect screening in the first injection and target MS/MS confirmation in the second injection	Bovine, rabbit, poultry, goat and pork feeds	116	40% of the analytes detected and 10% confirmed at 0.02 mg/kg in all 5 matrices 75% of the analytes detected and 55% confirmed at 0.2 mg/kg in all 5 matrices	[47]
HPLC-ESI-Orbitrap with full-scan MS	Pig, poultry, cattle lamb and fish feed	48	50% of the analytes with LOD $\leq$ 10 µg/kg 94% of the analytes with LOD $\leq$ 20 µg/kg 79% of the analytes with LOQ $\leq$ 50 µg/kg 98% of the analytes with LOQ $\leq$ 100 µg/kg	[48]

Table 2. *Cont.*

Instrument and Scanning Technique	Matrix	Number of Analytes	Method Sensitivity	Reference
Nanoflow LC-ESI-Q-Orbitrap with simultaneous full-scan MS and MS/MS scans	Honey, veal muscle, egg and milk	87	38% of the analytes with LOQ $\leq$ 0.1 µg/kg in all 4 matrices 100% of the analytes with LOQ $\leq$ 1 µg/kg in all 4 matrices	[49]
UHPLC-ESI-Q-Orbitrap with simultaneous full-scan MS and MS/MS scans	Pork meat	37	LOD between 0.8 and 3.5 µg/kg LOQ between 2.4 and 10.5 µg/kg	[50]
UHPLC-ESI-Q-Orbitrap with full-scan MS	Bovine, chicken and porcine meat	164	10.9% of the analytes confirmed at 1 µg/kg 32.3% of the analytes confirmed at 10 µg/kg 83.5% of the analytes confirmed at 100 µg/kg	[51]
UHPLC-ESI-Q-Orbitrap with simultaneous full-scan MS and MS/MS scans	Milk	105	58% of the analytes detected and 51% confirmed at 1 µg/kg 96% of the analytes detected and 84% confirmed at 10 µg/kg	[52]

4. Analysis of Natural Toxins

By contrast with pesticides and veterinary drugs, toxins are not man-made. They are metabolites produced by living organisms that are typically not harmful to the organisms themselves but can adversely affect human or animal health when consumed [55]. Natural toxins have multiple sources, including plants (phytotoxins or plant toxins), fungi (mycotoxins), algae (phycotoxins or biotoxins) and bacteria (bacterial toxins). The diversity of these biological systems and the disparate properties of toxins present challenges to analytical chemists and wide-ranging food safety implications. Moreover, the chemical structures of the toxins can be altered by the metabolism of the organisms as part of their defence against xenobiotics, increasing the wide spectrum of possible occurring contaminants [56]. For the most prevalent and potent toxins in both animal feed and human food, regulatory limits have been set in European Union legislation [3,57–59].

To ensure food and feed safety and compliance with European legislation, several detection and screening methods have been developed for the analysis of natural toxins. Methods based on chromatographic analysis coupled to fluorescence or ultraviolet (UV) detection are progressively replaced by chromatographic separation coupled to tandem mass spectrometric analysers, such as triple quadrupoles. This technique is, however, limited to targeted analysis, making necessary the use of an analytical standard, which is a critical issue for modified toxin analysis [56]. In recent years, thanks to technological advances and improved affordability, HRMS-based methods have been developed to circumvent this issue. These methods allow screening for and quantification of hundreds of parent toxins and associated metabolites in food and feed samples.

Mycotoxins are the most studied toxins, for which multiple HRMS-based methods have been proposed in recent years. Zachariasova and co-workers developed, for example, methods to analyse multiple mycotoxins in cereals [60] and beer [61]. Depending on the matrix, different strategies for mycotoxin quantification were suggested. The use of isotopically labelled surrogates and analytes in pure solvent standards for the construction of calibration curves provided better recovery and repeatability of results for cereals, whereas a matrix-matched calibration was preferred for beer. In both studies, the authors insisted on the importance of the resolving power of the instrument used for the analysis, improving the selectivity of the detection in the presence of abundant co-eluting matrix components. Another method was developed by Lattanzio and co-workers [62] for the simultaneous determination of multiple mycotoxins in wheat flour, barley flour and crisp bread. A critical comparison between the HRMS method and a validated method based on triple quadrupole mass spectrometry showed similar performance in terms of detection limits (in the range of 0.1 to 2.9 µg/kg with the HRMS method, adequate to assess mycotoxin contamination in cereal foods at regulatory levels), recoveries, repeatability and matrix effects. In their conclusion, the authors described HRMS-based methods as a reliable and robust alternative tool for the routine analysis of major mycotoxins in foods, with the additional advantage of the possibility to perform retrospective analysis, and to search for mycotoxin metabolites. Ates and co-workers [63] developed a method combining an automated on-line sample clean-up and LC-HRMS for the analysis of multiple mycotoxins in maize, wheat and animal feed. With this approach, interfering matrix compounds with different chemical properties and macromolecules such as fats and proteins can be removed. The method was validated with good repeatability, and quantification limits were all acceptable with respect to legislative limits. Certified reference materials, which have been analysed as representative samples of maize, wheat and animal feed for the target compounds, demonstrated a high method accuracy. Moreover, the developed method was successfully employed in proficiency testing of animal feed samples, confirming its applicability for routine analysis. The same approach was utilized by authors to screen plant and fungal metabolites in wheat, maize and animal feed, using an empirical database of over 600 metabolites [64]. The wide applicability of the method was first demonstrated by the validation of 15 fungal and plant metabolites in maize, wheat and animal feed samples. The method was then applied to market samples. In addition to regulated and known secondary metabolites, 3 other mycotoxin metabolites were identified for the first time, demonstrating the capacity of HRMS full-scan analysis. The applicability of HRMS-based methods, in routine analysis, for the determination of multiple mycotoxins in complex matrices was also demonstrated in several studies [65–67].

Besides mycotoxins, biotoxins also pose a significant food safety risk, for which dedicated HRMS methods were proposed. Blay and co-workers [68] developed, for instance, an LC-MS platform for the non-targeted screening of two major classes (hydrophilic and lipophilic) of biotoxins commonly found in shellfish. Although two different modes of separation were employed for the two classes, authors insisted on the minimum required MS method development time and the possibility to easily extend the approach to other toxins or toxin analogues. Rúbies and co-workers [69] developed a high-throughput confirmatory quantitative method for the analysis of regulated biotoxins in fresh and canned bivalves. Thanks to the high resolving power of the Q-Orbitrap instrument, accurate mass data were obtained for each analyte—both the molecular ion and the selected fragment. The different compounds were, therefore, identified with high confidence and the risk of false positive results is limited. Using a matrix-matched calibration curve and a HRMS/MS acquisition mode, the method provided reliable quantitative results at the regulated concentration levels. This method is currently used in a routine laboratory in Spain. Finally, HRMS based methods were also developed for the analysis of phytotoxins, such as tropane alkaloids in animal feed [70] or teas [71] or alkenylbenzenes in pepper [72]. These studies, summarized in Table 3, demonstrate, once again, the versatility of HRMS and its suitability for routine analysis.

Table 3. Selected studies on natural toxins analysis by HRMS.

Instrument and Scanning Technique	Matrix	Number of Analytes	Method Sensitivity	Reference
UHPLC-ESI-Orbitrap and UHPLC-ESI-ToF with full-scan MS	Maize, barley and wheat	11	5 analytes with LOD $\leq$ 10 µg/kg and with ToF 2 analytes with LOD $\leq$ 10 µg/kg with Orbitrap 9 analytes with LOD $\leq$ 25 µg/kg with ToF 9 analytes with LOD $\leq$ 25 µg/kg with Orbitrap 5 analytes with LOQ $\leq$ 25 µg/kg with ToF 3 analytes with LOQ $\leq$ 25 µg/kg with Orbitrap 9 analytes with LOQ $\leq$ 50 µg/kg with ToF 10 analytes with LOQ $\leq$ 50 µg/kg with Orbitrap	[60]
UHPLC-Atmospheric pressure chemical ionisation-Orbitrap with full-scan MS	Pale lager, non-alcoholic, and black lager beers	32	16% of the analytes with the lowest calibration level $\leq$2 µg/L 87% of the analytes with the lowest calibration level $\leq$10 µg/L	[61]
HPLC-ESI- High-energy collision dissociation–Orbitrap with full-scan MS	Wheat flour, barley flour, wheat crisp bread and rye crisp bread	9	Analytes detected in 0.1–1.6 µg/kg range and confirmed in 0.1–3.4 µg/kg range in all 4 matrices	[62]
HPLC-ESI-High-energy collision dissociation–Orbitrap with full-scan MS	Maize, wheat and animal feed	6	LOD between 2 and 150 µg/kg in all 3 matrices LOQ between 5 and 375 µg/kg in all 3 matrices	[63]
HPLC-ESI-High-energy collision dissociation–Orbitrap with full-scan MS	Maize, wheat and animal feed	15	99% identification rate in multiple replicates at 250 µg/kg in all 3 matrices	[64]
HPLC-Atmospheric pressure chemical ionisation-Q-Orbitrap with simultaneous full-scan MS and MS/MS scans	Forage maize and maize silage	8	Analytes detected in 11–88 µg/kg range and confirmed in 20–141 µg/kg range in both matrices	[65]
UHPLC-ESI-TOF with full-scan MS	Maize	9	5 analytes with LOD $\leq$ 1 µg/kg 7 analytes with LOD $\leq$ 25 µg/kg 5 analytes with LOQ $\leq$ 2 µg/kg 7 analytes with LOQ $\leq$ 50 µg/kg	[66]
UHPLC-ESI-Q-Orbitrap with simultaneous full-scan MS and MS/MS scans	Corn, rice, wheat, almond, peanut and pistachio	26	46% of the analytes with LOD $\leq$ 0.1 µg/kg 76% of the analytes with LOD $\leq$ 1 µg/kg 54% of the analytes with LOQ $\leq$ 0.5 µg/kg 81% of the analytes with LOQ $\leq$ 5 µg/kg	[67]
HPLC-ESI-Orbitrap with full-scan MS and separated LC methods for lipophilic and hydrophilic toxins	Mussel tissue	10 lipophilic 12 hydrophilic	LOD between 0.041 and 5.1 µg/L for lipophilic toxins LOD between 3.4 and 14 µg/L for hydrophilic toxins	[68]
UHPLC-ESI-Q-Orbitrap with simultaneous full-scan MS and MS/MS scans	Fresh and canned bivalves	10	LOQ at 25 µg/kg	[69]
HPLC-ESI-High-energy collision dissociation–Orbitrap with full-scan MS	Chicken feed	12	LOQ between 5 and 25 µg/kg	[70]
HPLC-ESI-High-energy collision dissociation–Orbitrap with full-scan MS	Tea and herbal teas	13	LOQ between 5 and 20 µg/kg	[71]
GC–Electron Ionization-Q-Orbitrap with full-scan MS	Pepper	8	LOD between 0.01 and 0.02 mg/kg LOQ at 0.2 mg/kg	[72]

5. Multi-Class Analysis

The previous sections concerned HRMS analysis methods dedicated to the determination of a single class of contaminants, including pesticides, veterinary residues and toxins. However, in full-scan mode, high-resolution instruments are able to screen for a theoretically unlimited number of compounds in a single analysis. The simultaneous analysis of multi-class compounds is, therefore, possible with the HRMS approach. Food and feed can, indeed, be contaminated with different types of undesired compounds. Pesticides and mycotoxins can, for example, be found in crops, just as pesticides and veterinary drug residues can be found in milk. Sample preparation is not addressed in this review but is of the utmost importance in non-targeted methods and is even more challenging in this context, with the high variability of physicochemical properties of the screened compounds. The impact that sample preparation and instrument parameters on data quality and chemical coverage is illustrated in the study of Knolhoff and co-workers [73].

In a feasibility study, Pérez-Ortega and co-workers [74] demonstrated that a time-of-flight instrument was suitable for the screening of 625 multiclass food contaminants in baby-food samples (containing meat and vegetables). Due to the resolving power of the instrument, retention time, isotopic profile and fragment ions, almost all the components, including pesticides, veterinary drugs, mycotoxins and other contaminants of concern, were successfully resolved. Only three pairs of compounds, out of the 76% of component involved in isobaric groups, were not resolved using these tools. Highlighted weaknesses of this approach were chromatographic issues with highly polar species, low sensitivities for selected compounds, which does not map well against electrospray ionization, and quantitation of compounds affected by signal suppression effects due to co-eluting matrix components or analytes. However, these issues are not specific to the HRMS approach and also affect the traditional triple quadrupole-based targeted approach. In contrast, only the HRMS approach is capable of screening for such a large number of contaminants. Residue analysis of infant food are is of high importance because this specific age group is more sensitive to several chemicals due to their high food intake/body weight ratio and the immaturity of their defence systems against chemical stressors [75]. A specific directive was, therefore, defined by the European Commission for such foodstuffs [76]. It prohibits the use of certain very toxic pesticides in production and requires that infant formula and follow-on formula contain no detectable levels of pesticide residues (meaning <0.01 mg/kg). The detection of residues at such a challengingly low level was partially met in another study from Gómez-Pérez and co-workers [77]. More than 300 pesticide and veterinary drug residues were quantified with a validated method presenting limits of detection from 0.5 to 50 mg/kg and limits of quantification between 10 and 100 mg/kg.

A study by Cotton and co-workers [78] provides a relevant example of the potential of HRMS for the analysis of water. They developed and validated (repeatability, selectivity, linearity and matrix effect) a multi-residue targeted method for the analysis of more than 500 pesticides and drugs in water. More than 30 different compounds were detected in 20 tap water samples collected in and around Paris but at level lower than 0.1 µg/L, the European Union limit for pesticides in drinking water. In another study, Albergamo and co-workers [79] developed an HRMS-based method for the identification and quantification of 33 polar micropollutants, representative for several classes of emerging contaminants (herbicides, sweeteners, pharmaceutically active compounds, anticorrosive agents and industrial chemicals), in natural drinking water sources. The authors argued that most polar micropollutants are overlooked by the current regulatory actions, resulting in the need to use accurate, sensitive and robust analytical tools to efficiently monitor source waters. The method detection limit for the 33 considered micropollutants in riverbank filtrate water ranged between 8 and 83 ng/L and was lower than 20 ng/L for 27 of them. The authors also highlighted that the developed method is suitable for a larger number of compounds, such as analogues and metabolites of the micropollutants, and the database of target compounds is extendable and could be used for retrospective suspect screening.

However, the use of routine non-target analysis is still uncommon for most environmental monitoring agencies and environmental scientists [80]. Compound identification in targeted and suspect-screening analysis (using a non-targeted data-acquisition approach) rely on reference standards, databases containing compounds structure, isotope pattern, presence of additional adducts, chromatographic retention behaviour, fragmentation information, and other experimental evidence. Purely non-targeted analysis aims at identifying compounds present in the sample without prior information. Data processing in such an approach is laborious, and the achievement of quantitative results remains difficult. These statements are well illustrated in a study by Schymanski and co-workers, who reviewed a collaborative trial on water analysis using non-targeted screening with HRMS. It was observed that the analytical methods are already reasonably well harmonised, contrary to processing workflow and used databases. Targets from some laboratories were found to be suspects or unknowns in other laboratories. Participants in the trial expressed the need to harmonise information sources. Authors insisted on the fact that enhancing this by uploading mass spectra of target compounds to an open access database would help improve the success of target, suspect and even non-target screening immensely. However, it is likely that several degradation products of industrial contaminants and pharmaceuticals would not be found in any current libraries. Improvement of the entire data treatment process is necessary to more extensively characterize water samples for a non-targeted approach and answer questions regarding the origin of the contamination or the dynamics of the contaminants [80].

Several other HRMS-based methods were developed for multi-class residue analysis, such as pesticides, veterinary drugs and mycotoxins, and quantitative determination in both bakery raw materials and finished products [81]. The authors highlighted that a more satisfactory performance may still be seen by way of a robust triple quadrupole MRM analysis. HRMS-generated data, however, allow for additional flexibility in post-acquisition processing, with the consequent advantage of the possibility to execute retrospective data mining. Several recent studies were identified for the determination of various undesired residues in multiple matrices, including feed, honey, vegetables, cereals, tea, botanical nutraceuticals or edible insects, with a growing interest in alternatives to the increasing food demand [82–88]. These numerous studies, summarized in Table 4, confirmed the suitability of HRMS for multi-class residue analysis, in particular, with hundreds of compounds screened in a single run, which was possible due to full-scan mode.

Table 4. Selected studies on HRMS analysis of multi-class contaminants.

Instrument and Scanning Technique	Matrix	Number of Analytes	Classes of Analytes	Method Sensitivity	Reference
UHPLC-ESI-Collision induced dissociation-Q-ToF with full-scan MS	Orange, tomato and baby food	625	Pesticides, veterinary drugs, mycotoxins, food-packaging contaminants, perfluoroalkyl substances or nitrosamines	80–85% of the analytes detected at 50 µg/kg	[74]
UHPLC-ESI-Q-Orbitrap with simultaneous full-scan MS and MS/MS scans	Water	539	Pesticides and drugs	44% of the analytes with LOD ≤ 0.001 µg/L 84% of the analytes with LOD ≤ 0.01 µg/L 99% of the analytes with LOD ≤ 0.1 µg/L	[78]
UHPLC-ESI-Collision induced dissociation-Q-TOF with full-scan MS	Surface water and groundwater	33	Herbicides, sweeteners, drugs, anticorrosive agents and chemicals	LOD between 0.009 and 0.093 µg/L	[79]
UHPLC-ESI-Orbitrap with full-scan MS	Milk, flours and minicakes	36	Pesticides, antibiotics and mycotoxins	17% of the analytes with LOD ≤ 1 µg/L 72% of the analytes with LOD ≤ 10 µg/L 83% of the analytes with LOD ≤ 100 µg/L	[81]
UHPLC-ESI-Q-Orbitrap with simultaneous full-scan MS and MS/MS scans	Botanical Nutraceuticals	16	Pesticides and mycotoxins	LOQ between 0.2 and 6.25 µg/kg	[82]
UHPLC-ESI-Q-Orbitrap with full-scan MS	Edible insects	77	Pesticides, (veterinary) drugs and mycotoxins	75 analytes detected in 1–100 µg/kg range	[83]

Table 4. *Cont.*

Instrument and Scanning Technique	Matrix	Number of Analytes	Classes of Analytes	Method Sensitivity	Reference
UHPLC-ESI-ToF with full-scan MS	Tea brew and tea leaves	32	Pesticides, mycotoxins, process-induced toxicants and packaging contaminants	81% of the analytes detected at 10 µg/kg 63% of the analytes quantified at 10 µg/kg	[84]
HPLC-ESI-High-energy collision dissociation–Orbitrap with full-scan MS	Nutraceutical products (green tea and royal jelly)	260	Pesticides and mycotoxins	LOD between 0.5 and 10 µg/kg LOQ between 1 and 20 µg/kg	[85]
UHPLC-ESI-High-energy collision dissociation–Orbitrap with full-scan MS	Cattle feed	77	Veterinary drugs, ergot alkaloids, plant toxins and other undesirable substances	52% of the analytes with LOQ ≤ 5 µg/kg 87% of the analytes with LOQ ≤ 25 µg/kg	[86]
HPLC-ESI-Q-Orbitrap with simultaneous full-scan MS and MS/MS scans	Leek, wheat, and tea	389	Pesticide, mycotoxins, and pyrrolizidine alkaloids	82% of the analytes with LOQ≤10 µg/kg in leek 81% of the analytes with LOQ≤10 µg/kg in wheat 61% of the analytes with LOQ≤10 µg/kg in tea	[87]
HPLC-ESI-High-energy collision dissociation–Orbitrap with full-scan MS	Honey	350	Pesticides and veterinary drugs	95% of the analytes with LOD ≤ 10 µg/kg	[88]

6. Food Authenticity

Besides analysis aiming to detect contaminants and residues in food and feed, HRMS has been used to assess food authenticity and detect fraud or adulteration. Food fraud is motivated by economic gain but can represent a serious health risk for consumers. By assessing food authenticity, consumers are protected from purchasing products of inferior quality or with incorrect descriptions and honest traders are defended from unfair competition. Wine, spirits, olive oil, fish, meat, cheese, honey and herbs and spices represent the most commonly reported adulterated foods [89].

HRMS analysis with untargeted data acquisition and chemometrics tools, such as principal component analysis, are a powerful combination for the evaluation of food authenticity. Among the wide range of analysed compounds, chemometric tools can decrease the number of detected features remarkably and suggest characteristic markers responsible for different types of authenticity issues, such as adulteration, variety or geographical origin discrimination, organoleptic profiles, ripening and method production.

Rubert and co-workers used metabolic fingerprinting for wine authentication according to the grape varieties. The validated discriminant analysis models based on the acquired data were able to correctly classify 95% of over 300 wine samples. Using online libraries, the markers used for wine authentication were tentatively identified and corresponded to different flavanol glucosides and polyphenols. Hrbek and co-workers [90] used a similar approach to identify garlic origin, which is known to influence its organoleptic properties. The data generated by an HPLC-HRMS analysis of 47 samples of garlic from different geographical origins were used to construct statistical models to authenticate garlic origin. A number of robust targets, including free amino acids and characteristic sulphur compounds, were identified as the most suitable markers. The last selected example was a study by Fiorino and co-workers [91], aiming to assess fish authenticity and discriminate between wild-type and farmed salmon. Wild salmon is, indeed, known to be richer in the more valuable omega-3 fatty acids. A fast HRMS analysis method, using an Orbitrap instrument, was developed and combined with data integration via principal component analysis. The analysis of wild-type and farmed salmon samples from different origins led to the conclusion that saturated and polyunsaturated fatty acids with 20 or 22 carbon atoms on their side chains were the most suitable markers to discriminate between the two types of salmons. The developed method was successfully applied to commercial samples.

These few examples demonstrate the power of HRMS analysis combined with chemometrics tools in food authenticity assessment. Identification is based on specific markers from the theoretically unlimited number of components detected during the analysis. Chemometrics tools enable the highlighting of these specific markers among the massive amount of data. If present in libraries, these specific markers can be identified to achieve a more robust authentication method. Due to the great amount of time and number of reference samples required to develop a model, this approach is currently limited to a few food products, mostly with high added value. However, associations such as the Association of Official Analytical Collaboration (AOAC) International are currently working on guidelines for method development and validation, including untargeted approaches [92].

7. Conclusions

In recent years, food and feed analysis based on HRMS coupled to chromatographic separation techniques has become increasingly common, and today, some developed methods are used in routine analysis. Contrary to targeted methods using low-resolution instruments, such as highly popular triple quadrupole instruments, HRMS-based methods using untargeted data acquisition are able to record a theoretically unlimited number of compounds in full-scan mode. The development of HRMS acquisition methods is relatively simple compared to the time-consuming and standard-requiring development and optimisation of MRM methods. Moreover, due to untargeted data acquisition, retrospective analysis is possible and could be relevant when new contaminants or residues are discovered.

Triple quadrupole instruments operating in MRM mode generally demonstrate higher sensitivities than HRMS instruments operating in full-scan mode. However, using the quadrupole of a Q-TOF or Q-Orbitrap to isolate a narrower mass range frequently improves HRMS sensitivity [15]. Moreover, components producing many fragments grant superior HRMS sensitivity since the compounds are preferably detected as unfragmented precursor ions. The sensitivity gap between the two technologies has likely narrowed over the last decade, and this process will probably continue. Sensitivity and quantification issues due to the poor ionization of certain targets and coeluting matrix components can also be encountered. However, these issues are not specific to HRMS-based methods and are also encountered with a triple quadrupole instrument. Nevertheless, the numerous studies and developed methods presented in this review demonstrate the capability of HRMS to detect several types of components at levels in compliance with the current relevant legislation. Moreover, alternative and innovative approaches using HRMS have recently been developed, such as untargeted metabolomics, allowing screening for banned compounds.

Currently, HRMS-based analysis is limited to components characterized in databases. Purely non-targeted screening without any prior information on the compounds remains challenging, and efforts towards developing computational tools are necessary to enable the use of this approach in the future. Therefore, the current priority is the expanding and the disseminating of libraries and databases for a wide range of contaminants, residues and associated transformation products. This will extend the scope of HRMS analysis for food and feed samples and make this approach essential.

Author Contributions: Conceptualization, M.G. and N.G.; methodology, M.G.; writing—original draft preparation, M.G.; writing—review and editing, M.G. and N.G.; supervision, P.D. and N.G. All authors have read and agreed to the published version of the manuscript.

Funding: This research received no external funding.

Conflicts of Interest: The authors declare no conflict of interest.

References

1. Steiner, D.; Malachová, A.; Sulyok, M.; Krska, R. Challenges and future directions in LC-MS-based multiclass method development for the quantification of food contaminants. *Anal. Bioanal. Chem.* **2020**, *413*, 25–34. [CrossRef] [PubMed]
2. European Communities. Treaty on European Union and of the Treaty Establishing the European Community. *Off. J. Eur. Comm.* **2002**, *45*, 1–184.
3. Commission Regulation (EC). Comission Regulation EC No 1881/2006 of 19 December 2006 setting maximum levels for certain contaminants in foodstuffs. *Off. J. Eur. Union* **2006**, *L 364/5*, 5–24.
4. Commission Regulation (EC). Comission Regulation EC No 37/2010 of 22 December 2009 on pharmacologically active substances and their classification regarding maximum residue limits in foodstuffs of animal origin. *Off. J. Eur. Union* **2009**, *L 15/1*, 1–72.
5. Commission Regulation (EC). Comission Regulation EC No 396/2005 of the European Parliament and of the Council of 23 February 2005 on maximum residue levels of pesticides in or on food and feed of plant and animal origin and amending Council Directive 91/414/EEC. *Off. J. Eur. Union* **2005**, *L 70/1*, 1–16.
6. Magnuson, B.; Munro, I.; Abbot, P.; Baldwin, N.; Lopez-Garcia, R.; Ly, K.; McGirr, L.; Roberts, A.; Socolovsky, S. Review of the regulation and safety assessment of food substances in various countries and jurisdictions. *Food Addit. Contam. Part A Chem. Anal. Control. Expo. Risk Assess.* **2013**, *30*, 1147–1220. [CrossRef]
7. Belet, A.; Wolfs, C.; Mahy, J.G.; Poelman, D.; Vreuls, C.; Gillard, N.; Lambert, S.D. Sol-gel syntheses of photocatalysts for the removal of pharmaceutical products in water. *Nanomaterials* **2019**, *9*, 126. [CrossRef] [PubMed]
8. Moreno, J.L.F.; Frenich, A.G.; Bolaños, P.P.; Vidal, J.L.M. Multiresidue method for the analysis of more than 140 pesticide residues in fruits and vegetables by gas chromatography coupled to triple quadrupole mass spectrometry. *J. Mass Spectrom.* **2008**, *43*, 1235–1254. [CrossRef] [PubMed]
9. Robert, C.; Brasseur, P.Y.; Dubois, M.; Delahaut, P.; Gillard, N. Development and validation of rapid multiresidue and multi-class analysis for antibiotics and anthelmintics in feed by ultra-high-performance liquid chromatography coupled to tandem mass spectrometry. *Food Addit. Contam. Part A Chem. Anal. Control. Expo. Risk Assess.* **2016**, *33*, 1312–1323. [CrossRef]
10. Socas-Rodríguez, B.; Herrera-Herrera, A.V.; Hernández-Borges, J.; Rodríguez-Delgado, M.Á. Multiresidue determination of estrogens in different dairy products by ultra-high-performance liquid chromatography triple quadrupole mass spectrometry. *J. Chromatogr. A* **2017**, *1496*, 58–67. [CrossRef] [PubMed]
11. Hakme, E.; Lozano, A.; Uclés, S.; Fernández-Alba, A.R. Further improvements in pesticide residue analysis in food by applying gas chromatography triple quadrupole mass spectrometry (GC-QqQ-MS/MS) technologies. *Anal. Bioanal. Chem.* **2018**, *410*, 5491–5506. [CrossRef] [PubMed]
12. Georgiou, C.A.; Danezis, G.P. Elemental and Isotopic Mass Spectrometry. *Compr. Anal. Chem.* **2015**, *68*, 131–243. [CrossRef]
13. Hermes, N.; Jewell, K.S.; Wick, A.; Ternes, T.A. Quantification of more than 150 micropollutants including transformation products in aqueous samples by liquid chromatography-tandem mass spectrometry using scheduled multiple reaction monitoring. *J. Chromatogr. A* **2018**, *1531*, 64–73. [CrossRef] [PubMed]
14. Wong, J.W.; Wang, J.; Chow, W.; Carlson, R.; Jia, Z.; Zhang, K.; Hayward, D.G.; Chang, J.S. Perspectives on liquid chromatography-high-resolution mass spectrometry for pesticide screening in foods. *J. Agric. Food Chem.* **2018**, *66*, 9573–9581. [CrossRef]
15. Kaufmann, A. High-resolution mass spectrometry for bioanalytical applications: Is this the new gold standard? *J. Mass Spectrom.* **2020**, *55*. [CrossRef] [PubMed]
16. Anumol, T.; Lehotay, S.J.; Stevens, J.; Zweigenbaum, J. Comparison of veterinary drug residue results in animal tissues by ultrahigh-performance liquid chromatography coupled to triple quadrupole or quadrupole–time-of-flight tandem mass spectrometry after different sample preparation methods, including use of. *Anal. Bioanal. Chem.* **2017**, *409*, 2639–2653. [CrossRef]
17. Cavaliere, C.; Antonelli, M.; Capriotti, A.L.; La Barbera, G.; Montone, C.M.; Piovesana, S.; Laganà, A. A Triple Quadrupole and a Hybrid Quadrupole Orbitrap Mass Spectrometer in Comparison for Polyphenol Quantitation. *J. Agric. Food Chem.* **2019**, *67*, 4885–4896. [CrossRef]
18. Morin, L.P.; Mess, J.N.; Garofolo, F. Large-molecule quantification: Sensitivity and selectivity head-to-head comparison of triple quadrupole with Q-TOF. *Bioanalysis* **2013**, *5*, 1181–1193. [CrossRef]
19. Hernández-Mesa, M.; Escourrou, A.; Monteau, F.; Le Bizec, B.; Dervilly-Pinel, G. Current applications and perspectives of ion mobility spectrometry to answer chemical food safety issues. *TrAC—Trends Anal. Chem.* **2017**, *94*, 39–53. [CrossRef]
20. Yan, X.; Zhang, Y.; Zhou, Y.; Li, G.; Feng, X. Technical Overview of Orbitrap High Resolution Mass Spectrometry and Its Application to the Detection of Small Molecules in Food (Update Since 2012). *Crit. Rev. Anal. Chem.* **2020**. [CrossRef] [PubMed]
21. Kalogiouri, N.P.; Aalizadeh, R.; Dasenaki, M.E.; Thomaidis, N.S. Application of High Resolution Mass Spectrometric methods coupled with chemometric techniques in olive oil authenticity studies—A review. *Anal. Chim. Acta* **2020**, *1134*, 150–173. [CrossRef] [PubMed]
22. Williamson, C. *FAPAS®—Food Chemistry Proficiency Test Report 02381—Avermectins & Anthelmintics in Bovine Liver*; FAPAS: York, UK, 2019.
23. Fernández-Alba, A.R. *Screening Methods 11 (EUPT-FV-SM11) Pesticide Residues in Red Cabbage Homogenate*; European Union Reference Laboratory For Pesticide Residues in Fruits and Vegetables in the University of Almeria: Almería, Spain, 2019.
24. Fernández-Alba, A.R. *Screening Methods 10 (EUPT-FV-SM10) Pesticide Residues in Green Bean Homogenate*; European Union Reference Laboratory for Pesticide Residues in Fruits and Vegetables in the University of Almeria: Almería, Spain, 2018.
25. Zhang, W. Global pesticide use: Profile, trend, cost / benefit and more. *Proc. Int. Acad. Ecol. Environ. Sci.* **2018**, *8*, 1–27.

26. Klein, J.; Alder, L. Applicability of gradient liquid chromatography with tandem mass spectrometry to the simultaneous screening for about 100 pesticides in crops. *J. AOAC Int.* **2003**, *86*, 1015–1037. [CrossRef]
27. Wang, Z.; Chang, Q.; Kang, J.; Cao, Y.; Ge, N.; Fan, C.; Pang, G.F. Screening and identification strategy for 317 pesticides in fruits and vegetables by liquid chromatography-quadrupole time-of-flight high resolution mass spectrometry. *Anal. Methods* **2015**, *7*, 6385–6402. [CrossRef]
28. Wang, J.; Chow, W.; Leung, D.; Chang, J. Application of ultrahigh-performance liquid chromatography and electrospray ionization quadrupole orbitrap high-resolution mass spectrometry for determination of 166 pesticides in fruits and vegetables. *J. Agric. Food Chem.* **2012**, *60*, 12088–12104. [CrossRef] [PubMed]
29. Wang, J.; Chow, W.; Chang, J.; Wong, J.W. Ultrahigh-performance liquid chromatography electrospray ionization Q-orbitrap mass spectrometry for the analysis of 451 pesticide residues in fruits and vegetables: Method development and validation. *J. Agric. Food Chem.* **2014**, *62*, 10375–10391. [CrossRef]
30. Del Mar Gómez-Ramos, M.; Rajski, Ł.; Heinzen, H.; Fernández-Alba, A.R. Liquid chromatography Orbitrap mass spectrometry with simultaneous full scan and tandem MS/MS for highly selective pesticide residue analysis. *Anal. Bioanal. Chem.* **2015**, *407*, 6317–6326. [CrossRef]
31. Kiefer, K.; Müller, A.; Singer, H.; Hollender, J. New relevant pesticide transformation products in groundwater detected using target and suspect screening for agricultural and urban micropollutants with LC-HRMS. *Water Res.* **2019**, *165*, 114972. [CrossRef]
32. Gasparini, M.; Angelone, B.; Ferretti, E. Glyphosate and other highly polar pesticides in fruit, vegetables and honey using ion chromatography coupled with high resolution mass spectrometry: Method validation and its applicability in an official laboratory. *J. Mass Spectrom.* **2020**, *55*, e4624. [CrossRef] [PubMed]
33. He, P.; Aga, D.S. Comparison of GC-MS/MS and LC-MS/MS for the analysis of hormones and pesticides in surface waters: Advantages and pitfalls. *Anal. Methods* **2019**, *11*, 1436–1448. [CrossRef]
34. Besil, N.; Uclés, S.; Mezcúa, M.; Heinzen, H.; Fernández-Alba, A.R. Negative chemical ionization gas chromatography coupled to hybrid quadrupole time-of-flight mass spectrometry and automated accurate mass data processing for determination of pesticides in fruit and vegetables. *Anal. Bioanal. Chem.* **2015**, *407*, 6327–6343. [CrossRef]
35. Vargas-Pérez, M.; Domínguez, I.; González, F.J.E.; Frenich, A.G. Application of full scan gas chromatography high resolution mass spectrometry data to quantify targeted-pesticide residues and to screen for additional substances of concern in fresh-food commodities. *J. Chromatogr. A* **2020**, *1622*, 461118. [CrossRef]
36. Beyene, T. Veterinary Drug Residues in Food-animal Products: Its Risk Factors and Potential Effects on Public Health. *J. Vet. Sci. Technol.* **2015**, *7*. [CrossRef]
37. Baynes, R.E.; Dedonder, K.; Kissell, L.; Mzyk, D.; Marmulak, T.; Smith, G.; Tell, L.; Gehring, R.; Davis, J.; Riviere, J.E. Health concerns and management of select veterinary drug residues. *Food Chem. Toxicol.* **2016**, *88*, 112–122. [CrossRef]
38. Comission Regulation (EC) Comission Regulation EC No 2377/90 of 26 June 1990 laying down a community procedure for the establishment of maximum residue limits of veterinary medicinal products in foodstuffs of animal origin. *Off. J. Eur. Communities* **1990**, *224*, 1–8.
39. Daeseleire, E.; Van Pamel, E.; Van Poucke, C.; Croubels, S. Veterinary Drug Residues in Foods. In *Chemical Contaminants and Residues in Food*, 2nd ed.; Woodhead Publishing: Cambridge, UK, 2017; pp. 117–153. ISBN 9780081006740.
40. Lees, P.; Toutain, P.L. The role of pharmacokinetics in veterinary drug residues. *Drug Test. Anal.* **2012**, *4*, 34–39. [CrossRef] [PubMed]
41. Delatour, T.; Racault, L.; Bessaire, T.; Desmarchelier, A. Screening of veterinary drug residues in food by LC-MS/MS. Background and challenges. *Food Addit. Contam. Part A Chem. Anal. Control. Expo. Risk Assess.* **2018**, *35*, 632–645. [CrossRef] [PubMed]
42. Katragadda, S.; Mahmoud, S.; Ramanathan, D.M. UHPLC-ESI-HRMS Quantitation of Metabolites without Using Reference Standards: Impact of LC Flow Rate and Mobile Phase Composition on MS Responses. *Am. J. Anal. Chem.* **2013**, *04*, 36–46. [CrossRef]
43. Romero-González, R.; Aguilera-Luiz, M.M.; Plaza-Bolaños, P.; Frenich, A.G.; Vidal, J.L.M. Food contaminant analysis at high resolution mass spectrometry: Application for the determination of veterinary drugs in milk. *J. Chromatogr. A* **2011**, *1218*, 9353–9365. [CrossRef] [PubMed]
44. Berendsen, B.J.A.; Meijer, T.; Mol, H.G.J.; van Ginkel, L.; Nielen, M.W.F. A global inter-laboratory study to assess acquisition modes for multi-compound confirmatory analysis of veterinary drugs using liquid chromatography coupled to triple quadrupole, time of flight and orbitrap mass spectrometry. *Anal. Chim. Acta* **2017**, *962*, 60–72. [CrossRef]
45. León, N.; Roca, M.; Igualada, C.; Martins, C.P.B.; Pastor, A.; Yusá, V. Wide-range screening of banned veterinary drugs in urine by ultra high liquid chromatography coupled to high-resolution mass spectrometry. *J. Chromatogr. A* **2012**, *1258*, 55–65. [CrossRef]
46. Staub Spörri, A.; Jan, P.; Cognard, E.; Ortelli, D.; Edder, P. Comprehensive screening of veterinary drugs in honey by ultra-high-performance liquid chromatography coupled to mass spectrometry. *Food Addit. Contam. Part A Chem. Anal. Control. Expo. Risk Assess.* **2014**, *31*, 806–816. [CrossRef]
47. Boix, C.; Ibáñez, M.; Sancho, J.V.; León, N.; Yusá, V.; Hernández, F. Qualitative screening of 116 veterinary drugs in feed by liquid chromatography-high resolution mass spectrometry: Potential application to quantitative analysis. *Food Chem.* **2014**, *160*, 313–320. [CrossRef] [PubMed]
48. George, K.; Vincent, U.; von Holst, C. Analysis of antimicrobial agents in pig feed by liquid chromatography coupled to orbitrap mass spectrometry. *J. Chromatogr. A* **2013**, *1293*, 60–74. [CrossRef]

49. Alcántara-Durán, J.; Moreno-González, D.; Gilbert-López, B.; Molina-Díaz, A.; García-Reyes, J.F. Matrix-effect free multi-residue analysis of veterinary drugs in food samples of animal origin by nanoflow liquid chromatography high resolution mass spectrometry. *Food Chem.* **2018**, *245*, 29–38. [CrossRef] [PubMed]
50. Chen, Q.; Pan, X.D.; Huang, B.F.; Han, J.L.; Zhou, B. Screening of multi-class antibiotics in pork meat by LC-Orbitrap-MS with modified QuEChERS extraction. *RSC Adv.* **2019**, *9*, 28119–28125. [CrossRef]
51. Pugajeva, I.; Ikkere, L.E.; Judjallo, E.; Bartkevics, V. Determination of residues and metabolites of more than 140 pharmacologically active substances in meat by liquid chromatography coupled to high resolution Orbitrap mass spectrometry. *J. Pharm. Biomed. Anal.* **2019**, *166*, 252–263. [CrossRef] [PubMed]
52. Wang, J.; Leung, D.; Chow, W.; Chang, J.; Wong, J.W. Target screening of 105 veterinary drug residues in milk using UHPLC/ESI Q-Orbitrap multiplexing data independent acquisition. *Anal. Bioanal. Chem.* **2018**, *410*, 5373–5389. [CrossRef] [PubMed]
53. Dervilly-Pinel, G.; Chereau, S.; Cesbron, N.; Monteau, F.; Le Bizec, B. LC-HRMS based metabolomics screening model to detect various β-agonists treatments in bovines. *Metabolomics* **2015**, *11*, 403–411. [CrossRef]
54. Dervilly-Pinel, G.; Royer, A.L.; Bozzetta, E.; Pezzolato, M.; Herpin, L.; Prevost, S.; Le Bizec, B. When LC-HRMS metabolomics gets ISO17025 accredited and ready for official controls–application to the screening of forbidden compounds in livestock. *Food Addit. Contam. Part A Chem. Anal. Control. Expo. Risk Assess.* **2018**, *35*, 1948–1958. [CrossRef] [PubMed]
55. Fletcher, M.T.; Netzel, G. Food safety and natural toxins. *Toxins* **2020**, *12*, 236. [CrossRef] [PubMed]
56. Righetti, L.; Paglia, G.; Galaverna, G.; Dall'Asta, C. Recent advances and future challenges in modified mycotoxin analysis: Why HRMS has become a key instrument in food contaminant research. *Toxins* **2016**, *8*, 361. [CrossRef]
57. Comission Regulation (EC) Commission Regulation (EC) No 853/2004 of 29 April 2004 laying down specific hygiene rules for food of animal origin. *Off. J. Eur. Union* **2004**, *L 269*, 1–15.
58. Comission Regulation (EC). Commission Regulation (EU) 2016/239 of 19 February 2016 Amending Regulation (EC) No 1881/2006 as Regards Maximum Levels of Tropane Alkaloids in Certain Cereal-Based Foods for Infants and Young Children. *Off. J. Eur. Union* **2016**, *L 45/3*, 1–3.
59. Commission Regulation (EC) Comission Regulation EC No 32/2002 of the European Parliament and of the Council of 7 May 2002 on undesirable substances in animal feed. *Off. J. Eur. Communities* **2002**, *L 269*, 1–15.
60. Zachariasova, M.; Lacina, O.; Malachova, A.; Kostelanska, M.; Poustka, J.; Godula, M.; Hajslova, J. Novel approaches in analysis of Fusarium mycotoxins in cereals employing ultra performance liquid chromatography coupled with high resolution mass spectrometry. *Anal. Chim. Acta* **2010**, *662*, 51–61. [CrossRef] [PubMed]
61. Zachariasova, M.; Cajka, T.; Godula, M.; Malachova, A.; Veprikova, Z.; Hajslova, J. Analysis of multiple mycotoxins in beer employing (ultra)-high-resolution mass spectrometry. *Rapid Commun. Mass Spectrom.* **2010**, *24*, 3357–3367. [CrossRef] [PubMed]
62. Lattanzio, V.M.T.; Gatta, S.D.; Godula, M.; Visconti, A. Quantitative analysis of mycotoxins in cereal foods by collision cell fragmentation-high-resolution mass spectrometry: Performance and comparison with triple-stage quadrupole detection. *Food Addit. Contam. Part A Chem. Anal. Control. Expo. Risk Assess.* **2011**, *28*, 1424–1437. [CrossRef] [PubMed]
63. Ates, E.; Mittendorf, K.; Stroka, J.; Senyuva, H. Determination of fusarium mycotoxins in wheat, maize and animal feed using on-line clean-up with high resolution mass spectrometry. *Food Addit. Contam. Part A Chem. Anal. Control. Expo. Risk Assess.* **2013**, *30*, 156–165. [CrossRef] [PubMed]
64. Ates, E.; Godula, M.; Stroka, J.; Senyuva, H. Screening of plant and fungal metabolites in wheat, maize and animal feed using automated on-line clean-up coupled to high resolution mass spectrometry. *Food Chem.* **2014**, *142*, 276–284. [CrossRef]
65. Jensen, T.; De Boevre, M.; Preußke, N.; De Saeger, S.; Birr, T.; Verreet, J.A.; Sönnichsen, F.D. Evaluation of high-resolution mass spectrometry for the quantitative analysis of mycotoxins in complex feed matrices. *Toxins* **2019**, *11*, 531. [CrossRef]
66. Silva, A.S.; Brites, C.; Pouca, A.V.; Barbosa, J.; Freitas, A. UHPLC-ToF-MS method for determination of multi-mycotoxins in maize: Development and validation. *Curr. Res. Food Sci.* **2019**, *1*, 1–7. [CrossRef] [PubMed]
67. Liao, C.D.; Wong, J.W.; Zhang, K.; Hayward, D.G.; Lee, N.S.; Trucksess, M.W. Multi-mycotoxin analysis of finished grain and nut products using high-performance liquid chromatography-triple-quadrupole mass spectrometry. *J. Agric. Food Chem.* **2013**, *61*, 4771–4782. [CrossRef] [PubMed]
68. Blay, P.; Hui, J.P.M.; Chang, J.; Melanson, J.E. Screening for multiple classes of marine biotoxins by liquid chromatography-high-resolution mass spectrometry. *Anal. Bioanal. Chem.* **2011**, *400*, 577–585. [CrossRef]
69. Rúbies, A.; Muñoz, E.; Gibert, D.; Cortés-Francisco, N.; Granados, M.; Caixach, J.; Centrich, F. New method for the analysis of lipophilic marine biotoxins in fresh and canned bivalves by liquid chromatography coupled to high resolution mass spectrometry: A quick, easy, cheap, efficient, rugged, safe approach. *J. Chromatogr. A* **2015**, *1386*, 62–73. [CrossRef]
70. Romera-Torres, A.; Romero-González, R.; Martínez Vidal, J.L.; Frenich, A.G. Study of the occurrence of tropane alkaloids in animal feed using LC-HRMS. *Anal. Methods* **2018**, *10*, 3340–3346. [CrossRef]
71. Romera-Torres, A.; Romero-González, R.; Martínez Vidal, J.L.; Garrido Frenich, A. Simultaneous analysis of tropane alkaloids in teas and herbal teas by liquid chromatography coupled to high-resolution mass spectrometry (Orbitrap). *J. Sep. Sci.* **2018**, *41*, 1938–1946. [CrossRef] [PubMed]
72. Rivera-Pérez, A.; López-Ruiz, R.; Romero-González, R.; Garrido Frenich, A. A new strategy based on gas chromatography–high resolution mass spectrometry (GC–HRMS-Q-Orbitrap) for the determination of alkenylbenzenes in pepper and its varieties. *Food Chem.* **2020**, *321*, 126727. [CrossRef] [PubMed]

73. Knolhoff, A.M.; Kneapler, C.N.; Croley, T.R. Optimized chemical coverage and data quality for non-targeted screening applications using liquid chromatography/high-resolution mass spectrometry. *Anal. Chim. Acta* **2019**, *1066*, 93–101. [CrossRef] [PubMed]

74. Pérez-Ortega, P.; Lara-Ortega, F.J.; García-Reyes, J.F.; Gilbert-López, B.; Trojanowicz, M.; Molina-Díaz, A. A feasibility study of UHPLC-HRMS accurate-mass screening methods for multiclass testing of organic contaminants in food. *Talanta* **2016**, *160*, 704–712. [CrossRef]

75. Nougadère, A.; Sirot, V.; Cravedi, J.P.; Vasseur, P.; Feidt, C.; Fussell, R.J.; Hu, R.; Leblanc, J.C.; Jean, J.; Rivière, G.; et al. Dietary exposure to pesticide residues and associated health risks in infants and young children Results of the French infant total diet study. *Environ. Int.* **2020**, *137*, 105529. [CrossRef] [PubMed]

76. Comission Regulation (EC). Commission Directive 2006/141/EC of 22 December on infant formulae and follow-on formulae and amending Directive 1999/21/EC. *Off. J. Eur. Union* **2006**, *L 401*, 1–33.

77. Luz Gómez-Pérez, M.; Romero-González, R.; José Luis Martínez, V.; Garrido Frenich, A. Analysis of pesticide and veterinary drug residues in baby food by liquid chromatography coupled to Orbitrap high resolution mass spectrometry. *Talanta* **2015**, *131*, 1–7. [CrossRef] [PubMed]

78. Cotton, J.; Leroux, F.; Broudin, S.; Poirel, M.; Corman, B.; Junot, C.; Ducruix, C. Development and validation of a multiresidue method for the analysis of more than 500 pesticides and drugs in water based on on-line and liquid chromatography coupled to high resolution mass spectrometry. *Water Res.* **2016**, *104*, 20–27. [CrossRef] [PubMed]

79. Albergamo, V.; Helmus, R.; de Voogt, P. Direct injection analysis of polar micropollutants in natural drinking water sources with biphenyl liquid chromatography coupled to high-resolution time-of-flight mass spectrometry. *J. Chromatogr. A* **2018**, *1569*, 53–61. [CrossRef]

80. Kruve, A. Semi-quantitative non-target analysis of water with liquid chromatography/high-resolution mass spectrometry: How far are we? *Rapid Commun. Mass Spectrom.* **2019**, *33*, 54–63. [CrossRef] [PubMed]

81. De Dominicis, E.; Commissati, I.; Gritti, E.; Catellani, D.; Suman, M. Quantitative targeted and retrospective data analysis of relevant pesticides, antibiotics and mycotoxins in bakery products by liquid chromatography-single-stage Orbitrap mass spectrometry. *Food Addit. Contam. Part A Chem. Anal. Control. Expo. Risk Assess.* **2015**, *32*, 1617–1627. [CrossRef]

82. Narváez, A.; Rodríguez-Carrasco, Y.; Castaldo, L.; Izzo, L.; Ritieni, A. Ultra-high-performance liquid chromatography coupled with quadrupole Orbitrap high-resolution mass spectrometry for multi-residue analysis of mycotoxins and pesticides in botanical nutraceuticals. *Toxins* **2020**, *12*, 114. [CrossRef] [PubMed]

83. De Paepe, E.; Wauters, J.; Van Der Borght, M.; Claes, J.; Huysman, S.; Croubels, S.; Vanhaecke, L. Ultra-high-performance liquid chromatography coupled to quadrupole orbitrap high-resolution mass spectrometry for multi-residue screening of pesticides, (veterinary)drugs and mycotoxins in edible insects. *Food Chem.* **2019**, *293*, 187–196. [CrossRef]

84. Cladière, M.; Delaporte, G.; Le Roux, E.; Camel, V. Multi-class analysis for simultaneous determination of pesticides, mycotoxins, process-induced toxicants and packaging contaminants in tea. *Food Chem.* **2018**, *242*, 113–121. [CrossRef] [PubMed]

85. Martínez-Domínguez, G.; Romero-González, R.; Garrido Frenich, A. Multi-class methodology to determine pesticides and mycotoxins in green tea and royal jelly supplements by liquid chromatography coupled to Orbitrap high resolution mass spectrometry. *Food Chem.* **2016**, *197*, 907–915. [CrossRef] [PubMed]

86. León, N.; Pastor, A.; Yusà, V. Target analysis and retrospective screening of veterinary drugs, ergot alkaloids, plant toxins and other undesirable substances in feed using liquid chromatography-high resolution mass spectrometry. *Talanta* **2016**, *149*, 43–52. [CrossRef] [PubMed]

87. Dzuman, Z.; Zachariasova, M.; Veprikova, Z.; Godula, M.; Hajslova, J. Multi-analyte high performance liquid chromatography coupled to high resolution tandem mass spectrometry method for control of pesticide residues, mycotoxins, and pyrrolizidine alkaloids. *Anal. Chim. Acta* **2015**, *863*, 29–40. [CrossRef] [PubMed]

88. Gómez-Pérez, M.L.; Plaza-Bolaños, P.; Romero-González, R.; Martínez-Vidal, J.L.; Garrido-Frenich, A. Comprehensive qualitative and quantitative determination of pesticides and veterinary drugs in honey using liquid chromatography-Orbitrap high resolution mass spectrometry. *J. Chromatogr. A* **2012**, *1248*, 130–138. [CrossRef] [PubMed]

89. Kendall, H.; Clark, B.; Rhymer, C.; Kuznesof, S.; Hajslova, J.; Tomaniova, M.; Brereton, P.; Frewer, L. A systematic review of consumer perceptions of food fraud and authenticity: A European perspective. *Trends Food Sci. Technol.* **2019**, *94*, 79–90. [CrossRef]

90. Hrbek, V.; Rektorisova, M.; Chmelarova, H.; Ovesna, J.; Hajslova, J. Authenticity assessment of garlic using a metabolomic approach based on high resolution mass spectrometry. *J. Food Compos. Anal.* **2018**, *67*, 19–28. [CrossRef]

91. Fiorino, G.M.; Losito, I.; De Angelis, E.; Arlorio, M.; Logrieco, A.F.; Monaci, L. Assessing fish authenticity by direct analysis in real time-high resolution mass spectrometry and multivariate analysis: Discrimination between wild-type and farmed salmon. *Food Res. Int.* **2019**, *116*, 1258–1265. [CrossRef] [PubMed]

92. Godefroy, S.B.; Barrere, V.; Théolier, J.; Baker, R.C.; Zhang, G.; Hamilton, M.; Pellegrino, M.; Byrne, P.; Ben Embarek, P. Summary of the AOAC-Sponsored Workshop Series Related to the Global Understanding of Food Fraud (GUFF): Mobilization of Resources for Food Authenticity Assurance and Food Fraud Prevention and Mitigation. *J. AOAC Int.* **2020**, *103*, 470–479. [CrossRef] [PubMed]

 *foods*

MDPI

Review

Optical Screening Methods for Pesticide Residue Detection in Food Matrices: Advances and Emerging Analytical Trends

Aristeidis S. Tsagkaris *, **Jana Pulkrabova and Jana Hajslova**

Department of Food Analysis and Nutrition, Faculty of Food and Biochemical Technology, University of Chemistry and Technology Prague, Technická 5, Prague 6—Dejvice, 166 28 Prague, Czech Republic; pulkrabj@vscht.cz (J.P.); hajslovj@vscht.cz (J.H.)
* Correspondence: tsagkara@vscht.cz

Abstract: Pesticides have been extensively used in agriculture to protect crops and enhance their yields, indicating the need to monitor for their toxic residues in foodstuff. To achieve that, chromatographic methods coupled to mass spectrometry is the common analytical approach, combining low limits of detection, wide linear ranges, and high accuracy. However, these methods are also quite expensive, time-consuming, and require highly skilled personnel, indicating the need to seek for alternatives providing simple, low-cost, rapid, and on-site results. In this study, we critically review the available screening methods for pesticide residues on the basis of optical detection during the period 2016–2020. Optical biosensors are commonly miniaturized analytical platforms introducing the point-of-care (POC) era in the field. Various optical detection principles have been utilized, namely, colorimetry, fluorescence (FL), surface plasmon resonance (SPR), and surface enhanced Raman spectroscopy (SERS). Nanomaterials can significantly enhance optical detection performance and handheld platforms, for example, handheld SERS devices can revolutionize testing. The hyphenation of optical assays to smartphones is also underlined as it enables unprecedented features such as one-click results using smartphone apps or online result communication. All in all, despite being in an early stage facing several challenges, i.e., long sample preparation protocols or interphone variation results, such POC diagnostics pave a new road into the food safety field in which analysis cost will be reduced and a more intensive testing will be achieved.

Keywords: pesticide residues; optical detection; screening methods; point-of-care diagnostics; smartphones; biosensors; bioassays; food

Citation: Tsagkaris, A.S.; Pulkrabova, J.; Hajslova, J. Optical Screening Methods for Pesticide Residue Detection in Food Matrices: Advances and Emerging Analytical Trends. *Foods* **2021**, *10*, 88. https://doi.org/10.3390/foods10010088

Received: 15 November 2020
Accepted: 25 December 2020
Published: 5 January 2021

Publisher's Note: MDPI stays neutral with regard to jurisdictional claims in published maps and institutional affiliations.

1. Introduction

The ever-increasing demand for food production unfortunately still requires a widespread use of pesticides. According to the European Commission (EC), pesticides "prevent, destroy, or control a harmful organism ("pest") or disease, or protect plants or plant products during production, storage, and transport". Pesticides can be clustered on the basis of the target pest (Table 1), for example, compounds combating insects are called insecticides [1]. Another useful classification was proposed by the World Health Organization (WHO) and is based on hazard expressed as lethal dose (LD) in rat specimen (Table 1) [2]. Alternatively, pesticides can be classified focusing on how they enter into the target pest, for instance, systemic pesticides are absorbed by tissues (leaves, roots, etc.) (Table 1) [3].

Table 1. Summary of various classification systems for pesticides.

a. Based on Target Pest	
Pesticide Type	**Pest**
Algicide	Algae
Avicide	Birds
Bactericide	Bacteria
Fungicide	Fungi
Herbicide	Weeds
Insecticide	Insects
Miticide	Mites
Molluscicide	Snails
Nematicide	Nematodes
Piscicide	Fish
Rodenticide	Rodents

b. Based on Toxicity			
Type	**Toxicity Level**	**LD$_{50}$ for Rats (mg kg^{-1} Body Weight)**	
		Oral	**Dermal**
Ia	extremely hazardous	<5	<50
Ib	highly hazardous	5 to 50	50–200
II	moderately hazardous	50–2000	200–2000
U	unlikely to present acute hazard	>5000	

c. Based on the Way of Entry into a Pest	
Ways of Entry	**Details**
Systemic	Absorption by tissues such as leaves, stems, and roots
Non-systemic	Physical contact between the pesticides and the target organism
Stomach poisoning	Pesticide digestion
Fumigants	Target organism killing through vapors
Repellents	Inhibit the ability of pests to localize in crops

Regardless their classification, pesticide residues are related to toxicity issues, which can be either acute or chronic. The various pesticide classes can potentially affect their targets in different ways, including humans. In the case of organochlorine (OC) pesticides, which were extensively used during the 20th century, nervous system stimulation has been noticed. For example, lindane inhibits the calcium ion influx and Ca- and Mg-ATPase, causing release of neurotransmitters [4] and acting as a hormone disruptor causing both acute and chronic adverse effects ranging from dermal irritation or headache to cancer, Parkinson's disease, or deficit immune system [5]. In the case of carbamate (CM) and organophosphate (OP) insecticides, their toxicity is related to the inhibition of acetylcholinesterase (AChE), a vital enzyme in the neural system of insects or mammals, including humans. Normally, AChE hydrolyzes the neurotransmitter acetylcholine into choline and acetic acid, an essential reaction that enables the cholinergic neuron to return to its resting state after activation. However, AChE activity is reduced in the presence of CMs and OPs due to carbamylation or phosphorylation of the serine hydroxyl group in the enzyme active cite [6], respectively. This results in acetylcholine accumulation, which can lead to serious health problems, including respiratory and myocardial malfunctions [7]. Another example of pesticide toxicity it is the class of pyrethroid pesticides. Pyrethroids cause neuronal hyperexcitation, resulting in repetitive synaptic firing and persistent depolarization. Their molecular targets are similar in mammals and insects, and include voltage-gated sodium, chloride, and calcium channels; nicotinic acetylcholine receptors; and intercellular gap junctions [8]. Therefore, it is obvious that the presence of pesticide residues in food has to be strictly regulated and monitored to protect consumer health.

To achieve that, accurate, sensitive, and robust analytical methods are of indispensable importance to assure that pesticide residues in food matrices are efficiently controlled. Liquid chromatography–tandem mass spectrometry (LC–MS/MS) and gas chromatography–tandem mass spectrometry (GC–MS/MS) are commonly applied [9,10] in various matrices, e.g., fruits and vegetables [11], honey [12], rice [13], and food of animal origin [14], enabling wide linear ranges and limits of detection (LODs) down to the $\mu g\ kg^{-1}$ level. The use of triple quadrupole (QqQ) as the mass analyzer operating in the selected reaction monitoring (SRM) mode is the common way to detect for pesticide residues. However, at least two product ions are necessary for a compound identification while the ion ratio from sample extracts should be within $\pm 30\%$ of calibration standards from the same sequence (SANTE/12682/2019 guideline). Therefore, this requirement highlights a major drawback of SRM mode as the more pesticides included in the method, the more the necessary ion transitions that have to be measured. Thus, there is an increased chance of common or overlapped transitions affecting the method detectability [15]. To counter this problem, high-resolution mass spectrometry (HRMS) targeted methods have been proposed as an alternative [16–18]. Orbitrap, time-of-flight (TOF), and hybrid analyzers such as quadrupole-Orbirtap (q-Orbitrap) and quadrupole-TOF (qTOF) are used as the mass detectors, providing accurate mass measurement (<5 ppm), high resolution (more than 20,000 full width at half maximum (FWHM)), structural elucidation, and full MS scan capabilities (usually for the range 100–1000 Da). HRMS detectors resolve SRM-related problems, but there is still controversy on their quantification capabilities in comparison to QqQ methods. In any case, although chromatographic methods coupled to MS detectors provide the aforementioned merits, they are also time-consuming, laborious, and expensive methods that cannot be applicable by any laboratory around the world. Consequently, it is necessary to seek for alternatives able to combine sufficient detectability with cost-efficiency, simplicity, and applicability at the point of need.

In this way, screening methods have been introduced in food contaminant analysis featuring a great potential [9]. According to the Decision 2002/657/EC, "screening methods are used to detect the presence of a substance or class of substances at the level of interest". There are several methods fitting within this concept aiming to achieve rapid, selective, cost-efficient, and sensitive screening in the food safety field [19]. Such methods are usually based on bio-affinity interactions between selective biomolecules, e.g., antibodies [20] or enzymes [21], and pesticide residues, while biorecognition events are typically monitored by either optical or electrochemical transducers [22]. In fact, optical transduction systems correlate biorecognition events to a color development/change, indicating their user-friendliness. The potential of such optical screening methods can be enhanced by coupling them with smartphones to achieve ubiquitous biosensing [23]. As we comprehensively discussed in our recent study [24], unprecedented characteristics have been introduced into chemical analysis due to smartphones, such as online results or end-user implementation, and this can obviously impact pesticide residue analysis as well.

In this study, a comprehensive overview on optical screening methods used in pesticide residue analysis is presented, focusing on the period 2016–2020. To identify the analytical performance that screening methods need to attain, we provide a critical discussion on EU regulatory framework. In fact, pesticide residues set two great challenges that need to be urgently faced. Firstly, pesticide regulatory limits are quite low (see Section 3), meaning that the developed screening methods need to demonstrate sufficient detectability into food extracts. Secondly, multi-step sample preparation protocols are commonly utilized (see Section 4.1), increasing the total analysis time and eliminating the advantage of rapid analysis provided by screening methods. Last but not least, the emergence of smartphones as analytical detectors is discussed, highlighting the novel capabilities brought by this technology in the field.

2. Pesticide Residue Occurrence in Food Distributed in the EU

The European Food Safety Authority (EFSA) compiles yearly the EU report on pesticide residues in food, which contains data from the EU countries as well as Iceland and Norway. Therefore, pesticide residue monitoring is systematically performed, and a clear view of the applied testing is available. On the basis of the latest available data from the official EU reports [25–29], the vast majority of tested samples (always more than 95% of the samples, Figure 1) fell below the maximum residue levels (MRLs). However, although the tested samples were complied with regulatory requirements, there was a minor tendency of more samples be non-compliant during the last five reported years. In fact, the number of samples with non-quantifiable residues or contained residues within the legally permitted levels dropped from 97.1% in 2014 to 95.5% in 2018. This is likely related to (i) the slightly increased tested samples (about 83,000 samples were tested in 2014 while 91,000 samples were tested in 2018) and (ii) the globalization of food market, resulting in increased food imports from countries with different regulatory requirements. Worth noticing is that samples containing non-quantifiable amounts of pesticide residues are transferred to the labs and analyzed by expensive and time-consuming chromatographic methods underpinning the importance to implement screening methods into residue controlling. Obviously, the use of screening methods aims to assist instrumental analysis, resulting in rapid results and a better utilization of available recourses. Significantly, CM and OP residues have been commonly detected or even exceeded the MRLs. In fact, chlorpyrifos, carbofuran, dimethoate, acephate, profenofos, methomyl, methamidophos, and ethephon (all CM and OP insecticides) residues were among the compounds with the most frequent MRL exceedances [25–29]. Chlorpyrifos, an OP compound, was steadily within the top five pesticide residues with the most exceedances (except in 2017, when it was reported in ninth place), whilst in the latest report, chlorpyrifos was the compound with the most exceedances of its acute reference dose (ARfD). In this way, an official ban has been recently applied in the EU due to concerns predominantly related to neurotoxicity issues [30]. This fact can also explain why there is a variety of screening methods measuring CM and OP residues (see Section 4.2).

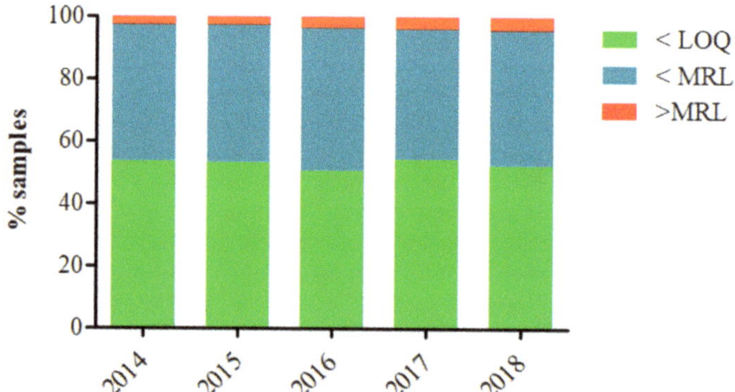

Figure 1. Temporal evaluation of the percentage samples that contained (i) no quantifiable residues (<limit of quantification, LOQ), (ii) residues at or below maximum residue levels (MRLs), and (iii) residues at a higher concentration than MRLs. The depicted data are extracted from the official EU reports on pesticide residues in food [25–29].

3. EU regulatory Requirements on Pesticide Residues

The EU regulatory framework related to pesticide residues is comprehensively set. In detail, MRLs for about 1100 pesticides in 300 different matrices has been established according to the EC Regulation 396/2005. To navigate and find the regulatory limits for a selected analyte, an online database has been developed permitting regulatory levels export in an excel file format (https://ec.europa.eu/food/plant/pesticides/eu-pesticides-database/mrls/?event=search.pr, last accessed 23 December 2020). However, although EU MRLs are established for unprocessed food, there are no EU MRLs for processed or composite foodstuffs. The Food and Agriculture Organization of the United Nations and the World Health Organization (FAO/WHO) have included MRLs for selected processed food in the Codex Alimentarius (http://www.fao.org/fao-who-codexalimentarius/codex-texts/dbs/pestres/commodities/en/, last accessed 18 August 2020). A similar approach has also been followed by the German Federal Institute for Risk Assessment (BfR), which provides an online tool for MRL calculation in processed food (https://www.bfr.bund.de/cm/349/bfr-compilation-of-processing-factors.xlsx, last accessed 18 May 2020). In case that there is no MRL for a pesticide, then a default 0.010 mg kg^{-1} limit is set; moreover, the default MRL is also used for infant food according to the Directive 2006/141. Infants (up to 12 months old) and young children (1 to 3 years old) are quite sensitive towards residues since their body weight is low and they face a greater risk when consuming a contaminant compared to an adult individual. Regarding the cumulative risk assessment, this is a major issue since the MRLs are prescribed for single residues, but food may be contaminated with multiple pesticide residues. In this context, a large amount of effort has been devoted to establish guidelines and a step towards this direction was an online tool called "Acropolis" developed by the National Institute for Public Health and the Environment for the Netherlands (RIVM) [31]. It is noteworthy that although the EFSA cannot set any regulatory requirements, its opinion is highly anticipated by the European Commission to prescribe any regulations. Undoubtedly, the legislation application is directly linked to the analytical capabilities and the quality assurance of the provided results.

4. Pesticide Residue Optical Screening in Food Matrices

The detection of pesticide residues is a great analytical challenge considering their diverse physicochemical characteristics and the numerous combinations of analyte-matrix. In addition, using optical screening methods pose further challenges, as in contrast to instrumental analysis, such methods sometimes face specificity, sensitivity, or robustness problems. In the following paragraphs, a critical discussion on sample preparation, optical screening methods, and their coupling to smartphones is provided to monitor the readiness of this upcoming technology in the pesticide residue analysis.

4.1. Sample Preparation

Sample preparation is a key step towards specific, sensitive, and accurate detection of pesticide residues. In the case of screening methods, high-throughput (in terms of tested samples) and short analysis duration need to be achieved while detectability should also be satisfactory (attained LODs lower than MRLs). Nevertheless, pesticide residues are commonly extracted using organic solvents and long sample preparation protocols. This is a major challenge for screening methods as they usually exploit selective biomolecules that have certain tolerance towards organic solvents (typically used as pesticide residue extractants). In fact, after a certain organic solvent content (commonly 20–30%) biomolecules are denaturized and lose their functionality, for example catalytic activity in the case of enzymes. Therefore, there have been efforts to extract pesticide residues using aqueous buffers, e.g., phosphate-buffered saline (PBS), since such solutions can adjust the pH value, which is vital for the proper biomolecule function. Sample incubation or mixing with a buffer, followed by a filtration to reduce matrix interferent compounds is a simple procedure that can be applied when using screening methods. Obviously, accuracy and/or detectability can be affected by such simplified sample preparation (due to co-isolated matrix compounds), underlying the need for highly selective recognition elements. It is worth noting that the emergence of paper analytical devices can provide a solution in this problem. Paper matrix can be used as an evaporation platform due to its large specific surface enabling air–liquid contact, which speeds up organic solvent evaporation easily [32] (Figure 2a). Therefore, extraction using organic solvents followed by paper-based solvent evaporation and then addition of the recognition element can be applied to face this challenge. Another practical and cost-efficient solution was recently published [33], in which adhesive tape (Figure 2b) was stuck to a vegetable surface, peeled off, and dipped into a water–methanol solution achieving a LOD around 0.20 µM (0.066 mg kg^{-1}) for malathion depending the tested matrix. In any case, there are still screening methods that use sample preparation protocols commonly applied in instrumental analysis, for example, quick easy cheap effective rugged and safe (QuEChERS) extraction [34,35] to achieve a better analytical performance. Unfortunately, the use of multi-step sample preparation protocols in pesticide residue screening methods remains a bottleneck.

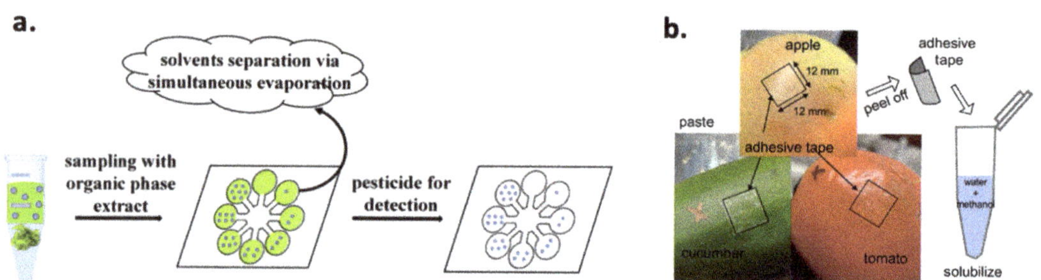

Figure 2. (**a**) Paper-based organic solvent evaporation for pesticide residue screening using enzymatic recognition. Reproduced with permission from [32]. (**b**) A simple and cost-efficient sample preparation protocol using an adhesive tape and a water–methanol solution to extract pesticides from fruit and vegetable peels. Reprinted with permission from [33]. Copyright 2020 American Chemical Society.

4.2. Optical Screening Methods

4.2.1. Biochemical Assays

Biochemical assays using antibodies or enzymes as recognition elements have been traditionally used in a microplate format, which provides high-throughput, simplicity, good sensitivity, and ease of operation. The enzyme-linked immunosorbent assay (ELISA) is a striking example of such bioassays. ELISA is based on the specific interaction between an enzyme-labelled analyte-specific antibody and its antigen. Owing to the labelling of the antibody with an enzyme, upon the addition of a substrate, a measurable color change is initiated. A recent review by Wu et al. [36] is recommended for a deeper understanding of the ELISA mechanism, various types (Figure 3a), as well as recent advances. ELISAs have been developed for the screening of various pesticide residues in food matrices, for example, OPs [37,38], CMs [39], neonicotinoids [40], or fungicides [41]. In terms of cholinesterase microplate assays, cholinesterases have been employed as recognition elements (both AChE [42] and butyrylcholinesterase, BChE [43]) to screen for CM and OP. Considering that, in vitro, cholinesterases hydrolase colorless substrates to colored products, the presence of CMs and OPs can be correlated to a color decrease similarly to competitive ELISAs. A great variety of substrates, resulting in different colored products (Figure 3b), have been used including acetylthiocholine and butyrylthiocholine halides for AChE and BChE, respectively; indoxyl acetate; α-naphthyl acetate; 2,6-dichloroindophenol acetate; and others [44]. Importantly, reduced sample and reagent consumption (typically less than 100 µL) as well as low LODs at the µg kg^{-1} level [42,45,46], depending on the matrix, were achieved by cholinesterase microplate assays. However, biochemical assays are still applicable in laboratories as they require certain apparatus and well-trained operators (commonly such assays contain multiple steps).

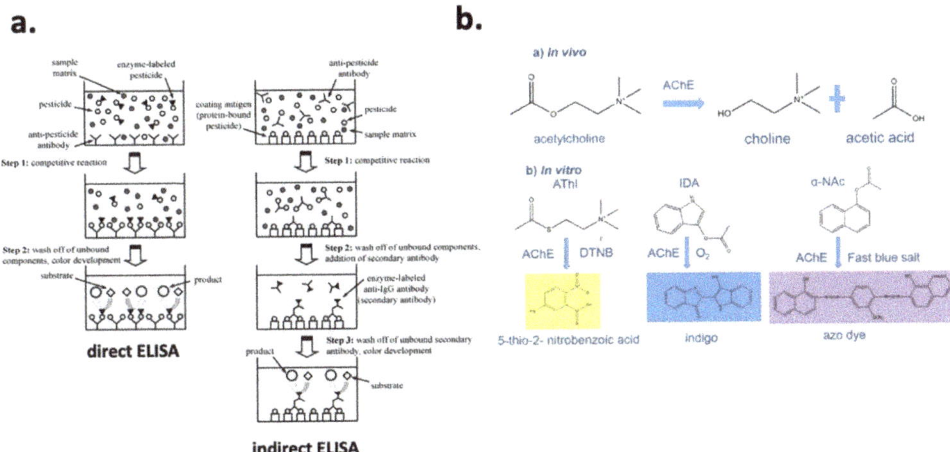

Figure 3. (**a**) Multistep direct and indirect ELISA protocols for pesticide residues screening. Reprinted with permission from [47]. Copyright 2013 American Chemical Society. (**b**) In vivo and in vitro acetylcholinesterase hydrolytic activity producing, in vitro, various colored products depending the catalyzed substrate. Reprinted from [42] under CC BY 4.0.

4.2.2. Biosensors

Biosensors are analytical platforms that convert a biological response into a quantifiable and processable signal. Besides the described attractive characteristics of biochemical assays, biosensors can be miniaturized and automated, indicating their potential for on-site testing. On the basis of the biorecognition element, we can distinguish three main groups of biosensors, i.e., immunosensors [20], cholinesterase [21] and lipase sensors [48] (enzymatic recognition), and aptasensors [49,50]. It is of note that aptamers emerge as an alternative to counter problems related to antibodies, such as the challenge to trigger an immune response for small molecules or their higher temperature stability, a problem related to biomolecules [51]. Biomolecules can be negatively affected by organic solvents (e.g., denaturation problems resulting in decreased activity), certain pH values (commonly neutral pH values are the optimum for antibodies and enzymes), or hydrostatic and osmotic pressure. Nevertheless, increased stability can be accomplished by immobilizing biomolecules on surfaces as in the case of biosensors [52]. For instance, the immobilization of AChE on cellulose strips resulted in retained enzyme activity over a two-month period [34]. Other less used recognition elements include, but are not limited to, molecularly imprinted polymers (MIPs, synthetic molecules), cells, and DNA probes. In the following paragraphs, further discussion on various biosensors is provided on the basis of the detection principle used, and tables summarizing interesting publications in the field during the period 2016–2020 are presented.

Colorimetric Biosensors

Colorimetry is probably the simplest approach as a biorecognition event is related to a color development. This fact significantly increases colorimetric platforms potential for on-site analysis as colorimetric signals can be monitored even by the naked eye or they can be easily coupled to a smartphone readout (see Section 4.3). On the downside, colorimetric signals are vulnerable to minor lighting variations while most of the food extracts are colored, which negatively effects method detectability. Of importance is the ever-increased use of analytical platforms commonly based on colorimetric responses such as membrane-based assays (lateral flow (LF) or paper-based assays), microfluidic chips, or lab-on-a-chip (LOC) devices (Table 2). LF assays are membrane tests consisting of various polymeric zones on which various substances can be accommodated and react with an analyte [53].

Liquid samples or extracts containing an analyte move through this lateral device due to capillary forces. Two different formats of LF assays can be distinguished, namely, competitive and sandwich formats. Competitive assays are used for low molecular weight analytes, i.e., pesticide residues, and a positive result is related to the absence of a test line due to the blocking of antibody binding sites to protein conjugates by the analyte. In terms of big molecules, for example, allergens, the sandwich format is used, and the analyte is immobilized between two complementary antibodies. Besides research studies using LF assays for pesticide residue screening [54,55], LF assays are one of the few cases that have reached the commercialization stage [19]. Regarding microfluidics, this is a relatively new field that was established in 2006 following the publication of G.M Whitesides in the prestigious *Nature* journal [56]. In this way, microfluidics are related to the manipulation of fluids in channels with dimensions of tens of micrometers. Fluidic behavior under these micro-level confined regions significantly differs from fluidic behavior in the macroscale. In this context, essential parameters such as viscosity, density, and pressure need to be strictly controlled to reach optimum microfluidic performances [57]. Although no strict criteria have been proposed to define microfluidic systems, the length and internal size of the channels is considered of critical importance. Microfluidic channels are combined to LOC devices to develop fully portable and autonomous analytical platforms. In fact, LOC systems are able to mimic different apparatus such as reactors and pumps to carry out injection, filtration, dilution, and detection in a reduced portion, eliminating handling errors and enhancing robustness while retaining the analysis cost low [58]. Regarding the application of colorimetric microfluidic and LOC platforms, paper-based microfluidics can combat problems related to intolerance towards organic solvents that are used to extract pesticide residues by spontaneous evaporation on the paper-platform before loading an enzyme solution for pesticide recognition [32]. However, overall, such platforms are still in an early stage, with the majority of the studies focusing on proof-of-concept applications [59]. Unfortunately, the majority of colorimetric analytical platforms utilize traditional sample preparation protocols, highlighting the need to automate and simplify sample pretreatment to increase the applicability of such methods in the field.

Table 2. Selected studies on pesticide residue screening using colorimetric biosensors.

Analyte	Matrix	Analytical Platform	Sample Preparation	LOD	EU MRL	Reference
Methyl-paraoxon and chlorpyrifos-oxon	cabbage and dried mussel	paper-based device coated with nanoceria using an enzyme inhibition assay with AChE and ChOX	methanol vortex extraction, centrifugation, PSA clean-up, centrifugation, evaporation	0.040 mg kg^{-1}	0.010 mg kg^{-1}	[60]
Carbofuran and carbofuran-3-hydroxy	water	LF immunoassay	none	7 µg L^{-1} (carbofuran) and 10 µg L^{-1} (carbofuran-3-hydroxy)	0.1 µg L^{-1}	[54]
Malathion	apple	aptasensor employing gold nanoparticles	methanol extraction, filtered and evaporation	5.2 pM (or 0.001 µg kg^{-1})	0.02 mg kg^{-1}	[61]
Paraoxon	vegetable irrigation water	enzyme cascade and iodine starch color reaction	filtration	10 µg L^{-1}	n.a.	[62]

Table 2. *Cont.*

Analyte	Matrix	Analytical Platform	Sample Preparation	LOD	EU MRL	Reference
Ethoprophos	tap water	gold nanoparticle aggregation combined to adenosine triphosphate	no	4 µM (or 0.96 mg L^{-1})	0.1 µg L^{-1}	[63]
Paraoxon	rice and cabbage	AChE assay coupled to carbon dots	acetonitrile ultrasonic extraction, centrifugation, filtration through sodium sulfate and evaporation	0.005 mg kg^{-1}	0.01 mg kg^{-1} (cabbage) and 0.02 kg^{-1} (rice)	[64]
Acetamiprid	spinach	aptamer with DNA probe	ethanol ultrasonic extraction, centrifugation, filtration, and 20-times dilution	0.1 nM (or 0.022 µg kg^{-1})	0.6 mg kg^{-1}	[65]

Fluorescent Biosensors

Biosensors with fluorescent detection combine the selectivity provided by the recognition part to the sensitivity of fluorescence (FL), as it is a zero-background method and only specific compounds (based on their structure) are able to fluoresce. Fluorescent biosensors (Table 3) are based on the principle that the interaction of a fluorescent probe (chemical or physical) with an analyte leads to either fluorescence enhancement or quenching [66], which is also known as analyte-induced "on–off" fluorescent behavior [67]. A great variety of fluorescent probes have been used, namely, fluorescent dyes, nanocomposite materials, rare earth elements, or semiconductors [68]. The great advancements in nanomaterial field have further improved fluorescent detection, as they have countered, at a certain extent, bottlenecks related to dyes, e.g., high photobleaching. Quantum dots, which are semiconductor crystalline nanomaterials with unique optical properties due to quantum confinement effects, are an example of nanocomposite probes that have enhanced fluorescent detection for pesticide residue screening [66]. This was recently demonstrated for the detection of four OP pesticides, namely, paraoxon, dichlorvos, malathion, and triazophos, using CdTe quantum dots as the fluorescent probe coupled to an AChE-choline oxidase enzyme system [69]. In this case, when AChE was active (resulting in choline production), H_2O_2 was produced by choline oxidase, which in turn "turned off" the FL of the CdTe quantum dots. However, in the presence of an OP, the FL induced by CdTe quantum dots was retained and a correlation between OP concentration and FL signal was feasible. Impressively, a LOD of 0.5 ng mL^{-1} was achieved in water, tomato juice, and apple juice, while the fluorescent biosensor could be regenerated using pyridine oximate. In another study, an "off–on–off" strategy was applied by using AChE as the recognition element and lanthanide-doped upconversion nanoparticles (UCNPs) with Cu^{+2} as the fluorescent probe [70]. This analytical platform achieved an LOD of 0.005 mg kg^{-1} for diazinon detection in apple and tea powder and, importantly, the results were cross-confirmed to GC–MS. It should be kept in mind that although it is necessary to benchmark the results attained using screening methods, this practice is commonly omitted in the published literature as it is comprehensively discussed in our previous study [9]. In conclusion, FL biosensors can attain sensitive results, which is extremely important in the food safety field. However, their principles and analytical configuration are commonly more complicated than colorimetric platforms that may influence their applicability within the point-of-care (POC) testing concept.

Table 3. Selected studies on pesticide residue screening using fluorescent biosensors.

Analyte	Matrix	Analytical Platform	Sample Preparation	LOD	EU MRL	Reference
Acetamiprid	tea	aptasensor	methylene chloride extraction, filtration, and evaporation	0.002 mg kg^{-1}	0.05 mg kg^{-1}	[71]
Dichlorvos	cabbage and fruit juice	carbon dots–Cu(II) system	PBS extraction	0.84 ng mL^{-1}	n.a.	[72]
Paraoxon	water	BChE assay	no	0.25 µg L^{-1}	0.1 µg L^{-1}	[73]
Imidacloprid	Chinese leek, sweet potato, and potato	LF immunoassay	PBS extraction and supernatant dilution with PBS	0.5 ng g^{-1}	0.5 mg kg^{-1}	[74]
Diazinon	cucumber and apple	aptasensor	Dilution with water, water-heated bath, centrifugation	0.13 nM (0.039 µg kg^{-1})	0.01 mg kg^{-1}	[75]
Aldicarb	ginger	AChE-based assay	QuEChERS	100 µg kg^{-1}	0.05 mg kg^{-1}	[76]
Eight rodenticides	wheat	LF immunoassay combined with quantum dots	acetonitrile ultrasonic extraction, centrifugation, filtration, and filtrate 10-times dilution in PBS	1–100 µg kg^{-1} depending the analyte	0.01 mg kg^{-1}	[77]

Surface Plasmon Resonance Biosensors

Surface plasmon resonance (SPR) biosensors are based on an optical phenomenon that happens on a thin conducting film at the interface between media of different refractive index [78]. SPR provides label-free sensing, which is a great advantage as labeling procedures are omitted, resulting in reduced cost and prevention against false positive signals related to labeling. Moreover, SPR is especially useful to calculate association (or dissociation) kinetics and affinity constants or bounded analyte content in the case of immunorecognition [79]. Interestingly, only a few enzyme-based biosensors have employed SPR detection [80]. Detecting pesticide residues in trace amounts is a challenging task as it is difficult to attain a measurable change in the refractive index due to their low molecular mass. To face this problem, sensor surface modification using nanoparticles is commonly applied since nanomaterials can enhance SPR signals due to their high refractive index. Furthermore, nanomaterials are also preferred because of their facile synthesis, high surface to volume ratio, and high biocompatibility and photostability [81]. The nanomaterials commonly utilized in such analytical platforms include, but are not limited to, metal nanoparticles, i.e., Au or Ag; carbon nanoparticles; and quantum dots. Besides signal enhancement using nanomaterials, SPR phase-measurement instead of amplitude (which is the case in conventional SPR systems) is an alternative approach that is based on the topological nature of the phase of a system. Considering that our study focuses on the analytical developments and applications in pesticide residue analysis, no further discussion on the physics behind phase sensitive SPR measurement is provided, and two studies [82,83] are recommended for a deeper understanding of the phenomenon. In any case, SPR biosensors have found several applications in pesticide residue analysis based mainly on immunorecognition (Table 4). It can be noticed that the problem of laborious sample preparation when analyzing solid food matrices was also the case for SPR-based biosensors. In addition, the low molecular weight of pesticides set a great challenge in terms of detectability and compliance to regulatory limits for SPR-based analytical platforms. More effort is definitely needed to further improve such platforms, considering the miniaturization potential (handheld SPR systems or coupling to smartphones) [84] that can be highly beneficial for the field.

Table 4. Selected studies on pesticide residue screening using surface plasmon resonance (SPR) biosensors.

Analyte	Matrix	Analytical Platform	Sample Preparation	LOD	EU MRL	Reference
Parathion	cabbage washing solutions	AChE + SPR	The spiked cabbage sample was washed with 30 mL of distilled water twice	0.069 mg L^{-1}	n.a.	[85]
Profenofos	water	fiber optic sensor based on MIP recognition	No sample preparation	0.02 μg L^{-1}	0.1 μg L^{-1}	[86]
Triazophos	cabbage, cucumber, apple	immunosensor	QuEChERS, 10-times dilution for cabbage and cucumber 20-times dilution for apple	0.1 μg kg^{-1} (cabbage and cucumber) and 0.4 μg kg^{-1}	0.01 mg kg^{-1}	[87]
Carbendazim	medlar	immunosensor with Au/Fe$_3$O$_4$ nanocomposite probe for SPR signal enhancement	80% methanol extraction, centrifugation, dilution with PBS to 5% methanol	5 ng mL^{-1} in the extract (there is no information about sample weight)	0.01 mg kg^{-1}	[88]
Chlorothalonil	lettuce, cabbage, onion	immunosensor	Methanol extraction, centrifugation, 8.5 times dilution to 10% methanol	1 mg kg^{-1}	0.6 mg kg^{-1} (cabbage) and 0.01 mg kg^{-1} (lettuce, onion)	[89]
Chlorpyrifos	maize, apple, cabbage, medlar	immunosensor	80% methanol extraction, supernatant diluted 10-times with PBS	0.0025 mg kg^{-1}	0.01 mg kg^{-1} (apple, cabbage, medlar) and 0.05 mg kg^{-1} (maize)	[90]

Surface-Enhanced Raman Spectroscopy

Although some consider surface-enhanced Raman spectroscopy (SERS) as an optical biosensor due to its coupling to biorecognition events [20], SERS is in principle a spectroscopic method based on light scattering, specifically to inelastic collisions occurring between a sample and incident photons emitted by a monochromatic light source, such as a laser beam [91]. Combining biorecognition events to SERS can significantly enhance the analytical performance of such methods, but also it increases method complexity and cost. For example, a multiplexed immunochromatographic assay for the simultaneous detection of cypermethrin and esfenvalerate (pyrethroid pesticides) achieved impressive results in milk matrix [92]. Specifically, the acquired LOD was at the parts per trillion level (LOD = 0.005 ng mL^{-1}), a performance that would not be possible without using SERS-based detection considering that immunochromatographic assays mostly provide qualitative results. Regarding direct SERS screening, this is feasible as molecules provide specific Raman spectra due to their unique structure, which is also called "Raman fingerprint". However, Raman signals are not strong enough, with only 1 out of 10 million of the scattered photons experiencing Raman scattering when incident light interacts with an analyte [93]. Therefore, it is necessary to enhance such signals by employing nanocomposite substrates resulting in electromagnetic and chemical enhancement [94]. Two different types of substrates can be distinguished, namely, colloidal and solid substrates. Although the synthesis of colloidal substrates such as Ag or Au nanoparticles is quite facile and cost-effective, poor reproducibility of signals remains a problem [95]. In terms of solid substrates, these provide more robust signals and counter the risk of nanoparticle aggregation, which is a problem for colloidal substrates. Solid substrates can be immobilized on various surfaces for example paper [96] or hydrogels [97]. In fact, paper-based SERS substrates can further increase the method potential to be applied on-site as such substrates can be used to swab the surface of a sample and then screen using a portable Raman spectrometer. In this way, paper SERS substrate coated with a monolayer of core-shell nanospheres was recently developed and was successfully used for the detection of thiram in orange juice [98]. This simple and non-destructive method achieved a LOD of 0.25 μM or 0.060 mg L^{-1} by using 4-methylthiobenzoic acid (4-MBA) as the internal standard (IS) to attain quantitative results. Similarly, in another study, 4-MBA was accommodated in Au@Ag nanocubes and exploited as the IS [99]. Moreover, it was noticed that water molecules can be used as a IS since their Raman scattering signal is quite stable [100]. Alternatively, the use of anisotropic nanoparticles, e.g., nanocubes, nanorods, and nanostars, positively affected SERS quantification capabilities by achieving more stable signals [101]. Nevertheless, SERS can mostly detect analytes on the surface of food, which does not correspond to the whole amount of a pesticide in a food matrix. Pesticide residues depending their polarity can be found in the non-polar peel or the polar-aquatic inner part of a fruit. Moreover, LODs have been mostly expressed using the "ng cm^{-2}" unit [102] because pesticide residues were measured on a surface. Nevertheless, such a concentration expression is not in line to the regulated MRL units (mg kg^{-1}). There were also cases in which QuEChERS extraction [103] or other long sample preparation protocols (Table 5) were used prior to SERS screening, an approach that comes in contrast to the non-destructive and direct measurements than can be acquired using SERS. In conclusion, SERS can highly improve the current status of pesticide residue screening at the point of need due to the discussed merits and the ever-decreased price of such portable platforms (approximately EUR 35,000 to 50,000 at the moment).

Table 5. Selected studies on pesticide residue screening using SERS methods.

Analyte	Matrix	Analytical Platform	Sample Preparation	LOD	EU MRL	Reference
Methyl parathion	apple	portable SERS	none	0.011 μg cm^{-2}	0.010 mg kg^{-1}	[102]
Prometryn and simetryn	wheat and rice	MIP-SERS	QuEChERS	20 μg·kg^{-1}	0.010 mg kg^{-1}	[103]
Thiram	lemon	SERS with nanowire Si paper as a substrate	none	72 ng cm^{-2}	0.100 mg kg^{-1}	[104]

Table 5. *Cont.*

Analyte	Matrix	Analytical Platform	Sample Preparation	LOD	EU MRL	Reference
Difenoconazole	pak choi	portable SERS	acetonitrile extraction, centrifugation, dSPE clean-up, evaporation, and reconstitution to ethyl acetate	0.41 mg kg^{-1}	2.0 mg kg^{-1}	[105]
Paraquat	apple and grape juice	portable SERS	none	100 nM (0.025 mg L^{-1})	n.a.	[106]
Dimethoate	olive leaves	portable SERS	none	5×10^{-7} M	n.a.	[107]
Edifenphos	rice	SERS	two times acetone extraction, centrifugation; six times pre-concentration	0.1 mg kg^{-1}	0.01 mg kg^{-1}	[108]
Thiram	apple, pear, and grape	"drop-wipe-test" using portable SERS	none	5 ng cm^{-2}	5 mg kg^{-1} (apple and pear) and 0.1 mg kg^{-1} (grape)	[109]

4.3. Coupling Optical Screening Methods to Smartphones

As already discussed in the previous paragraphs, the analytical signal of optical screening methods, especially in the case of colorimetry, is a simple and user-friendly indication of pesticide residue presence in food matrices. In terms of biochemical assays, such signals are commonly monitored using benchtop instruments, for example, absorbance readers, to acquire semi-quantitative or quantitative data. Regarding biosensors, these analytical platforms can also be handheld, providing on-site results, which can be extremely useful for detecting pesticide residues in imported foodstuff at the control point, i.e., border controls or at the field testing. Nevertheless, optical biosensors usually attain either qualitative results on the basis of visual inspection of the tested assay or semiquantitative results using readers, e.g., readers for LF assays, which significantly decrease the portability potential of such analytical platforms.

To face this challenge and introduce further unprecedented characteristics, smartphones have emerged as an alternative analytical detector combined to bioassays [23,110]. In principle, smartphone camera can be used as an optical biosensor to record images or videos containing the analytical useful information, enabling result semi-quantitation. Moreover, on-site one-click results exploiting smartphone computing power are feasible using smartphone apps. Interestingly, these results can be instantly communicated due to the online connectivity provided by smartphones as well as geo-located, potentially creating heatmaps during an outbreak situation. Such an option could be extremely useful during the fipronil insecticide scandal in 2017 (https://edition.cnn.com/2017/08/10/health/europe-egg-scandal-contamination-arrests/index.html, last accessed 8 November 2020), when egg farms in the Netherlands violated the regulatory limits and supplied contaminated eggs in the EU market. Actually, the available analytical scheme posed itself a key challenge during the fipronil scandal. In detail, samples needed to be collected; transported; marked with a unique laboratory code to assure traceability; and finally analyzed using instrumental analysis, in this case chromatographic methods [111]. The response in this health threat for the EU consumers would be totally different if smartphone assays were available at that moment. Smartphone assays could be used for an initial on-site screening, omitting the collection and transportation steps, generating instantly a sample ID, and providing a screening result with a certain false positive/false negative rate. In other words, smartphone-based analysis can assist the current analytical scheme by accelerating processes and sending only suspected samples to the lab.

Unfortunately, smartphone-based analysis has not yet reached such a technology readiness level (TRL) to be actively implemented into the analytical scheme. The majority of studies focus on proof-of-concept results (Table 6), with insufficient application on food matrices, especially in the case of solid food [24]. This is mostly related to the laborious sample preparation protocols that are necessary to extract pesticides from food matrices, mostly fruits and vegetables. Obviously, combining pocket-sized analytical platforms to laboratory protocols minimizes their actual portability potential and drives the field to the so called "chip-in-a-lab" era [112]. Chip-in-a-lab is a term used to describe the development of POC platforms that are unable to operate without the complementary use of certain laboratory equipment. In our view, the development of micro total analysis systems (μTAS) enabling integrated sample preparation is a necessity for field-ready and consumer-focused diagnostics [113]. To date, there is a lack of such systems, especially in the case of solid food matrices for the vast majority of analytes. Recently, a smartphone-based platform providing a sampling-to-result solution was developed for multiplex allergen detection in cookies [114]. This platform integrates a completed analytical protocol on the device, which can be even applied by non-experts following simple instructions. Undoubtedly, such an approach paves the road for smartphone diagnostics in food analysis. Additionally, the use of prototype 3D-printed apparatus pinpoints the significance of implementing 3D printing into chemical analysis. Another significant bottleneck is result ruggedness when using different smartphone models. Indeed, smartphone-based analytical platforms are mostly coupled to a specific device questioning whether comparable results can be obtained with a different smartphone model [115]. In terms of the analytical signal used in smartphone-based optical assays, various approaches have been utilized, specifically the RGB color space [43], other color spaces (i.e., HSV or CIE-Lab) [116], and random combination of color spaces based on algorithms [117] or barcodes [118]. In general, there has not been a clear conclusion on which is the most useful approach, but RGB is the smartphone primary color space and thus can be directly used without the need of mathematical transformation as in the case of other color spaces. It is also unclear as to whether it is necessary to use auxiliary attachable parts such as 3D-printed elements [34] to standardize optical conditions or record under ambient light using correction algorithms [119]. Overall, smartphone-based pesticide residue analysis is at an early stage and further developments are definitely expected, indicating this technology potential to revolutionize the field.

Table 6. Selected studies on pesticide residue screening using smartphone-based methods.

Analyte	Matrix	Analytical Platform	Sample Preparation	LOD	EU MRL	Reference
Chlorpyrifos, diazinon, and malathion	spinach, lettuce, and cabbage	LF multiplex aptasensor	homogenization and homogenate filtration	0.010 mg kg^{-1}	0.01 to 0.5 mg kg^{-1}, depending the analyte matrix	[120]
Carbofuran	apple	hybrid paper-LOC prototype	QuEChERS and evaporation	0.050 mg kg^{-1}	0.001 mg kg^{-1}	[34]
Chlorpyrifos methyl	cabbage	chemiluminescent enzyme origami paper-based biosensor	mixing with water and centrifugation	0.6 mM (193 mg kg^{-1})	0.01 mg kg^{-1}	[121]

Table 6. Cont.

Analyte	Matrix	Analytical Platform	Sample Preparation	LOD	EU MRL	Reference
Acetochlor and fenpropathrin	corn, apple, and cabbage	multiplex LF immunoassay	PBS 0.05% Tween-20 and 10% methanol extraction, centrifugation, dilution	6.3 ng g^{-1} (acetochlor) and 2.4 ng g^{-1} (fenpropathin)	0.010 mg kg^{-1}	[122]
Chlorpyrifos	fruit and vegetable wash water	lipase paper-based device		65 ng mL^{-1}	n.a.	[55]
Methyl paraoxon	pear	nanoceria-based assay	ethyl acetate ultrasonic extraction, centrifugation, and evaporation	0.060 mg kg^{-1}	0.010 mg kg^{-1}	[123]
2,4-Dichlorophenoxyacetic acid	water	ELISA in 3D-printed device	no	1 µg L^{-1}	0.1 µg L^{-1}	[124]

5. Conclusions

A critical overview of the developed optical methods for pesticide residue screening is comprehensively presented. Importantly, yearly reports on the occurrence of pesticide residues in the food chain as well as well-established regulation are available in the EU. Pesticide residue monitoring and control are strictly related to the available analytical methods, which need to attain low LODs, high accuracy, and ruggedness. These performance characteristics are provided up to date by chromatographic methods coupled to MS detectors. However, there is an intensive research effort to establish more optical screening methods able to assist instrumental analysis and face challenges related to their high cost, laborious protocols, and necessity of highly trained users. Thus, various biochemical assays and biosensors based on optical detection have been developed during the last five years. Sample preparation using common laboratory protocols, for example, QuEChERS, remains a bottleneck that limits the current applicability of POC screening methods, indicating the need to develop fully integrated µTAS. Nevertheless, sometimes such protocols are the only way to satisfactory extract pesticide residues from complicated food matrices. Assay sensitivity and selectivity are critical performance characteristics that need to be always assessed. In this way, LODs must be attained in the tested food matrix and not in buffer solutions, which was the case in few cases. Acquiring LODs in buffer is useful during method optimization to monitor the optimum assay performance and test parameters, for example, enzyme substrate concentration. In terms of assay selectivity, this is also a crucial performance characteristic as biorecognition elements may be affected by other compounds with structure similar to analytes. A characteristic example of this is AChE, an enzyme widely utilized in bioanalytical methods for pesticide residue screening. Although both CM and OP pesticides inhibit AChE activity, their inhibitory potency highly varies depending on their structure. Therefore, cross-reactivity studies are of indispensable importance to monitor bio-affinity interactions and determine potential interfering compound effect on assay performance. Additionally, the absence of result confirmation using instrumental analysis is another challenge since screening results need to be verified. In terms of optical detection, colorimetry is the simplest and most user-friendly detection system, but FL, SPR, and SERS can usually provide more sensitive results due to their selectivity and combination to nanomaterials. In these cases, nanomaterials enhance the optical properties of detection systems proving their indispensable importance for POC diagnostics.

Portable handheld SERS devices can further improve on-site pesticide residue detection at the point of need without the need of sample preparation. On-site screening can also be achieved by hyphenating optical screening assays to smartphones for ubiquitous sensing. Smartphone-based pesticide residue analysis can be extremely useful at border controls, considering the ever-increased globalization of the food market or at the field testing. To achieve that, however, sufficient detectability and a minimum false negative rate need to be achieved. Moreover, interphone result variation is a key parameter that has to be investigated more as most of the smartphone-based studies are applicable on a specific smartphone. In any case, the hyphenation of screening methods to smartphones is a step towards the "democratization" of chemical analysis and the introduction of new era, in which sensing is not strictly related to laboratories.

Author Contributions: Conceptualization, A.S.T.; methodology, A.S.T.; resources, J.P. and J.H.; writing—original draft preparation, A.S.T.; writing—review and editing, A.S.T., J.P., and J.H.; supervision, J.H.; project administration, J.P. and J.H.; funding acquisition, J.P. and J.H. All authors have read and agreed to the published version of the manuscript.

Funding: This project has received funding from the European Union's Horizon 2020 research and innovation program under the Marie Sklodowska-Curie grant agreement no. 720325, FoodSmartphone.

Conflicts of Interest: The authors declare no conflict of interest. The funders had no role in the design of the study; in the collection, analyses, or interpretation of data; in the writing of the manuscript; or in the decision to publish the results.

References

1. Kim, K.-H.; Kabir, E.; Jahan, S.A. Exposure to pesticides and the associated human health effects. *Sci. Total Environ.* **2017**, *575*, 525–535. [CrossRef] [PubMed]
2. World Health Organization. *The WHO Recommended Classification of Pesticides by Hazard and Guidelines to Classification 2009*; World Health Organization: Geneva, Switzerland, 2010.
3. Sulaiman, N.S.; Rovina, K.; Joseph, V.M. Classification, extraction and current analytical approaches for detection of pesticides in various food products. *J. Consum. Prot. Food Saf.* **2019**, *14*, 209–221. [CrossRef]
4. Jayaraj, R.; Megha, P.; Sreedev, P. Organochlorine pesticides, their toxic effects on living organisms and their fate in the environment. *Interdiscip. Toxicol.* **2016**, *9*, 90–100. [CrossRef] [PubMed]
5. Chopra, A.K.; Sharma, M.K.; Chamoli, S. Bioaccumulation of organochlorine pesticides in aquatic system—An overview. *Environ. Monit. Assess.* **2011**, *173*, 905–916. [CrossRef] [PubMed]
6. Patočka, J.; Cabal, J.; Kuča, K.; Jun, D. Oxime reactivation of acetylcholinesterase inhibited by toxic phosphorus esters: In vitro kinetics and thermodynamics. *J. Appl. Biomed.* **2005**, *3*, 91–99. [CrossRef]
7. Lin, J.-N.; Lin, C.-L.; Lin, M.-C.; Lai, C.-H.; Lin, H.-H.; Yang, C.-H.; Kao, C.-H. Increased risk of dementia in patients with acute organophosphate and carbamate poisoning: A nationwide population-based cohort study. *Medicine* **2015**, *94*, e1187. [CrossRef] [PubMed]
8. Gupta, R.C.; Crissman, J.W. Chapter 42—Agricultural Chemicals. In *Haschek and Rousseaux's Handbook of Toxicologic Pathology*, 3rd ed.; Haschek, W.M., Rousseaux, C.G., Wallig, M.A., Eds.; Academic Press: Boston, MA, USA, 2013; pp. 1349–1372. ISBN 978-0-12-415759-0.
9. Tsagkaris, A.S.; Nelis, J.L.D.; Ross, G.M.S.; Jafari, S.; Guercetti, J.; Kopper, K.; Zhao, Y.; Rafferty, K.; Salvador, J.P.; Migliorelli, D.; et al. Critical assessment of recent trends related to screening and confirmatory analytical methods for selected food contaminants and allergens. *TrAC Trends Anal. Chem.* **2019**, *121*, 115688. [CrossRef]
10. Stachniuk, A.; Fornal, E. Liquid Chromatography-Mass Spectrometry in the Analysis of Pesticide Residues in Food. *Food Anal. Methods* **2016**, *9*, 1654–1665. [CrossRef]
11. Hakme, E.; Lozano, A.; Uclés, S.; Fernández-Alba, A.R. Further improvements in pesticide residue analysis in food by applying gas chromatography triple quadrupole mass spectrometry (GC-QqQ-MS/MS) technologies. *Anal. Bioanal. Chem.* **2018**, *410*, 5491–5506. [CrossRef]
12. Mrzlikar, M.; Heath, D.; Heath, E.; Markelj, J.; Borovšak, A.K.; Prosen, H. Investigation of neonicotinoid pesticides in Slovenian honey by LC-MS/MS. *LWT* **2019**, *104*, 45–52. [CrossRef]
13. Hou, X.; Han, M.; Dai, X.; Yang, X.; Yi, S. A multi-residue method for the determination of 124 pesticides in rice by modified QuEChERS extraction and gas chromatography-tandem mass spectrometry. *Food Chem.* **2013**, *138*, 1198–1205. [CrossRef] [PubMed]
14. Hamamoto, K.; Iwatsuki, K.; Akama, R.; Koike, R. Rapid multiresidue determination of pesticides in livestock muscle and liver tissue via modified QuEChERS sample preparation and LC-MS/MS. *Food Addit. Contam. Part A* **2017**, *34*, 1162–1171. [CrossRef] [PubMed]

15. Mezcua, M.; Malato, O.; García-Reyes, J.F.; Molina-Díaz, A.; Fernández-Alba, A.R. Accurate-Mass Databases for Comprehensive Screening of Pesticide Residues in Food by Fast Liquid Chromatography Time-of-Flight Mass Spectrometry. *Anal. Chem.* **2009**, *81*, 913–929. [CrossRef] [PubMed]
16. Sun, F.; Tan, H.; Li, Y.; De Boevre, M.; Zhang, H.; Zhou, J.; Li, Y.; Yang, S. An integrated data-dependent and data-independent acquisition method for hazardous compounds screening in foods using a single UHPLC-Q-Orbitrap run. *J. Hazard. Mater.* **2021**, *401*, 123266. [CrossRef]
17. Musatadi, M.; González-Gaya, B.; Irazola, M.; Prieto, A.; Etxebarria, N.; Olivares, M.; Zuloaga, O. Focused ultrasound-based extraction for target analysis and suspect screening of organic xenobiotics in fish muscle. *Sci. Total Environ.* **2020**, *740*, 139894. [CrossRef]
18. Vargas-Pérez, M.; Domínguez, I.; González, F.J.E.; Frenich, A.G. Application of full scan gas chromatography high resolution mass spectrometry data to quantify targeted-pesticide residues and to screen for additional substances of concern in fresh-food commodities. *J. Chromatogr. A* **2020**, *1622*, 461118. [CrossRef]
19. Nelis, J.L.D.; Tsagkaris, A.S.; Zhao, Y.; Lou-Franco, J.; Nolan, P.; Zhou, H.; Cao, C.; Rafferty, K.; Hajslova, J.; Elliott, C.T.; et al. The end user sensor tree: An end-user friendly sensor database. *Biosens. Bioelectron.* **2019**, *130*, 245–253. [CrossRef]
20. Fang, L.; Liao, X.; Jia, B.; Shi, L.; Kang, L.; Zhou, L.; Kong, W. Recent progress in immunosensors for pesticides. *Biosens. Bioelectron.* **2020**, *164*, 112255. [CrossRef]
21. Cao, J.; Wang, M.; Yu, H.; She, Y.; Cao, Z.; Ye, J.; Abd El-Aty, A.M.; Hacımüftüoğlu, A.; Wang, J.; Lao, S. An Overview on the Mechanisms and Applications of Enzyme Inhibition-Based Methods for Determination of Organophosphate and Carbamate Pesticides. *J. Agric. Food Chem.* **2020**, *68*, 7298–7315. [CrossRef]
22. Capoferri, D.; Della Pelle, F.; Del Carlo, M.; Compagnone, D. Affinity sensing strategies for the detection of pesticides in food. *Foods* **2018**, *7*, 148. [CrossRef]
23. Nelis, J.; Elliott, C.; Campbell, K. "The smartphone's guide to the galaxy": In situ analysis in space. *Biosensors* **2018**, *8*, 96. [CrossRef] [PubMed]
24. Nelis, J.L.D.; Tsagkaris, A.S.; Dillon, M.J.; Hajslova, J.; Elliott, C.T. Smartphone-based optical assays in the food safety field. *TrAC Trends Anal. Chem.* **2020**, *129*, 115934. [CrossRef]
25. Authority, E.F.S. The 2014 European Union report on pesticide residues in food. *EFSA J.* **2016**, *14*, e04611. [CrossRef]
26. Authority, E.F.S. The 2015 European Union report on pesticide residues in food. *EFSA J.* **2017**, *15*, e04791.
27. Authority, E.F.S. European Food Safety Authority The 2016 European Union report on pesticide residues in food. *EFSA J.* **2018**, *16*, e05348.
28. Authority, E.F.S. The 2017 European Union report on pesticide residues in food. *EFSA J.* **2019**, *17*, e05743.
29. Medina-Pastor, P.; Triacchini, G. The 2018 European Union report on pesticide residues in food. *EFSA J.* **2020**, *18*, e06057. [PubMed]
30. EFSA. Statement on the available outcomes of the human health assessment in the context of the pesticides peer review of the active substance chlorpyrifos. *EFSA J.* **2019**, *17*, e05809.
31. Van Klaveren, J.D.; Kennedy, M.C.; Moretto, A.; Verbeke, W.; van der Voet, H.; Boon, P.E. The ACROPOLIS project: Its aims, achievements, and way forward. *Food Chem. Toxicol.* **2015**, *79*, 1–4. [CrossRef]
32. Jin, L.; Hao, Z.; Zheng, Q.; Chen, H.; Zhu, L.; Wang, C.; Liu, X.; Lu, C. A facile microfluidic paper-based analytical device for acetylcholinesterase inhibition assay utilizing organic solvent extraction in rapid detection of pesticide residues in food. *Anal. Chim. Acta* **2020**, *1100*, 215–224. [CrossRef]
33. Jang, I.; Carrão, D.B.; Menger, R.F.; de Oliveira, A.R.M.; Henry, C.S. Pump-Free Microfluidic Rapid Mixer Combined with a Paper-Based Channel. *ACS Sens.* **2020**, *5*, 2230–2238. [CrossRef] [PubMed]
34. Tsagkaris, A.S.; Pulkrabova, J.; Hajslova, J.; Filippini, D. A Hybrid Lab-on-a-Chip Injector System for Autonomous Carbofuran Screening. *Sensors* **2019**, *19*, 5579. [CrossRef] [PubMed]
35. Arduini, F.; Forchielli, M.; Scognamiglio, V.; Nikolaevna, K.A.; Moscone, D. Organophosphorous pesticide detection in olive oil by using a miniaturized, easy-to-use, and cost-effective biosensor combined with QuEChERS for sample clean-up. *Sensors* **2017**, *17*, 34. [CrossRef] [PubMed]
36. Wu, L.; Li, G.; Xu, X.; Zhu, L.; Huang, R.; Chen, X. Application of nano-ELISA in food analysis: Recent advances and challenges. *TrAC Trends Anal. Chem.* **2019**, *113*, 140–156. [CrossRef]
37. Hongsibsong, S.; Prapamontol, T.; Xu, T.; Hammock, B.D.; Wang, H.; Chen, Z.-J.; Xu, Z.-L. Monitoring of the organophosphate pesticide chlorpyrifos in vegetable samples from local markets in Northern Thailand by developed immunoassay. *Int. J. Environ. Res. Public Health* **2020**, *17*, 4723. [CrossRef]
38. Ivanov, Y. Development of MNPs based enzyme immuno-sorbent analysis for the determination of organophosphorus pesticides in milk. *Open Biotechnol. J.* **2019**, *13*, 146–154. [CrossRef]
39. He, J.; Tao, X.; Wang, K.; Ding, G.; Li, J.; Li, Q.X.; Gee, S.J.; Hammock, B.D.; Xu, T. One-step immunoassay for the insecticide carbaryl using a chicken single-chain variable fragment (scFv) fused to alkaline phosphatase. *Anal. Biochem.* **2019**, *572*, 9–15. [CrossRef]
40. Watanabe, E.; Miyake, S. Direct determination of neonicotinoid insecticides in an analytically challenging crop such as Chinese chives using selective ELISAs. *J. Environ. Sci. Health Part B Pestic. Food Contam. Agric. Wastes* **2018**, *53*, 707–712. [CrossRef]

41. Esteve-Turrillas, F.A.; Agulló, C.; Abad-Somovilla, A.; Mercader, J.V.; Abad-Fuentes, A. Fungicide multiresidue monitoring in international wines by immunoassays. *Food Chem.* **2016**, *196*, 1279–1286. [CrossRef]

42. Tsagkaris, A.S.; Uttl, L.; Pulkrabova, J.; Hajslova, J. Screening of Carbamate and Organophosphate Pesticides in Food Matrices Using an Affordable and Simple Spectrophotometric Acetylcholinesterase Assay. *Appl. Sci.* **2020**, *10*, 565. [CrossRef]

43. Tsagkaris, A.S.; Migliorelli, D.; Uttl, L.; Filippini, D.; Pulkrabova, J.; Hajslova, J. A microfluidic paper-based analytical device (µPAD) with smartphone readout for chlorpyrifos-oxon screening in human serum. *Talanta* **2020**, *222*, 121535. [CrossRef] [PubMed]

44. Halámek, E.; Kobliha, Z.; Pitschmann, V. *Analysis of Chemical Warfare Agents*; Univerzita Obrany: Brno, Czech Republic, 2009; ISBN 8072316583.

45. Yang, X.; Dai, J.; Yang, L.; Ma, M.; Zhao, S.-J.; Chen, X.-G.; Xiao, H. Oxidation pretreatment by calcium hypochlorite to improve the sensitivity of enzyme inhibition-based detection of organophosphorus pesticides. *J. Sci. Food Agric.* **2018**, *98*, 2624–2631. [CrossRef] [PubMed]

46. Pohanka, M.; Karasova, J.Z.; Kuca, K.; Pikula, J. Multichannel spectrophotometry for analysis of organophosphate paraoxon in beverages. *Turk. J. Chem.* **2010**, *34*, 91–98.

47. Watanabe, E.; Miyake, S.; Yogo, Y. Review of Enzyme-Linked Immunosorbent Assays (ELISAs) for Analyses of Neonicotinoid Insecticides in Agro-environments. *J. Agric. Food Chem.* **2013**, *61*, 12459–12472. [CrossRef]

48. Pohanka, M. Biosensors and bioassays based on lipases, principles and applications, a review. *Molecules* **2019**, *24*, 616. [CrossRef]

49. Mishra, G.K.; Sharma, V.; Mishra, R.K. Electrochemical aptasensors for food and environmental safeguarding: A review. *Biosensors* **2018**, *8*, 28. [CrossRef]

50. Lan, L.; Yao, Y.; Ping, J.; Ying, Y. Recent Progress in Nanomaterial-Based Optical Aptamer Assay for the Detection of Food Chemical Contaminants. *ACS Appl. Mater. Interfaces* **2017**, *9*, 23287–23301. [CrossRef]

51. Augusto, F.; Hantao, L.W.; Mogollón, N.G.S.; Braga, S.C.G.N. New materials and trends in sorbents for solid-phase extraction. *TrAC Trends Anal. Chem.* **2013**, *43*, 14–23. [CrossRef]

52. Nery, E.W.; Kubota, L.T. Evaluation of enzyme immobilization methods for paper-based devices—A glucose oxidase study. *J. Pharm. Biomed. Anal.* **2016**, *117*, 551–559. [CrossRef]

53. Koczula, K.M.; Gallotta, A. Lateral flow assays. *Essays Biochem.* **2016**, *60*, 111–120.

54. Lan, J.; Sun, W.; Chen, L.; Zhou, H.; Fan, Y.; Diao, X.; Wang, B.; Zhao, H. Simultaneous and rapid detection of carbofuran and 3-hydroxy-carbofuran in water samples and pesticide preparations using lateral-flow immunochromatographic assay. *Food Agric. Immunol.* **2020**, *31*, 165–175. [CrossRef]

55. Sankar, K.; Lenisha, D.; Janaki, G.; Juliana, J.; Kumar, R.S.; Selvi, M.C.; Srinivasan, G. Digital image-based quantification of chlorpyrifos in water samples using a lipase embedded paper based device. *Talanta* **2020**, *208*, 120408. [CrossRef]

56. Whitesides, G.M. The origins and the future of microfluidics. *Nature* **2006**, *442*, 368–373. [CrossRef]

57. Rapp, B. *Microfluidics: Modeling, Mechanics and Mathematics*; Holt, S., Ed.; Elsevier: Amsterdam, The Netherlands, 2016; ISBN 978-1-4557-3141-1.

58. Suska, A.; Filippini, D. Autonomous lab-on-a-chip generic architecture for disposables with integrated actuation. *Sci. Rep.* **2019**, *9*, 20320. [CrossRef]

59. Xu, B.; Guo, J.; Fu, Y.; Chen, X.; Guo, J. A review on microfluidics in the detection of food pesticide residues. *Electrophoresis* **2020**, *41*, 821–832. [CrossRef]

60. Nouanthavong, S.; Nacapricha, D.; Henry, C.S.; Sameenoi, Y. Pesticide analysis using nanoceria-coated paper-based devices as a detection platform. *Analyst* **2016**, *141*, 1837–1846. [CrossRef]

61. Bala, R.; Kumar, M.; Bansal, K.; Sharma, R.K.; Wangoo, N. Ultrasensitive aptamer biosensor for malathion detection based on cationic polymer and gold nanoparticles. *Biosens. Bioelectron.* **2016**, *85*, 445–449. [CrossRef]

62. Guo, L.; Li, Z.; Chen, H.; Wu, Y.; Chen, L.; Song, Z.; Lin, T. Colorimetric biosensor for the assay of paraoxon in environmental water samples based on the iodine-starch color reaction. *Anal. Chim. Acta* **2017**, *967*, 59–63. [CrossRef]

63. Li, X.; Cui, H.; Zeng, Z. A simple colorimetric and fluorescent sensor to detect organophosphate pesticides based on adenosine triphosphate-modified gold nanoparticles. *Sensors* **2018**, *18*, 4302. [CrossRef]

64. Li, H.; Yan, X.; Lu, G.; Su, X. Carbon dot-based bioplatform for dual colorimetric and fluorometric sensing of organophosphate pesticides. *Sens. Actuators B Chem.* **2018**, *260*, 563–570. [CrossRef]

65. Zeng, L.; Zhou, D.; Wu, J.; Liu, C.; Chen, J. A target-induced and equipment-free biosensor for amplified visual detection of pesticide acetamiprid with high sensitivity and selectivity. *Anal. Methods* **2019**, *11*, 1168–1173. [CrossRef]

66. Nsibande, S.A.; Forbes, P.B.C. Fluorescence detection of pesticides using quantum dot materials—A review. *Anal. Chim. Acta* **2016**, *945*, 9–22. [CrossRef]

67. Liu, B.; Zhuang, J.; Wei, G. Recent advances in the design of colorimetric sensors for environmental monitoring. *Environ. Sci. Nano* **2020**, *7*, 2195–2213. [CrossRef]

68. Kalyani, N.; Goel, S.; Jaiswal, S. On-site sensing of pesticides using point-of-care biosensors: A review. *Environ. Chem. Lett.* **2020**, 1–10. [CrossRef]

69. Korram, J.; Dewangan, L.; Karbhal, I.; Nagwanshi, R.; Vaishanav, S.K.; Ghosh, K.K.; Satnami, M.L. CdTe QD-based inhibition and reactivation assay of acetylcholinesterase for the detection of organophosphorus pesticides. *RSC Adv.* **2020**, *10*, 24190–24202. [CrossRef]

70. Wang, P.; Li, H.; Hassan, M.M.; Guo, Z.; Zhang, Z.-Z.; Chen, Q. Fabricating an Acetylcholinesterase Modulated UCNPs-Cu^{2+} Fluorescence Biosensor for Ultrasensitive Detection of Organophosphorus Pesticides-Diazinon in Food. *J. Agric. Food Chem.* **2019**, *67*, 4071–4079. [CrossRef]

71. Hu, W.; Chen, Q.; Li, H.; Ouyang, Q.; Zhao, J. Fabricating a novel label-free aptasensor for acetamiprid by fluorescence resonance energy transfer between NH$_2$-NaYF$_4$: Yb, Ho@SiO$_2$ and Au nanoparticles. *Biosens. Bioelectron.* **2016**, *80*, 398–404. [CrossRef]

72. Hou, J.; Dong, G.; Tian, Z.; Lu, J.; Wang, Q.; Ai, S.; Wang, M. A sensitive fluorescent sensor for selective determination of dichlorvos based on the recovered fluorescence of carbon dots-Cu(II) system. *Food Chem.* **2016**, *202*, 81–87. [CrossRef]

73. Wu, X.; Song, Y.; Yan, X.; Zhu, C.; Ma, Y.; Du, D.; Lin, Y. Carbon quantum dots as fluorescence resonance energy transfer sensors for organophosphate pesticides determination. *Biosens. Bioelectron.* **2017**, *94*, 292–297. [CrossRef]

74. Tan, G.; Zhao, Y.; Wang, M.; Chen, X.; Wang, B.; Li, Q.X. Ultrasensitive quantitation of imidacloprid in vegetables by colloidal gold and time-resolved fluorescent nanobead traced lateral flow immunoassays. *Food Chem.* **2020**, *311*, 126055. [CrossRef]

75. Arvand, M.; Mirroshandel, A.A. An efficient fluorescence resonance energy transfer system from quantum dots to graphene oxide nano sheets: Application in a photoluminescence aptasensing probe for the sensitive detection of diazinon. *Food Chem.* **2019**, *280*, 115–122. [CrossRef]

76. Wang, X.; Hou, T.; Dong, S.; Liu, X.; Li, F. Fluorescence biosensing strategy based on mercury ion-mediated DNA conformational switch and nicking enzyme-assisted cycling amplification for highly sensitive detection of carbamate pesticide. *Biosens. Bioelectron.* **2016**, *77*, 644–649. [CrossRef]

77. Li, H.; Dong, B.; Dou, L.; Yu, W.; Yu, X.; Wen, K.; Ke, Y.; Shen, J.; Wang, Z. Fluorescent lateral flow immunoassay for highly sensitive detection of eight anticoagulant rodenticides based on cadmium-free quantum dot-encapsulated nanospheres. *Sens. Actuators B Chem.* **2020**, *324*, 128771. [CrossRef]

78. Špačková, B.; Wrobel, P.; Bocková, M.; Homola, J. Optical Biosensors Based on Plasmonic Nanostructures: A Review. *Proc. IEEE* **2016**, *104*, 2380–2408. [CrossRef]

79. Ross, G.M.S.; Bremer, M.G.E.G.; Wichers, J.H.; Van Amerongen, A.; Nielen, M.W.F. Rapid antibody selection using surface plasmon resonance for high-speed and sensitive hazelnut lateral flow prototypes. *Biosensors* **2018**, *8*, 130. [CrossRef]

80. Pundir, C.S.; Malik, A. Preety Bio-sensing of organophosphorus pesticides: A review. *Biosens. Bioelectron.* **2019**, *140*, 111348. [CrossRef]

81. Mahmoudpour, M.; Dolatabadi, J.E.N.; Torbati, M.; Homayouni-Rad, A. Nanomaterials based surface plasmon resonance signal enhancement for detection of environmental pollutions. *Biosens. Bioelectron.* **2019**, *127*, 72–84. [CrossRef]

82. Huang, Y.H.; Ho, H.P.; Kong, S.K.; Kabashin, A. V Phase-sensitive surface plasmon resonance biosensors: Methodology, instrumentation and applications. *Ann. Phys.* **2012**, *524*, 637–662. [CrossRef]

83. Mohammadzadeh-Asl, S.; Keshtkar, A.; Dolatabadi, J.E.N.; de la Guardia, M. Nanomaterials and phase sensitive based signal enhancment in surface plasmon resonance. *Biosens. Bioelectron.* **2018**, *110*, 118–131. [CrossRef]

84. Soler, M.; Huertas, C.S.; Lechuga, L.M. Label-free plasmonic biosensors for point-of-care diagnostics: A review. *Expert Rev. Mol. Diagn.* **2019**, *19*, 71–81. [CrossRef]

85. Wu, S.; Li, D.; Gao, Z.; Wang, J. Controlled etching of gold nanorods by the Au(III)-CTAB complex, and its application to semi-quantitative visual determination of organophosphorus pesticides. *Microchim. Acta* **2017**, *184*, 4383–4391. [CrossRef]

86. Shrivastav, A.M.; Usha, S.P.; Gupta, B.D. Fiber optic profenofos sensor based on surface plasmon resonance technique and molecular imprinting. *Biosens. Bioelectron.* **2016**, *79*, 150–157. [CrossRef]

87. Guo, Y.; Liu, R.; Liu, Y.; Xiang, D.; Liu, Y.; Gui, W.; Li, M.; Zhu, G. A non-competitive surface plasmon resonance immunosensor for rapid detection of triazophos residue in environmental and agricultural samples. *Sci. Total Environ.* **2018**, *613–614*, 783–791. [CrossRef]

88. Li, Q.; Dou, X.; Zhao, X.; Zhang, L.; Luo, J.; Xing, X.; Yang, M. A gold/Fe$_3$O$_4$ nanocomposite for use in a surface plasmon resonance immunosensor for carbendazim. *Microchim. Acta* **2019**, *186*, 313. [CrossRef]

89. Hirakawa, Y.; Yamasaki, T.; Watanabe, E.; Okazaki, F.; Murakami-Yamaguchi, Y.; Oda, M.; Iwasa, S.; Narita, H.; Miyake, S. Development of an Immunosensor for Determination of the Fungicide Chlorothalonil in Vegetables, Using Surface Plasmon Resonance. *J. Agric. Food Chem.* **2015**, *63*, 6325–6330. [CrossRef]

90. Li, Q.; Dou, X.; Zhang, L.; Zhao, X.; Luo, J.; Yang, M. Oriented assembly of surface plasmon resonance biosensor through staphylococcal protein A for the chlorpyrifos detection. *Anal. Bioanal. Chem.* **2019**, *411*, 6057–6066. [CrossRef]

91. Boyaci, I.H.; Temiz, H.T.; Geniş, H.E.; Soykut, E.A.; Yazgan, N.N.; Güven, B.; Uysal, R.S.; Bozkurt, A.G.; İlaslan, K.; Torun, O. Dispersive and FT-Raman spectroscopic methods in food analysis. *RSC Adv.* **2015**, *5*, 56606–56624. [CrossRef]

92. Li, X.; Yang, T.; Song, Y.; Zhu, J.; Wang, D.; Li, W. Surface-enhanced Raman spectroscopy (SERS)-based immunochromatographic assay (ICA) for the simultaneous detection of two pyrethroid pesticides. *Sens. Actuators B Chem.* **2019**, *283*, 230–238. [CrossRef]

93. Gao, F.; Hu, Y.; Chen, D.; Li-Chan, E.C.Y.; Grant, E.; Lu, X. Determination of Sudan I in paprika powder by molecularly imprinted polymers–thin layer chromatography–surface enhanced Raman spectroscopic biosensor. *Talanta* **2015**, *143*, 344–352. [CrossRef]

94. Lin, Z.; He, L. Recent Advance in SERS techniques for food safety and quality analysis: A brief review. *Curr. Opin. Food Sci.* **2019**, *28*, 82–87. [CrossRef]

95. Yaseen, T.; Pu, H.; Sun, D.-W. Functionalization techniques for improving SERS substrates and their applications in food safety evaluation: A review of recent research trends. *Trends Food Sci. Technol.* **2018**, *72*, 162–174. [CrossRef]

96. Hoppmann, E.P.; Wei, W.Y.; White, I.M. Highly sensitive and flexible inkjet printed SERS sensors on paper. *Methods* **2013**, *63*, 219–224. [CrossRef] [PubMed]

97. Gong, Z.; Wang, C.; Pu, S.; Wang, C.; Cheng, F.; Wang, Y.; Fan, M. Rapid and direct detection of illicit dyes on tainted fruit peel using a PVA hydrogel surface enhanced Raman scattering substrate. *Anal. Methods* **2016**, *8*, 4816–4820. [CrossRef]

98. Lin, S.; Lin, X.; Han, S.; Liu, Y.; Hasi, W.; Wang, L. Flexible fabrication of a paper-fluidic SERS sensor coated with a monolayer of core–shell nanospheres for reliable quantitative SERS measurements. *Anal. Chim. Acta* **2020**, *1108*, 167–176. [CrossRef] [PubMed]

99. Lin, S.; Lin, X.; Liu, Y.; Zhao, H.; Hasi, W.; Wang, L. Self-assembly of Au@Ag core–shell nanocubes embedded with an internal standard for reliable quantitative SERS measurements. *Anal. Methods* **2018**, *10*, 4201–4208. [CrossRef]

100. Li, X.; Zhang, S.; Yu, Z.; Yang, T. Surface-enhanced Raman spectroscopic analysis of phorate and fenthion pesticide in apple skin using silver nanoparticles. *Appl. Spectrosc.* **2014**, *68*, 483–487. [CrossRef]

101. Fales, A.M.; Vo-Dinh, T. Silver embedded nanostars for SERS with internal reference (SENSIR). *J. Mater. Chem. C* **2015**, *3*, 7319–7324. [CrossRef]

102. Xie, J.; Li, L.; Khan, I.M.; Wang, Z.; Ma, X. Flexible paper-based SERS substrate strategy for rapid detection of methyl parathion on the surface of fruit. *Spectrochim. Acta Part A Mol. Biomol. Spectrosc.* **2020**, *231*, 118104. [CrossRef]

103. Yan, M.; She, Y.; Cao, X.; Ma, J.; Chen, G.; Hong, S.; Shao, Y.; Abd El-Aty, A.M.; Wang, M.; Wang, J. A molecularly imprinted polymer with integrated gold nanoparticles for surface enhanced Raman scattering based detection of the triazine herbicides, prometryn and simetryn. *Microchim. Acta* **2019**, *186*, 1–9. [CrossRef]

104. Cui, H.; Li, S.; Deng, S.; Chen, H.; Wang, C. Flexible, Transparent, and Free-Standing Silicon Nanowire SERS Platform for in Situ Food Inspection. *ACS Sens.* **2017**, *2*, 386–393. [CrossRef]

105. Huang, S.; Yan, W.; Liu, M.; Hu, J. Detection of difenoconazole pesticides in pak choi by surface-enhanced Raman scattering spectroscopy coupled with gold nanoparticles. *Anal. Methods* **2016**, *8*, 4755–4761. [CrossRef]

106. Wang, C.; Wu, X.; Dong, P.; Chen, J.; Xiao, R. Hotspots engineering by grafting Au@Ag core-shell nanoparticles on the Au film over slightly etched nanoparticles substrate for on-site paraquat sensing. *Biosens. Bioelectron.* **2016**, *86*, 944–950. [CrossRef] [PubMed]

107. Tognaccini, L.; Ricci, M.; Gellini, C.; Feis, A.; Smulevich, G.; Becucci, M. Surface enhanced Raman spectroscopy for in-field detection of pesticides: A test on dimethoate residues in water and on olive leaves. *Molecules* **2019**, *24*, 292. [CrossRef] [PubMed]

108. Weng, S.; Wang, F.; Dong, R.; Qiu, M.; Zhao, J.; Huang, L.; Zhang, D. Fast and Quantitative Analysis of Ediphenphos Residue in Rice Using Surface-Enhanced Raman Spectroscopy. *J. Food Sci.* **2018**, *83*, 1179–1185. [CrossRef]

109. Wang, K.; Huang, M.; Chen, J.; Lin, L.; Kong, L.; Liu, X.; Wang, H.; Lin, M. A "drop-wipe-test" SERS method for rapid detection of pesticide residues in fruits. *J. Raman Spectrosc.* **2018**, *49*, 493–498. [CrossRef]

110. Kanchi, S.; Sabela, M.I.; Mdluli, P.S.; Bisetty, K. Smartphone based bioanalytical and diagnosis applications: A review. *Biosens. Bioelectron.* **2018**, *102*, 136–149. [CrossRef]

111. Li, X.; Li, H.; Ma, W.; Guo, Z.; Li, X.; Song, S.; Tang, H.; Li, X.; Zhang, Q. Development of precise GC-EI-MS method to determine the residual fipronil and its metabolites in chicken egg. *Food Chem.* **2019**, *281*, 85–90. [CrossRef]

112. Dekker, S.; Isgor, P.K.; Feijten, T.; Segerink, L.I.; Odijk, M. From chip-in-a-lab to lab-on-a-chip: A portable Coulter counter using a modular platform. *Microsyst. Nanoeng.* **2018**, *4*, 1–8. [CrossRef]

113. Keçili, R.; Büyüktiryaki, S.; Hussain, C.M. 8—Micro total analysis systems with nanomaterials. In *Handbook of Nanomaterials in Analytical Chemistry*; Elsevier: Amsterdam, The Netherlands, 2020; pp. 185–198. ISBN 978-0-12-816699-4.

114. Ross, G.M.S.; Filippini, D.; Nielen, M.W.F.; Salentijn, G.I.J. Interconnectable solid-liquid protein extraction unit and chip-based dilution for multiplexed consumer immunodiagnostics. *Anal. Chim. Acta* **2020**, *1140*, 190–198. [CrossRef]

115. Nelis, J.L.D.; Bura, L.; Zhao, Y.; Burkin, K.M.; Rafferty, K.; Elliott, C.T.; Campbell, K. The Efficiency of Color Space Channels to Quantify Color and Color Intensity Change in Liquids, pH Strips, and Lateral Flow Assays with Smartphones. *Sensors* **2019**, *19*, 5104. [CrossRef]

116. Ross, G.; Salentijn, G.I.J.; Nielen, M.W.F. A critical comparison between flow-through and lateral flow immunoassay formats for visual and smartphone-based multiplex allergen detection. *Biosensors* **2019**, *9*, 143. [CrossRef] [PubMed]

117. Nelis, J.L.D.; Zhao, Y.; Bura, L.; Rafferty, K.; Elliott, C.T.; Campbell, K. A Randomized Combined Channel Approach for the Quantification of Color- And Intensity-Based Assays with Smartphones. *Anal. Chem.* **2020**, *92*, 7852–7860. [CrossRef] [PubMed]

118. Guo, J.; Wong, J.X.H.; Cui, C.; Li, X.; Yu, H.-Z. A smartphone-readable barcode assay for the detection and quantitation of pesticide residues. *Analyst* **2015**, *140*, 5518–5525. [CrossRef] [PubMed]

119. Zhao, Y.; Elliott, C.; Zhou, H.; Rafferty, K. Spectral Illumination Correction: Achieving Relative Color Constancy Under the Spectral Domain. In Proceedings of the 2018 IEEE International Symposium on Signal Processing and Information Technology (ISSPIT), Louisville, KY, USA, 6–8 December 2018; pp. 690–695.

120. Cheng, N.; Song, Y.; Fu, Q.; Du, D.; Luo, Y.; Wang, Y.; Xu, W.; Lin, Y. Aptasensor based on fluorophore-quencher nano-pair and smartphone spectrum reader for on-site quantification of multi-pesticides. *Biosens. Bioelectron.* **2018**, *117*, 75–83. [CrossRef] [PubMed]

121. Montali, L.; Calabretta, M.M.; Lopreside, A.; D'Elia, M.; Guardigli, M.; Michelini, E. Multienzyme chemiluminescent foldable biosensor for on-site detection of acetylcholinesterase inhibitors. *Biosens. Bioelectron.* **2020**, *162*, 112232. [CrossRef]

122. Cheng, N.; Shi, Q.; Zhu, C.; Li, S.; Lin, Y.; Du, D. Pt–Ni(OH)$_2$ nanosheets amplified two-way lateral flow immunoassays with smartphone readout for quantification of pesticides. *Biosens. Bioelectron.* **2019**, *142*, 111498. [CrossRef]

123. Wei, J.; Yang, L.; Luo, M.; Wang, Y.; Li, P. Nanozyme-assisted technique for dual mode detection of organophosphorus pesticide. *Ecotoxicol. Environ. Saf.* **2019**, *179*, 17–23. [CrossRef]
124. Wang, Y.; Zeinhom, M.M.A.; Yang, M.; Sun, R.; Wang, S.; Smith, J.N.; Timchalk, C.; Li, L.; Lin, Y.; Du, D. A 3D-Printed, Portable, Optical-Sensing Platform for Smartphones Capable of Detecting the Herbicide 2,4-Dichlorophenoxyacetic Acid. *Anal. Chem.* **2017**, *89*, 9339–9346. [CrossRef]

 *foods*

Article

A Novel Method for Antibiotic Detection in Milk Based on Competitive Magnetic Immunodetection

Jan Pietschmann [1,*], Dominik Dittmann [1], Holger Spiegel [1], Hans-Joachim Krause [2] and Florian Schröper [1]

1   Fraunhofer Institute for Molecular Biology and Applied Ecology IME, Forckenbeckstraße 6, 52074 Aachen, Germany; dominik.dittmann@ime.fraunhofer.de (D.D.); holger.spiegel@ime.fraunhofer.de (H.S.); florian.schroeper@ime.fraunhofer.de (F.S.)
2   Institute of Biological Information Processing, Bioelectronics IBI-3, Forschungszentrum Jülich, 52428 Jülich, Germany; h.-j.krause@fz-juelich.de
*   Correspondence: jan.pietschmann@ime.fraunhofer.de

Received: 30 October 2020; Accepted: 27 November 2020; Published: 30 November 2020

Abstract: The misuse of antibiotics as well as incorrect dosage or insufficient time for detoxification can result in the presence of pharmacologically active molecules in fresh milk. Hence, in many countries, commercially available milk has to be tested with immunological, chromatographic or microbiological analytical methods to avoid consumption of antibiotic residues. Here a novel, sensitive and portable assay setup for the detection and quantification of penicillin and kanamycin in whole fat milk (WFM) based on competitive magnetic immunodetection (cMID) is described and assay accuracy determined. For this, penicillin G and kanamycin-conjugates were generated and coated onto a matrix of immunofiltration columns (IFC). Biotinylated penicillin G or kanamycin-specific antibodies were pre-incubated with antibiotics-containing samples and subsequently applied onto IFC to determine the concentration of antibiotics through the competition of antibody-binding to the antibiotic-conjugate molecules. Bound antibodies were labeled with streptavidin-coated magnetic particles and quantified using frequency magnetic mixing technology. Based on calibration measurements in WFM with detection limits of 1.33 ng·mL^{-1} for penicillin G and 1.0 ng·mL^{-1} for kanamycin, spiked WFM samples were analyzed, revealing highly accurate recovery rates and assay precision. Our results demonstrate the suitability of cMID-based competition assay for reliable and easy on-site testing of milk.

Keywords: frequency mixing technology; immunofiltration; magnetic beads

1. Introduction

Antibiotics are small molecules either produced by different molds such *Penicillium* species or bacteria such as *Streptomyces* spp., or artificially synthesized, and can inhibit the growth of various pathogens [1,2]. Worldwide, annually more than 60,000 tons of antibiotics are used to treat bacterial infectious diseases in animal husbandry with numbers being expected to reach more than 100,000 tons by the year 2030 [3]. Antibiotics belonging to the classes of β-lactams such as penicillin (Figure 1A) and aminoglycosides such as kanamycin (Figure 1B) are primarily used for treatment of infectious diseases [4–8]. Different studies showed that a large part of the antibiotics used in animal husbandry can be released undegraded, with antimicrobial activity, into the environment [9]. Consumption of contaminated food leads to repeated and, consequently, long exposure times to these antibiotics, which poses a major threat for public health. The risks include development of bacterial resistances, allergies and hypersensitive reactions [3,6,8,10,11]. Mostly, antibiotic residues found in food samples are caused by injudicious usage, such as use as growth promotors, incorrect dosage or not maintaining proper detoxification times, e.g., affected by a lack of proper farmer education or awareness [7,12,13].

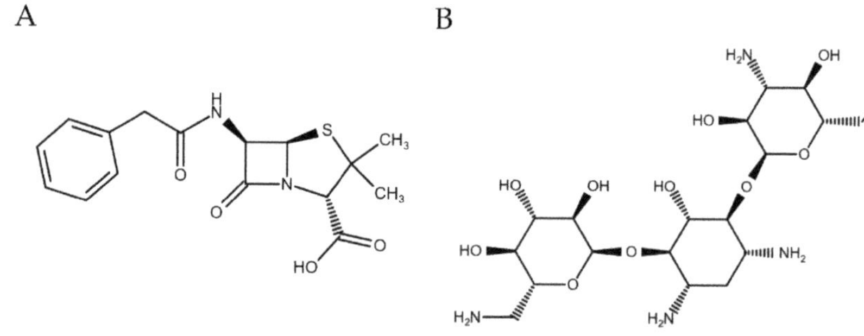

Figure 1. Chemical structures of (**A**) penicillin and (**B**) kanamycin.

Due to the growing risk of overexposure to antibiotics in animal derived foods such as meat, milk or eggs, countries of the European Union (EU) have defined residue limits which specify the acceptable dosage of an antibiotic that will probably not affect consumer health. For the EU, these maximum residue limits (MRL) are defined in the European Union Commission Regulation No. 37/2010 and are set at 4 ng·mL^{-1} for benzylpenicillin (penicillin G) and 150 ng·mL^{-1} for kanamycin in milk, which are comparable to those set in the US [14].

Currently used detection methods are mostly based on chromatographic, immunological and microbiological test procedures [7,15–17]. Especially in the field of chromatographic methods, LC-MS/MS-based analytical technologies enable a highly sensitive and simultaneous detection of multiple antibiotic residues within a single sample [18–20]. Using chromatographic methods, multiresidue analytics with detection limits lower than 1 ng·mL^{-1} for many antibiotics can be performed [21]. However, this method is restricted to analytical laboratories due to the need for highly trained staff and cost-intensive laboratory-based equipment [15,16,22]. Nowadays, immunological tests such as ELISA or lateral flow assays (LFAs) as well as microbial test kits are commonly used for monitoring of milk samples [7,15,17]. LFAs are simple, easy-to-handle and are usually performed within minutes, which makes them easy to use, even for untrained personnel [22]. However, such LFAs lack sensitivity and a quantitative measurement is typically not possible. In contrast, ELISAs have a high sensitivity and are at least semi-quantitative, but their dynamic range of detection is quite low and additionally they lack speed due to long incubation times [23]. Furthermore, experienced staff are needed for performing these assays. Microbiological tests such as the Brilliant Black Reduction Test are easy in procedure but also need laboratory-based equipment. By the application of a (probably) antibiotic-containing sample onto a reference microorganism, bacteria growth is inhibited and a colorimetric change cannot be seen. Although it is a quite simple procedure, it needs a few hours to enable bacteria growth and it lacks specificity since bacterial growth is inhibited by all kinds of antibiotics. Additionally, by analyzing low-contaminated samples, visual interpretation could be difficult, which increases the rate of false negative results [4,24,25]. However, if the milk samples seem to be above the MRL and the previously described tests show positive results, the suspect milk must be sent to analytical laboratories where it is retested for confirmation in accordance with regulatory requirements. Here, a quantitative detection of antibiotic residues is done, mainly using LC-MS/MS-based analytical methods. With those cost-intensive, quantitative results, farmers can calculate when the milk from treated husbandries has to be discarded.

In previous studies, magnetic immunodetection (MID) has been successfully employed for detection and quantification of human as well as plant pathogens and such bacterial toxins as cholera toxin B by sensing superparamagnetic particles [26–29]. In a recent study, a highly sensitive and quantitative detection of aflatoxin B1 has been demonstrated based on newly developed competitive magnetic immunodetection (cMID) [23]. Using frequency mixing magnetic detection (FMMD), a dose-depending measuring signal is obtained in a portable handheld measuring FMMD device [23,26,29,30]. A detailed explanation of FMMD is given in [23,29,31]. In a competitive MID setup, the obtained measuring signal is reciprocally correlated with the amount of analyte in the sample. Easy sample handling and, especially, the possibility for battery-driven operation of the handheld FMMD device, enables on-site analytics and readout without need for special laboratory equipment or electrical infrastructure. Due to several drawbacks of the currently used immunological or microbiological detection technologies and their mainly qualitative results with limited range of quantification, we developed a highly sensitive assay based on cMID for efficient detection, in combination with quantification of penicillin G and kanamycin, in milk samples. For this, penicillin-bovine serum albumin (BSA) and kanamycin-BSA conjugates were synthesized and affinity of monoclonal antibodies (mAb) were determined by dose-response analysis. Afterward, cMID assays were established and highly specific assay accuracy and quantification was demonstrated by spiking different concentrations of either penicillin G or kanamycin in whole fat milk (WFM) and evaluating recovery rates. With this demonstrated proof-of-concept assay, farmers can control their milk directly on-site and can estimate the needed detoxification time without cost-intensive analytics in regulatory laboratories.

2. Materials and Methods

2.1. Materials and Chemicals

Dimethyl sulfoxide (DMSO), 1-Ethyl-3-(3-dimethylaminopropyl)carbodiimide hydrochloride, EZ-Link™ NHS-PEG4 Biotinylation Kit, and NHS were acquired from Merck KGaA, Darmstadt, Germany. Albumin Fraction V (biotin-free), KCl, KH_2PO_4, NaCl, $Na_2(CO_3)$, $NaHCO_3$, $Na_2HPO_4 \times 12$ H_2O and Tween-20 were purchased from Carl Roth, Karlsruhe, Germany. Immuno-filtration columns (ABICAP HP columns) were purchased from Senova Gesellschaft für Biowissenschaft und Technik mbH, Weimar, Germany. Anti-penicillin G monoclonal antibody was kindly provided by the Milchprüfring Bayern e.V. (MPR), Wolnzach, Germany. Anti-kanamycin monoclonal antibody (article number CSB-MA000511I0m) was purchased from Cusabio, Wuhan, China. Detection antibody goat anti-mouse IgG coupled to HRPO (article number 115-035-008) was purchased from Jackson ImmunoResearch Europe Ltd., Ely, UK. Penicillin G and kanamycin A were acquired from Duchefa Biochemie, Haarlem, Netherlands. Magnetic particles with streptavidin-functionalized shell and a hydrodynamic diameter of 70 nm [synomag®-D, article number 104-19-701] were purchased from micromod Partikeltechnologie GmbH, Rostock, Germany.

Coupling buffer, phosphate buffered saline (PBS), PBS-Tween (PBS-T) as well as blocking solutions for ELISA and magnetic immunodetection were prepared as described in [23].

2.2. Generation of Antibiotic-BSA Conjugates

2.2.1. Penicillin-BSA

Penicillin-BSA conjugate was prepared according to an adapted protocol described by Venkataramana and colleagues (2015) [32]. For this, 5.6 mg NHS (43 µmol) and 8 mg EDC-HCl (41 µmol) were dissolved in 500 µL DMSO. The solution was then transferred to 5 mg penicillin G (15 µmol) and incubated for two hours at room temperature in dark surroundings, followed by an overnight incubation at 4 °C in dark surroundings. On the next day, the solution was added dropwise to 10 mg of BSA (0.15 µmol). Afterward, 2 mL of carbonate buffer (pH 9.6) was added to the solution, followed by an incubation for 2 h at room temperature (RT) in dark surroundings. The solution was then dialyzed for four days against daily exchanged 5 l PBS (pH 7.4). After sterile filtration through a filter with a pore diameter of 0.22 µm, the solution was stored at 4 °C until usage.

2.2.2. Kanamycin-BSA

Kanamycin-BSA conjugate was prepared according to a protocol described by Haasnoot et al. (1999) [33]. Firstly, 34 mg kanamycin A (701 µmol) and 10 mg BSA (0.15 µmol) were dissolved in 1 mL MilliQ-water. Afterward, 383 mg EDC-HCl (2 mmol) were dissolved in 1 mL MilliQ-water. This solution was added dropwise to the kanamycin and BSA solution and incubated for 2 h at RT while shaking. The solution was then dialyzed against daily exchanged 5 l PBS (pH 7.4) for four days at 4 °C in dark surroundings. After sterile filtration through a filter with a pore diameter of 0.22 µm, the solution was stored at 4 °C until usage.

2.3. Determination of Protein Concentration

Protein concentration of each antibiotic-BSA conjugate was determined after final sterile filtration with Bradford protein assay using the Roti®-Quant Bradford reagent (Carl Roth) against BSA as reference. The assay was performed according to the manufacturer's instructions. Briefly, a serial dilution of samples was prepared in a 96-well plate and 200 µL of Bradford reagent was added. After an incubation of 5 min at room temperature, the absorbance was measured at a wavelength of 595 nm.

2.4. Determination of Antibody Affinity

Affinity of the monoclonal antibodies to the antibiotic-BSA conjugates was determined with an ELISA. All incubation steps were performed for one hour at RT in dark surroundings. After each step, plates were washed by rinsing the wells three times with 200 µL PBS-T. A high-binding 96-well microtiter plate (Greiner Bio-One) was coated with 100 µL per well of 2 µg/mL penicillin-BSA or kanamycin-BSA. Afterward, remaining free binding sites were blocked with 200 µL per well of 5% (*w/v*) skimmed milk.

A serial dilution of anti-penicillin G or anti-kanamycin monoclonal antibodies with concentrations ranging from 9.76 ng·mL^{-1} up to 1250 ng·mL^{-1} was prepared and applied onto coated and blocked microtiter plate. Following incubation and washing, 100 µL of 80 ng·mL^{-1} detection antibody, diluted in PBS, was added to each well and incubated. Following a final washing step, 100 µL of 1 mg·mL^{-1} ABTS substrate in ABTS buffer was applied and absorption was measured at 405 nm wavelength after 10 min of incubation.

2.5. Preparation of Immunofiltration Columns

The equilibration of immunofiltration columns (IFCs) was performed as previously described [23,29]. For degassing of IFCs, the columns were placed in ethanol (96%) inside of a desiccator at a pressure of—0.8 bar for 20 min. Afterward, the columns were washed with each 750 µL ethanol-water (50/50), MilliQ-water and twice with carbonate buffer (pH 9.6). Matrices were coated by rinsing 500 µL of antibiotic-BSA conjugate solution (3.5 µg·mL^{-1}), diluted in coupling buffer, through each column. After an incubation of one hour at RT, columns were washed by rinsing 750 µL PBS through the matrix. Subsequently, remaining free binding sites inside the matrix were blocked by adding twice 750 µL of a 1% (*w/v*) PBS-BSA solution onto each column. After second application an incubation of 60 min was performed. After washing by applying twice 750 µL of PBS onto columns, the assay can be performed, or the columns can be stored at 4 °C in PBS for 14 days.

2.6. cMID Calibration Curve Analysis

For preparation of cMID calibration curve analysis, a pre-incubation of free antibiotics and biotinylated antibody was performed. Biotinylation of antibody was performed as described in [23]. For this, serially diluted penicillin G or kanamycin samples in PBS, with concentrations ranging from 0.011 ng·mL^{-1} to 3000 ng·mL^{-1} for penicillin G and 0.0057 ng·mL^{-1} to 1500 ng·mL^{-1} for kanamycin, were incubated with 1.2 µg·mL^{-1} biotinylated antibody, also diluted in PBS. As positive control, a sample of respective biotinylated antibody without the addition of antibiotic was prepared for determining the highest possible measuring signal, later called B0 signal. After incubating the sample for one hour at RT, 500 μL sample volume was applied on coated and blocked columns and also incubated for one hour at room temperature. Subsequently, columns were washed with 750 μL PBS and, afterward, 500 μL of 60 μg·mL^{-1} magnetic particles were rinsed through the column and incubated for one hour at room temperature. After final washing with 750 μL PBS through the matrix, the columns were measured using a portable FMMD magnetic reader.

2.7. Frequency Mixing Magnetic Detection (FMMD)

The magnetic nanoparticle markers were detected using a custom-made magnetic reader consisting of a measurement head with excitation and detection coils and an electronic readout [29]. In brief, the sample containing the magnetic particles is exposed to a magnetic field consisting of two distinct frequencies, a high frequency field of approximately a milli-tesla at $f_1 = 49$ kHz and a low-frequency field of about ten milli-tesla at $f_2 = 61$ Hz. Due to the particles' nonlinear superparamagnetic magnetization, intermodulation products are generated and picked up in the detection coil. The dominant mixing component at frequency $f_1 + 2 \cdot f_2$ is demodulated. Its amplitude is proportional to the particle concentration in the sample. Details of the measurement principle and of the setup are given in [23,28,29].

2.8. Sample Preparation and cMID in Milk

The calibration curve in milk was prepared as described above. For this purpose, unconjugated antibiotics were dissolved in whole milk (3.5% total fat) instead of PBS. For determination of assay accuracy, spiked milk samples were prepared in whole milk (3.5% total fat) at concentrations of 4 ng·mL^{-1}, 8 ng·mL^{-1}, 20 ng·mL^{-1}, 40 ng·mL^{-1} and 200 ng·mL^{-1} for penicillin G and 1 ng·mL^{-1}, 10 ng·mL^{-1} and 50 ng·mL^{-1} for kanamycin. The spiked samples were incubated with 1.2 μg·mL^{-1} of biotinylated antibody diluted in PBS. After application of samples onto pre-coated and blocked IFCs, the assay procedure was carried out as described above.

2.9. Data Analysis

For all data analysis and data fitting using Hill Fit, GraphPad Prism 8.3.1 and Hill parameters calculated by GraphPad Prism were used. To compare sample intensities in relation to highest maximum signal (sample without competitor), Equation (1) was used. Calculating limit of detection (LOD) or maximum of detection (MOD) on the signal scale, Equation (2) or (3), respectively, were used. For calculation of concentration of LOD, MOD as well as spiked samples, Equation (4) was used. Recovery rates of spiked samples were determined using Equation (5):

$$B/B0\ ratio = \frac{Measuring\ signal_{Sample}}{Average\ measuring\ signal_{Sample\ without\ antibiotic}} \tag{1}$$

$$Signal_{Limit\ of\ Detection} = Average\ B/B0\ signal_{Saturated\ samples} - 3 \times SD_{Saturated\ samples} \tag{2}$$

$$Signal_{Maximum\ of\ Detection} = Average\ B/B0\ signal_{Background\ samples} + 3 \times SD_{Background\ samples} \tag{3}$$

$$Concentration_{Sample} = \left[\frac{IC_{50}^h \times \left(B/B0\ Signal_{Sample} - B_{max} \right)}{B/B0\ Signal_{Sample}^{Hill\ Slope}} \right]^{-\frac{1}{Hill\ Slope}} \tag{4}$$

$$Recovery\ Rate = \frac{Concentration\ detected}{Concentration\ spiked} \times 100\ [\%] \tag{5}$$

3. Results and Discussion

3.1. Generation of Penicillin-BSA and Kanamycin-BSA Conjugates and Antibody Affinity Determination

In order to establish a reliable assay for highly sensitive detection and quantification of penicillin and kanamycin in milk samples, penicillin-BSA and kanamycin-BSA conjugates had to be prepared and tested regarding their binding capacity with respective monoclonal antibodies (mAb). For testing the affinity of corresponding mAb towards self-conjugated penicillin-BSA conjugate (Figure 2A) or kanamycin-BSA conjugate (Figure 2B) in an ELISA, antibodies were titrated ranging from 9.76 ng·mL^{-1} up to 1250 ng·mL^{-1} against coated antibiotic-conjugate. A high affinity of respective antibody against their antigen-conjugate could be detected with EC$_{50}$-values of 91.8 ng·mL^{-1} for penicillin-specific mAb and 177.7 ng·mL^{-1} for kanamycin-specific mAb. Although anti-penicillin antibody resulted in a lower EC$_{50}$-value, a higher kanamycin density on BSA compared to the density of penicillin on BSA could be concluded based on the higher absorbance of anti-kanamycin antibody dose response, achieved after 10 min of substrate incubation. Additionally, a later saturation of measuring signal could be detected with kanamycin-BSA and respective monoclonal antibody, which underlines the higher antigen-density on the carrier protein. However, on those high-affine dose responses, a successful conjugation of penicillin as well as kanamycin with BSA could be demonstrated.

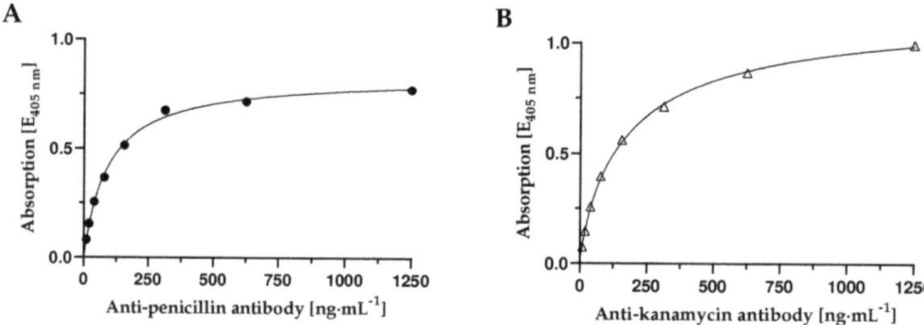

Figure 2. Affinity determination of (**A**) anti-penicillin G specific mAb against penicillin G-BSA conjugate and (**B**) anti-kanamycin specific mAb against kanamycin-BSA conjugate. For both, a microtiter plate was coated with respective antibiotic-BSA conjugate and remaining binding sites were blocked. Subsequently, antibody dilutions in the range from 9.76 ng·mL^{-1} to 1250 ng·mL^{-1} were applied and afterward labelled with mouse IgG-specific HRPO-conjugated secondary antibody. Absorbance was measured after 10 min of ABTS substrate incubation. $n = 1$.

3.2. Development of cMID for Detection of Penicllin and Kanamycin in Buffer

After successful generation of antibiotic-conjugates and the confirmation of high affinity binding of mAbs, first cMID experiments were performed. For this, biotinylated monoclonal antibodies are pre-incubated with antibiotics-containing samples. Afterward, the mixture is applied onto immunofiltration columns (IFC) containing antibiotic-carrier protein conjugate coated polyethylene matrices. While the sample is flushed through the IFC by gravity flow, a competitive binding reaction of biotinylated monoclonal antibodies between free soluble antigens in the sample and coated antigen-conjugates at the matrix takes place. Hence, non-saturated biotinylated antibodies are retained and subsequently can be magnetically labelled with streptavidin-functionalized superparamagnetic particles based on highly affine streptavidin-biotin reaction. Especially due to the comparable molecular weight and chemical properties of antibiotics and mycotoxins, coating concentration as well as antibody and magnetic particle concentration was adapted from intensive preliminary work of mycotoxin cMID assay development [23]. Additionally, each assay step was set to one hour, enabling full equilibrium of binding reactions. A schematic overview of cMID procedure is demonstrated in Figure 3A and assay times are presented in Figure 3B.

A

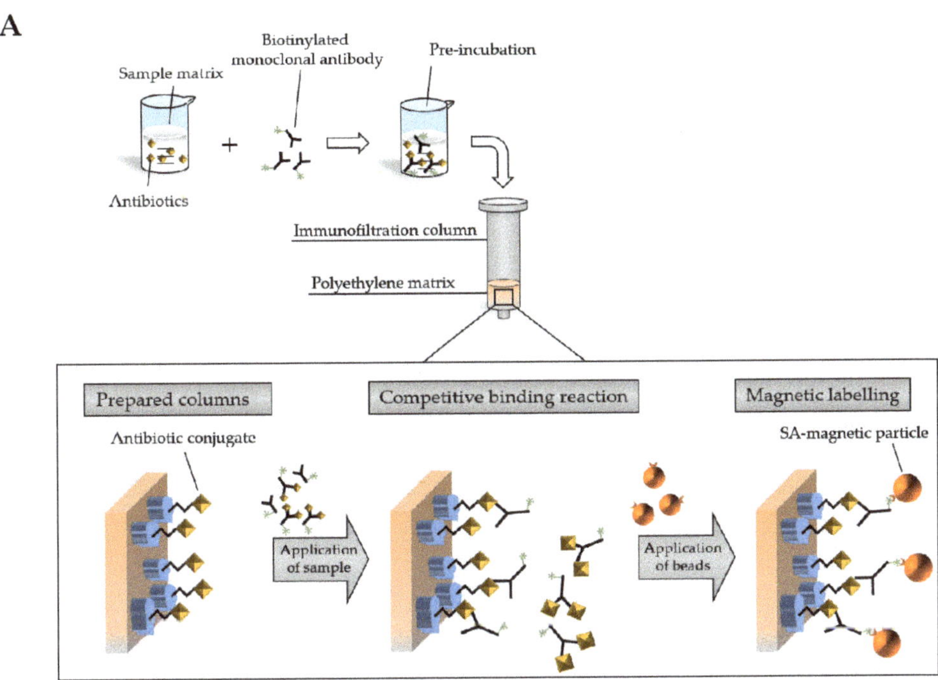

B

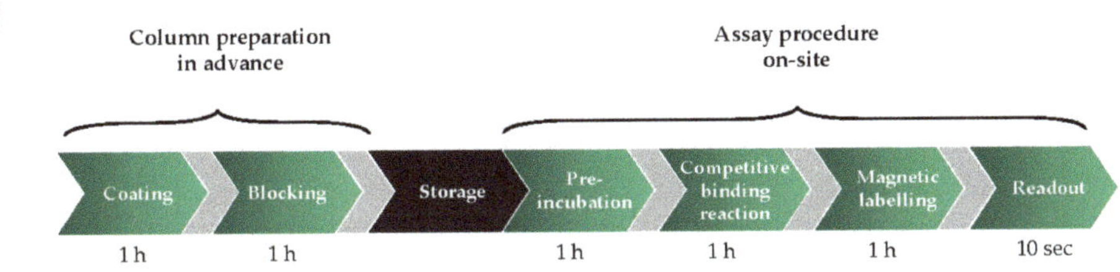

Figure 3. Schematic overview of competitive magnetic immunodetection assay procedure for detection of antibiotics in milk. (**A**) Biotinylated monoclonal antibodies are applied into the sample (either sample buffer or whole fat milk (WFM). Throughout pre-incubation, mAbs bind soluble antibiotic molecules. Afterward, the sample is applied onto antibiotic-conjugate coated and blocked polyethylene matrix of immunofiltration columns (IFCs). Here, a competitive binding reaction of monoclonal antibodies between coated antibiotic-conjugate and soluble antigen results in a retention of non-saturated antibodies within the IFC matrix. Subsequently, retained mAbs are magnetically labeled with streptavidin-functionalized magnetic particles (SA-magnetic particle). Finally, retained SA-magnetic particles can be detected and quantified by means of frequency mixing magnetic detection (FMMD). (**B**) Overview about assay steps with incubation times. Light grey arrows represent washing steps with 2 min of duration.

Calibration curve experiments with cMID for detection of penicillin G or kanamycin, respectively, were done by diluting penicillin G (Figure 4A) or kanamycin (Figure 4B) in range from 0.011 ng·mL^{-1} up to 1500 ng·mL^{-1} in phosphate-buffered saline (PBS), pre-incubation with respective monoclonal antibodies followed by application on pre-coated IFCs and subsequent magnetic labelling. Measuring signals were recorded using FMMD and normalized by calculating the B/B0 ratios using Equation (1). Calibration results are shown in Figure 4, LOD and MOD values were calculated using Equations (2) and (3) and are indicated.

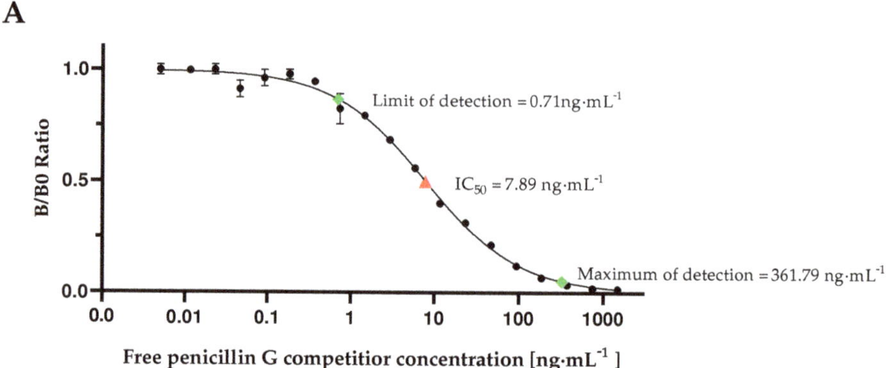

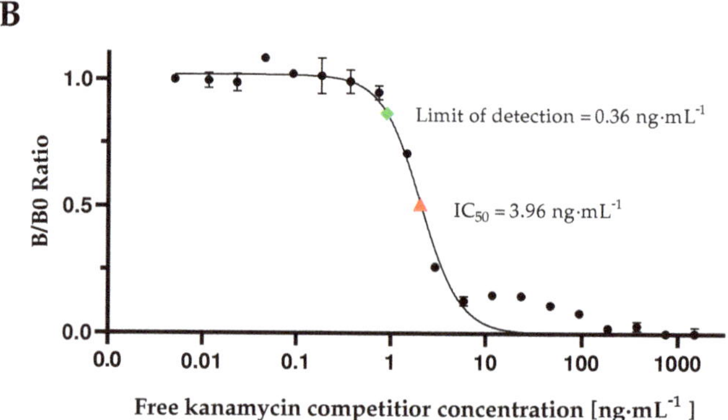

Figure 4. cMID calibration curves for detection of antibiotics in PBS sample buffer. cMID calibration curves for detection of antibiotics in PBS sample buffer. (**A**) Penicillin cMID calibration curve and (**B**) kanamycin cMID calibration curve. For both, columns were coated with respective antibiotic-BSA conjugate and subsequently blocked. After application of each antibiotic diluted from 0.011 ng·mL^{-1} up to 1500 ng·mL^{-1} pre-incubated with 1.2 µg·mL^{-1} of respective mAb, 60 µg·mL^{-1} streptavidin-functionalized superparamagnetic particles were applied and rinsed through the column. Readout was done using frequency mixing magnetic detection (FMMD). Each data point represents mean ± SD (n = 2). Limit of detection and, if possible, maximum of detection are indicated by green square. IC$_{50}$ (half maximal inhibitory concentration) is symbolized by red triangle.

In our B0 measuring, maximum signals of both calibration curves were at approximately 500 mV. Based on Equation (4), a limit of detection of 0.71 ng·mL^{-1} for penicillin G and a LOD of 0.36 ng·mL^{-1} for kanamycin could be determined. An approximately 2-fold lower IC$_{50}$-value for kanamycin with 3.96 ng·mL^{-1} in comparison to penicillin with 7.89 ng·mL^{-1} demonstrates a higher assay sensitivity. While for penicillin G a MOD of 361.79 ng·mL^{-1} was calculated according to Equation (3), this was not possible for kanamycin due to increasing B/B0 ratio of measuring signals in the range of 11.7 ng·mL^{-1} up to 93.7 ng·mL^{-1}. However, for penicillin G it could be demonstrated that the dynamic range (0.71 ng·mL^{-1} up to 361.79 ng·mL^{-1}) is greatly increased in comparison to commercially available test kits as EuroProxima Penicillin ELISA (r-biopharm, Darmstadt, Germany). For this ELISA-based assay, a calibration curve ranging from 0.125 ng·mL^{-1} up to 4 ng·mL^{-1} needs to be prepared which just covers the range of MRL for penicillin G. Hence, our results demonstrate an, in general, improved detection of penicillin G and kanamycin by employing our newly developed competitive magnetic immunodetection.

3.3. cMID Calibration Measurements in Whole Fat Milk (WFM)

With the previously shown calibration curve analysis in PBS, an efficient detection of antibiotics could be demonstrated. However, for performing the assay in WFM, studies of matrix interference, which could be reasoned by the inhibitory effects of fatty acids on antibody-binding or hydrophobic interaction of antigen, had to be done. To avoid such matrix effects, samples can be diluted in assay buffer, e.g., PBS, resulting in strong dilution of interfering substances. However, diluting the samples to be analyzed would also decrease the assay accuracy. Especially in the case of penicillin G with a low MRL of just 4 ng·mL^{-1}, too high dilution would increase the possibility of false negative results because penicillin G concentration might become lower than the detection limit of 0.71 ng·mL^{-1} in PBS. Hence, in this study, calibration curves were prepared in commercially available, controlled whole fat milk instead of PBS to analyze the influence of milk matrix effects on sensitivity and to determine cMID assay applicability (Figure 5). For this purpose, penicillin G (Figure 5A) and kanamycin (Figure 5B) were diluted in WFM in the same ranges as previously in PBS (Figure 4). For cMID analysis the samples were diluted twofold in PBS buffer, due to the addition of the respective antibody.

A

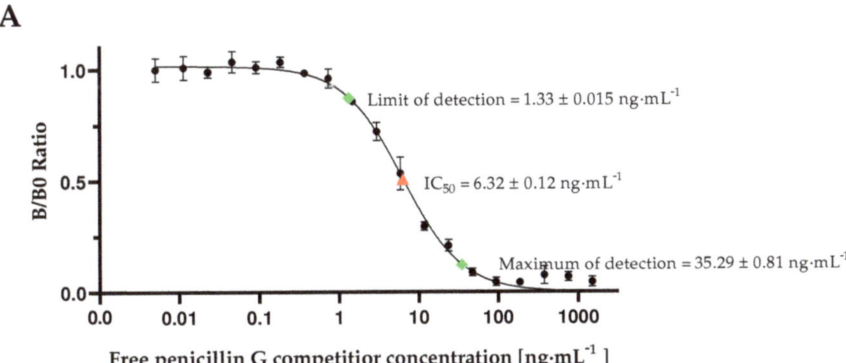

B

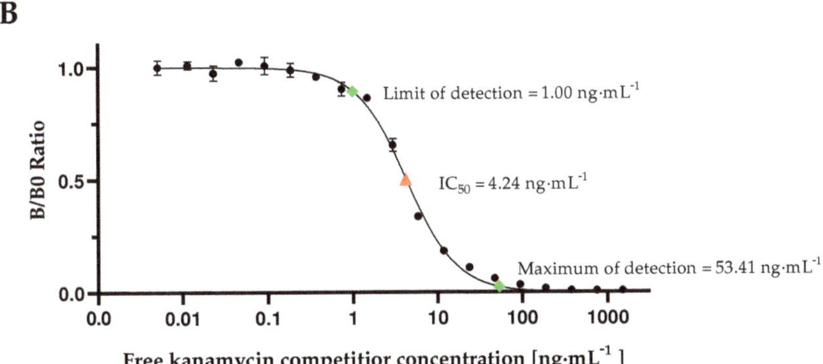

Figure 5. cMID calibration curves for detection of antibiotics spiked in whole fat milk. (**A**) Penicillin G calibration curve and (**B**) kanamycin calibration curve. Both antibiotics were diluted from 0.011 ng·mL^{-1}

up to 1500 ng·mL^{-1} in whole fat milk and were pre-incubated with 1.2 µg·mL^{-1} of respective biotinylated monoclonal antibodies (mAb). Afterward, the samples were applied onto 3.5 µg·mL^{-1} antibiotic-BSA conjugate coated and blocked IFC. Finally, 60 µg·mL^{-1} superparamagnetic particles functionalized with streptavidin were applied and rinsed through the column. Readout was done using FMMD. Each data point represents mean ± SD. For (**A**) penicillin, data points were averaged from two independent calibration curve experiments ($n = 4$) and for (**B**) kanamycin each data point represents the mean ± SD of $n = 2$. Limit of detection and maximum of detection are indicated by green square. IC$_{50}$ (half maximal inhibitory concentration) is symbolized by red triangle.

Especially in the case of penicillin G detection, an influence of WFM as matrix could be observed by the three-fold reduction of B0 measuring signal to a maximum of approximately 150 mV. Furthermore, an approximately ten-fold reduced dynamic detection range from 1.33 ± 0.015 ng·mL^{-1} up to 35.29 ± 0.81 ng·mL^{-1} compared to measurements in PBS was obtained. The reduced B0 signal as well as the reduced dynamic range of detection could be attributed to a lipid-mediated interference, which could inhibit the binding of monoclonal antibodies to coated antigen [34]. Compared to commercial ELISA kits such as EuroProxima Penicillin ELISA (r-biopharm, Darmstadt, Germany), in our approach a higher LOD of 1.33 ± 0.015 ng·mL^{-1} compared to 0.08 ng·mL^{-1} was obtained. However, the dynamic range obtained was still approximately ten-fold higher than in ELISA (up to 35.29 ± 0.81 ng·mL^{-1}). Currently further optimization experiments are planned to improve coating and antibody concentrations in a similar way as performed during our mycotoxin cMID assay development [23]. Hence, a further enhancement of assay sensitivity can be expected. Interestingly, in the case of IC$_{50}$-value, no relevant negative matrix interference in WFM could be detected for cMID. In contrast to penicillin G, for kanamycin cMID no reduction of the B0 signal was observed. A slight reduction of dynamic detection range from 1.00 ng·mL^{-1} up to 53.41 ng·mL^{-1} was determined, suggesting only a weak matrix effect. The IC$_{50}$-value of 4.24 ng·mL^{-1} was also comparable to the value from PBS calibration experiments. The obtained assay parameters were comparable to laboratory-based ELISA kits as e.g., Kanamycin ELISA kit (Creative Diagnostics, New York, USA) with a detection range from 0.5 ng·mL^{-1} up to 40.5 ng·mL^{-1}.

3.4. Spiked Sample Analysis and Determination of Recovery Rate

For determination of assay accuracy, milk samples were spiked with different concentrations of respective antibiotics and recovery rates were calculated using the fit function of previously obtained calibration curves (Figure 5). Penicillin was spiked at industrially relevant concentrations of 2 ng·mL^{-1}, 4 ng·mL^{-1}, 10 ng·mL^{-1} and 20 ng·mL^{-1}, and kanamycin at concentrations of 1 ng·mL^{-1}, 10 ng·mL^{-1} and 50 ng·mL^{-1} into milk samples. For penicillin G, recovery rates between 91.4% ± 14.9% up to 112.5% ± 13.1% could be achieved (Table 1). Compared to commercially available kits with maximally achieved recoveries of 86 ± 7% in milk (EuroProxima Penicillin ELISA, r-biopharm, Darmstadt, Germany), our newly developed assay demonstrated a superior assay performance. For kanamycin, recovery rates of 87.8% ± 4.3% for 1 ng·mL^{-1} and 94.1% ± 0.1% for 4 ng·mL^{-1} spiked samples were obtained (Table 1), which also demonstrate a higher assay accuracy compared to commercially available Kanamycin ELISA kit (Creative Diagnostics, New York, NY, USA) with an accuracy of 85% ± 10%. For the last spiked kanamycin sample with a concentration of 50 ng·mL^{-1}, a recovery rate of only 52.5% ± 0.7% was achieved. However, this result is not surprising since with this concentration the maximum detection limit of 53.41 ng·mL^{-1} is almost reached, making a reliable quantification difficult. Hence, these results reveal a high assay accuracy, a high sensitivity, and a good applicability of developed cMID for detecting antibiotics in milk samples.

Table 1. Detected concentration of spiked penicillin G or kanamycin in WFM using cMID with calculated recovery rates (penicillin G: $n = 4$; kanamycin: $n = 2$).

Analyte	Concentration Spiked [ng·mL^{-1}]	Concentration Detected [ng·mL^{-1}]	Recovery Rate [%]
Penicillin G	2	1.83 ± 0.30	91.4 ± 14.9
	4	3.90 ± 0.56	97.4 ± 14.0
	10	10.03 ± 0.67	100.3 ± 6.7
	20	22.6 ± 2.61	112.9 ± 13.1
Kanamycin	1	0.88 ± 0.04	87.8 ± 4.3
	10	9.40 ± 0.01	94.1 ± 0.1
	50	26.26 ± 0.36	52.5 ± 0.7

Hence, our results reveal a high assay accuracy, a high sensitivity and a good applicability of developed cMID for detecting and quantifying antibiotics in milk samples. Thus, the potential as a highly accurate detection and quantification method was demonstrated although just a limited number of spiked WFM samples were used. With this, a proof-of-concept approach was demonstrated. Before translating the method to commercial application, an extensive as well as a fit-for-purpose (FFP) validation process should be carried out to further confirm accuracy, reproducibility and robustness. A further applicability of this assay format could be a simpler qualitative detection of antibiotic residues in milk samples. In such a screening setup, the measured magnetic signal resulting from a contamination as low as the MRL could be defined as threshold. By this, a resulting measuring signal either defines a sample as contaminated above or below the MRL. To demonstrate this specific applicability, an additional FFP validation should be performed. Practically, additional spiking experiments with at least 20 different blank samples as well as 20 spiked samples containing relevant concentrations of respective antibiotics, at approximately 0.5 times the MRL, should be performed. According to this, for penicillin G a concentration of 2 ng·mL^{-1} should be used. In case of kanamycin an even lower concentrations of 10 ng·mL^{-1} would be recommended due to the higher assay accuracy of 94.1% ± 0.1% (Table 1). Such additional validation experiments would then be suitable to further confirm the practicability and overall assay accuracy of our method.

4. Conclusions

We demonstrate a newly developed proof-of-concept method for detection of penicillin G and kanamycin in milk. With our cMID approach, a detection of penicillin G in the range from 1.33 ± 0.015 ng·mL^{-1} up to 35.29 ± 0.81 ng·mL^{-1} in WFM with an accuracy ranging from 91.4% up to 112.5% and detection of kanamycin contaminations in WFM ranging from 1.00 ng·mL^{-1} up to 53.41 ng·mL^{-1}, with recovery rates between 87.8% up to 94.1% in the linear range of our calibration curve, is possible. Based on the MRL of 4 ng·mL^{-1} for penicillin, a direct analysis of milk can be performed following the addition of antibody, since no further preparation steps of milk are necessary. For kanamycin, samples should be at least diluted threefold, although no matrix effects could be noticed. This is reasoned by the high assay sensitivity. Milk samples with kanamycin concentrations lower than the MRL will be detected as positive, resulting in false positive assay outcome. However, compared to commercially ELISA kits, a higher range of detection of penicillin G and kanamycin was demonstrated. By spiking whole fat milk samples with different concentrations of these antibiotics and re-calculating concentrations using corresponding calibration measurements, a superior assay accuracy ranging from 91.4% up to 112.5% for penicillin G detection and 87.8% up to 94.1% in the linear range of calibration curve for kanamycin detection could furthermore be demonstrated.

Our easily applicable assay setup in combination with the handheld FMMD readout device could, in contrast to other laboratory based standard procedures, be suited for fast on-site testing and thus be applicable even for farmers. This easy assay setup with broad range of quantification might then lead to better estimation of necessary detoxification time after antibiotic treatment of milk cows. However, when using raw milk, an even stronger matrix interference as described in Section 3.3 when analyzing penicillin G contaminations could be expected, which further diminishes the dynamic range of detection. For this, further preparation steps, such as the usage of a detergent, should be tested, reducing the risk of lipid-mediated matrix interference by disrupting lipid interaction. Nevertheless, if a too strong reduction of dynamic range of detection is observed, the assay could still be used as qualitative screening approach after further, appropriate validation, as suggested in Section 3.4.

However, for fast on-site testing the overall assay time needs to be further reduced. Hence, in further studies a reduction of assay steps and time will be addressed by, e.g., testing optimized incubation times for pre-incubation, competitive binding reaction and magnetic labelling. Based on our experience, this should result in an approximately six-fold reduced assay time from 3 h (Figure 3B) to less than 30 min, as demonstrated for a sandwich based MID approach by Rettcher et al. (2015) [29]. Additionally, the assay setup can be easily adapted for detecting other antibiotics in milk, just by exchanging the coating antigen and using other specific antibodies. Our approach might thus be a platform-like assay system for antibiotic detection in general. By the distinction of different magnetic particle types by their characteristic measuring signals, as demonstrated by Achtsnicht et al. (2019) [35], even a multiplex detection of several antibiotics in one sample is feasible, which will be addressed in further studies. Hence, by pre-functionalizing different magnetic particles with monoclonal antibodies targeting several antibiotics, multiple residues could be detected in one sample. However, even in this setup, the newly established cMID assay demonstrates an easily applicable and powerful tool for on-site testing, which allows sensitive and reliable detection and quantification of penicillin G and kanamycin in WFM without laboratory-based cultivation equipment, absorbance measuring device or cost-intensive LC-MS/MS-based analysis methods.

Author Contributions: Conceptualization, J.P. and F.S.; methodology, J.P.; validation, J.P. and D.D.; formal analysis, J.P.; resources, F.S., H.S. and H.-J.K.; investigation, J.P. and D.D.; data curation, D.D. and J.P.; writing—original draft preparation, J.P.; writing—review and editing, F.S., H.-J.K. and H.S.; visualization, J.P.; supervision, H.S. and F.S.; project administration, F.S.; funding acquisition, F.S. All authors have read and agreed to the published version of the manuscript.

Funding: This research was funded by the land North Rhine-Westphalia and the European Regional Development Fund (ERDF; German: EFRE) under grant number EFRE-0801298, reference number LS-2-1-014 and the APC was funded by Fraunhofer IME.

Acknowledgments: The authors would like to thank Katrin Kloth-Everding and Christian Baumgartner from the Milchprüfring Bayern e.V. for kindly providing the anti-penicillin G monoclonal antibody. Additionally, the authors would like to thank Nadja Vöpel for her helpful advice given in discussion and critically reviewing the manuscript.

Conflicts of Interest: The authors declare no conflict of interest. The funders had no role in the design of the study; in the collection, analyses, or interpretation of data; in the writing of the manuscript, or in the decision to publish the results.

References

1. Fleming, A. On the Antibacterial Action of Cultures of a Penicillium, with Special Reference to their Use in the Isolation of B. influenzæ. *Br. J. Exp. Pathol.* **1929**, *10*, 226–236. [CrossRef]

2. Hotta, K.; Kondo, S. Kanamycin and its derivative, arbekacin: Significance and impact. *J. Antibiot. (Tokyo)* **2018**, *71*, 417–424. [CrossRef] [PubMed]

3. Van Boeckel, T.P.; Brower, C.; Gilbert, M.; Grenfell, B.T.; Levin, S.A.; Robinson, T.P.; Teillant, A.; Laxminarayan, R. Global trends in antimicrobial use in food animals. *Proc. Natl. Acad. Sci. USA* **2015**, *112*, 5649–5654. [CrossRef]

4. Fejzic, N.; Begagic, M.; Seric-Haracic, S.; Smajlovic, M. Beta lactam antibiotics residues in cow's milk: Comparison of efficacy of three screening tests used in Bosnia and Herzegovina. *Bosn. J. Basic Med. Sci.* **2014**, *14*, 155–159. [CrossRef]

5. Li, C.; Zhang, Y.; Eremin, S.A.; Yakup, O.; Yao, G.; Zhang, X. Detection of kanamycin and gentamicin residues in animal-derived food using IgY antibody based ic-ELISA and FPIA. *Food Chem.* **2017**, *227*, 48–54. [CrossRef]

6. Cuong, N.V.; Padungtod, P.; Thwaites, G.; Carrique-Mas, J.J. Antimicrobial Usage in Animal Production: A Review of the Literature with a Focus on Low- and Middle-Income Countries. *Antibiotics (Basel)* **2018**, *7*, 75. [CrossRef]

7. Sachi, S.; Ferdous, J.; Sikder, M.H.; Azizul Karim Hussani, S.M. Antibiotic residues in milk: Past, present, and future. *J. Adv. Vet. Anim. Res.* **2019**, *6*, 315–332. [CrossRef]

8. Payne, M.A.; Craigmill, A.; Riviere, J.E.; Webb, A.I. Extralabel use of penicillin in food animals. *J. Am. Vet. Med. Assoc.* **2006**, *229*, 1401–1403. [CrossRef]

9. Kummerer, K. Antibiotics in the aquatic environment–a review–part II. *Chemosphere* **2009**, *75*, 435–441. [CrossRef] [PubMed]

10. Farouk, F.; Azzazy, H.M.; Niessen, W.M. Challenges in the determination of aminoglycoside antibiotics, a review. *Anal. Chim. Acta* **2015**, *890*, 21–43. [CrossRef]

11. Blumenthal, K.G.; Peter, J.G.; Trubiano, J.A.; Phillips, E.J. Antibiotic allergy. *Lancet* **2019**, *393*, 183–198. [CrossRef]

12. Kanwar, N.; Scott, H.M.; Norby, B.; Loneragan, G.H.; Vinasco, J.; McGowan, M.; Cottell, J.L.; Chengappa, M.M.; Bai, J.; Boerlin, P. Effects of ceftiofur and chlortetracycline treatment strategies on antimicrobial susceptibility and on tet(A), tet(B), and bla CMY-2 resistance genes among E. coli isolated from the feces of feedlot cattle. *PLoS ONE* **2013**, *8*, e80575. [CrossRef] [PubMed]

13. Beyene, T. Veterinary Drug Residues in Food-animal Products: Its Risk Factors and Potential Effects on Public Health. *J. Vet. Sci. Technol.* **2016**, *7*. [CrossRef]

14. European Union. Commission Regulation (EU) No 37/2010 on Pharmacologically Active Substances and Their Classification Regarding Maximum Residue Limits in Foodstuffs of Animal Origin. 2009. Available online: https://eur-lex.europa.eu/legal-content/EN/TXT/HTML/?uri=CELEX:32010R0037&from=EN (accessed on 29 November 2020).

15. Shen, X.; Liu, L.; Xu, L.; Ma, W.; Wu, X.; Cui, G.; Kuang, H. Rapid detection of praziquantel using monoclonal antibody-based ic-ELISA and immunochromatographic strips. *Food Agric. Immunol.* **2019**, *30*, 913–923. [CrossRef]

16. Kloth, K.; Rye-Johnsen, M.; Didier, A.; Dietrich, R.; Martlbauer, E.; Niessner, R.; Seidel, M. A regenerable immunochip for the rapid determination of 13 different antibiotics in raw milk. *Analyst* **2009**, *134*, 1433–1439. [CrossRef]

17. Mukunzi, D.; Isanga, J.; Suryoprabowo, S.; Liu, L.; Kuang, H. Rapid and sensitive immunoassays for the detection of lomefloxacin and related drug residues in bovine milk samples. *Food Agric. Immunol.* **2017**, *28*, 599–611. [CrossRef]

18. Herrera-Herrera, A.V.; Hernández-Borges, J.; Rodríguez-Delgado, M.A.; Herrero, M.; Cifuentes, A. Determination of quinolone residues in infant and young children powdered milk combining solid-phase extraction and ultra-performance liquid chromatography–tandem mass spectrometry. *J. Chromatogr. A* **2011**, *1218*, 7608–7614. [CrossRef]

19. Junza, A.; Amatya, R.; Barrón, D.; Barbosa, J. Comparative study of the LC–MS/MS and UPLC–MS/MS for the multi-residue analysis of quinolones, penicillins and cephalosporins in cow milk, and validation according to the regulation 2002/657/EC. *J. Chromatogr. B* **2011**, *879*, 2601–2610. [CrossRef]

20. Freitas, A.; Barbosa, J.; Ramos, F. Development and validation of a multi-residue and multiclass ultra-high-pressure liquid chromatography-tandem mass spectrometry screening of antibiotics in milk. *Int. Dairy J.* **2013**, *33*, 38–43. [CrossRef]

21. Moreno-Gonzalez, D.; Hamed, A.M.; Gilbert-Lopez, B.; Gamiz-Gracia, L.; Garcia-Campana, A.M. Evaluation of a multiresidue capillary electrophoresis-quadrupole-time-of-flight mass spectrometry method for the determination of antibiotics in milk samples. *J. Chromatogr. A* **2017**, *1510*, 100–107. [CrossRef]

22. Wu, X.; Suryoprabowo, S.; Kuang, H.; Liu, L. Detection of aminophylline in serum using an immunochromatographic strip test. *Food Agric. Immunol.* **2020**, *31*, 33–44. [CrossRef]

23. Pietschmann, J.; Spiegel, H.; Krause, H.J.; Schillberg, S.; Schroper, F. Sensitive Aflatoxin B1 Detection Using Nanoparticle-Based Competitive Magnetic Immunodetection. *Toxins* **2020**, *12*, 337. [CrossRef] [PubMed]

24. Molina, M.P.; Althaus, R.L.; Molina, A.; Fernández, N. Antimicrobial agent detection in ewes' milk by the microbial inhibitor test brilliant black reduction test—BRT AiM®. *Int. Dairy J.* **2003**, *13*, 821–826. [CrossRef]

25. Romero, T.; Van Weyenberg, S.; Molina, M.P.; Reybroeck, W. Detection of antibiotics in goats' milk: Comparison of different commercial microbial inhibitor tests developed for the testing of cows' milk. *Int. Dairy J.* **2016**, *62*, 39–42. [CrossRef]

26. Achtsnicht, S.; Neuendorf, C.; Fassbender, T.; Nolke, G.; Offenhausser, A.; Krause, H.J.; Schroper, F. Sensitive and rapid detection of cholera toxin subunit B using magnetic frequency mixing detection. *PLoS ONE* **2019**, *14*, e0219356. [CrossRef] [PubMed]

27. Meyer, M.H.; Hartmann, M.; Krause, H.J.; Blankenstein, G.; Mueller-Chorus, B.; Oster, J.; Miethe, P.; Keusgen, M. CRP determination based on a novel magnetic biosensor. *Biosens Bioelectron* **2007**, *22*, 973–979. [CrossRef]

28. Meyer, M.H.; Stehr, M.; Bhuju, S.; Krause, H.J.; Hartmann, M.; Miethe, P.; Singh, M.; Keusgen, M. Magnetic biosensor for the detection of Yersinia pestis. *J. Microbiol. Methods* **2007**, *68*, 218–224. [CrossRef]

29. Rettcher, S.; Jungk, F.; Kuhn, C.; Krause, H.J.; Nolke, G.; Commandeur, U.; Fischer, R.; Schillberg, S.; Schroper, F. Simple and portable magnetic immunoassay for rapid detection and sensitive quantification of plant viruses. *Appl. Environ. Microbiol.* **2015**, *81*, 3039–3048. [CrossRef]

30. Achtsnicht, S.; Todter, J.; Niehues, J.; Teloken, M.; Offenhausser, A.; Krause, H.J.; Schroper, F. 3D Printed Modular Immunofiltration Columns for Frequency Mixing-Based Multiplex Magnetic Immunodetection. *Sensors* **2019**, *19*, 148. [CrossRef]

31. Krause, H.; Wolters, N.; Zhanga, Y.; Offenhäusser, A.; Miethe, P.; Meyer, M.H.F.; Hartmann, M.; Keusgen, M. Magnetic particle detection by frequency mixing for immunoassay applications. *J. Magn. Magn. Mater.* **2007**, *311*, 436–444. [CrossRef]

32. Venkataramana, M.; Rashmi, R.; Uppalapati, S.R.; Chandranayaka, S.; Balakrishna, K.; Radhika, M.; Gupta, V.K.; Batra, H.V. Development of sandwich dot-ELISA for specific detection of Ochratoxin A and its application on to contaminated cereal grains originating from India. *Front. Microbiol.* **2015**, *6*, 511. [CrossRef] [PubMed]

33. Haasnoot, W.; Stouten, P.; Cazemier, G.; Lommen, A.; Nouws, J.F.; Keukens, H.J. Immunochemical detection of aminoglycosides in milk and kidney. *Analyst* **1999**, *124*, 301–305. [CrossRef] [PubMed]

34. Tate, J.; Ward, G. Interferences in immunoassay. *Clin. Biochem. Rev.* **2004**, *25*, 105–120. [PubMed]

35. Achtsnicht, S.; Pourshahidi, A.M.; Offenhausser, A.; Krause, H.J. Multiplex Detection of Different Magnetic Beads Using Frequency Scanning in Magnetic Frequency Mixing Technique. *Sensors* **2019**, *19*, 2599. [CrossRef] [PubMed]

Publisher's Note: MDPI stays neutral with regard to jurisdictional claims in published maps and institutional affiliations.

Article

Vibrational Spectroscopy Coupled to a Multivariate Analysis Tiered Approach for Argentinean Honey Provenance Confirmation

Tito Damiani [1], Rosa M. Alonso-Salces [2,3], Inés Aubone [2], Vincent Baeten [4], Quentin Arnould [4], Chiara Dall'Asta [1,*], Sandra R. Fuselli [2,5] and Juan Antonio Fernández Pierna [4]

[1] Department of Food and Drugs, University of Parma, Viale delle Scienze 17/A, 43124 Parma, Italy; tito.damiani@studenti.unipr.it

[2] Grupo de Investigación Microbiología Aplicada, Centro de Investigación en Abejas Sociales, Facultad de Ciencias Exactas y Naturales, Universidad Nacional de Mar del Plata, Dean Funes B7602AYL, Mar del Plata 3350, Argentina; rosamaria.alonsosalces@gmail.com (R.M.A.-S.); inesaubone@gmail.com (I.A.); sandra.fuselli@gmail.com (S.R.F.)

[3] Departamento de Biología, CONICET, Facultad de Ciencias Exactas y Naturales, Universidad Nacional de Mar del Plata, Funes 3350, Mar del Plata 7600, Argentina

[4] Quality and Authentication of Products Unit, Knowledge and Valorization of Agricultural Products Department, Walloon Agricultural Research Centre (CRA-W), Chée de Namur, 24, 5030 Gembloux, Belgium; v.baeten@cra.wallonie.be (V.B.); q.arnould@cra.wallonie.be (Q.A.); j.fernandez@cra.wallonie.be (J.A.F.P.)

[5] Comisión de Investigaciones Científicas (CIC), La Plata, Argentina Camino General Belgrano 526, La Plata 1900, Argentina

[*] Correspondence: chiara.dallasta@unipr.it; Tel.: +39-0521-905406

Received: 15 September 2020; Accepted: 10 October 2020; Published: 13 October 2020

Abstract: In the present work, the provenance discrimination of Argentinian honeys was used as case study to compare the capabilities of three spectroscopic techniques as fast screening platforms for honey authentication purposes. Multifloral honeys were collected among three main honey-producing regions of Argentina over four harvesting seasons. Each sample was fingerprinted by FT-MIR, NIR and FT-Raman spectroscopy. The spectroscopic platforms were compared on the basis of the classification performance achieved under a supervised chemometric approach. Furthermore, low- mid- and high-level data fusion were attempted in order to enhance the classification results. Finally, the best-performing solution underwent to SIMCA modelling with the purpose of reproducing a food authentication scenario. All the developed classification models underwent to a "year-by-year" validation strategy, enabling a sound assessment of their long-term robustness and excluding any issue of model overfitting. Excellent classification scores were achieved by all the technologies and nearly perfect classification was provided by FT-MIR. All the data fusion strategies provided satisfying outcomes, with the mid- and high-level approaches outperforming the low-level data fusion. However, no significant advantage over the FT-MIR alone was obtained. SIMCA modelling of FT-MIR data produced highly sensitive and specific models and an overall prediction ability improvement was achieved when more harvesting seasons were used for the model calibration (86.7% sensitivity and 91.1% specificity). The results obtained in the present work suggested the major potential of FT-MIR for fingerprinting-based honey authentication and demonstrated that accuracy levels that may be commercially useful can be reached. On the other hand, the combination of multiple vibrational spectroscopic fingerprints represents a choice that should be carefully evaluated from a cost/benefit standpoint within the industrial context.

Keywords: honey; vibrational spectroscopy; geographical origin; chemometrics; data fusion

1. Introduction

According to the European Union Council Directive 2001/110/EC [1] and FAO/WHO Codex Alimentarius [2], honey is defined as the natural substance produced by *Apis mellifera* bees from plant nectar or excretions of plant-sucking insects. As a relative expensive food commodity, honey is known to be highly vulnerable to adulteration with the main concern historically being its dilution with cheaper sugars and/or syrups. Nowadays, the premium price usually commanded by mono-floral and mono-geographic products encourages other fraud practices such as false origin labelling or misdescription [3].

Reliable analytical methods for the honey authenticity assessment are highly claimed and lot of research has been undertaken in this field. Botanical origin is traditionally confirmed by melissopalynology, a microscopy-based qualitative and quantitative characterization of pollen [4]. This technique has been tested also for geographical discrimination purposes, but its application suffers from methodological shortages and limitations [3,5]. Consequently, novel alternative approaches have been proposed, including those based on mass spectrometry, vibrational spectroscopy and molecular biology [6–8]. The targeted quantification of specific compounds indicative for certain properties and/or origin would represent the most straightforward approach for food authentication; the comparison of the measured parameter with a control limit would empower the direct assessment of the product compliance and might also be used for forensic purposes [9]. However, finding reliable authenticity markers for honey's botanical/geographical origin proved to be a hard task due to the number of factors affecting its chemical composition (e.g., beekeeping technique, harvest and storage environmental conditions, etc.). In addition, the analytical output may strongly depend on the adopted sample preparation procedure, hindering the data comparison and interpretation [10].

Over the past few years, new food testing strategies based on the so-called fingerprinting approaches have been introduced. The intrinsic aim of food fingerprinting is the non-targeted detection of as many features as technically possible, by means of high-throughput techniques, to gain a comprehensive insight into the sample composition. The recorded output consists of multidimensional datasets which, beside relevant information, may also contain unintended systematic and random variation. For this reason, mathematical and statistical tools (multivariate analysis/chemometrics) constitute an integral part of the fingerprinting workflow for the extraction of meaningful information from the raw data [9]. A review of the main fingerprinting technologies has been published by Ellis et al., with particular interest toward vibrational spectroscopy techniques, namely Raman, near- and mid-infrared spectroscopy [11]. These platforms offer non-destructive and cost-effective solutions to get quick spectral information about the tested material; the easy-of-use and potential on/in-line implementation represent further advantages over traditional methods that contributed to their spread in virtually all branches of agricultural and food industries [12].

In the honey authenticity field, the potential of vibrational spectroscopy coupled to multivariate data analysis to confirm the product's claimed provenance [6,13–17] and/or botanical origin [18–20] has been widely investigated. Most of the published works are represented by truthful feasibility studies that demonstrated the capability of the employed technologies to capture differences between the analysed honey samples. To this end, discriminant analysis (DA) techniques have been used to develop supervised classification models that would correctly assign each sample to its belonging class. However, in real-world authentication contexts, no information is normally available about the alternative classes to which the tested item may belong. Indeed, the goal is typically to establish whether the analysed sample is compliant or not with a defined reference standard. For these reasons, DA methods have been defined inappropriate for solving food authenticity problems by several authors [21–23]. In contrast, one-class classifier (OCC) approaches should be preferred. Furthermore, the sample collection in the above-mentioned studies was most often limited to 1–2 years, thus hardly representative of the potential seasonal variability. This has certainly posed some limitations for a solid validation of the achieved classification results. As a matter of fact, the adaption of existing models to new harvests is a problem scarcely addressed in pilot studies, usually due to the limited samples and/or resources availability. Nevertheless, it represents an essential challenge to be faced for a relevant implementation of non-targeted fingerprinting approaches in routine analysis [24].

The present work deals with the geographical origin discrimination of Argentinian honeys. Multifloral honeys were collected from three main honey-producing Argentinian provinces (i.e., Buenos Aires, Catamarca, Misiones) and the sampling was repeated over four harvesting seasons, from 2014 to 2017. Each sample was fingerprinted by near-infrared (NIR), Fourier-transform mid-infrared (FT-MIR) and Raman (FT-Raman) spectroscopy. The main intention was not the development of a multivariate model able to correctly classify the analysed samples according to their provenance. Rather, the aim was to use this survey as a case study to compare the capabilities of the employed spectroscopic techniques as fast screening platforms for honey authentication purposes. In order to further improve the results obtained by the individual techniques, different data fusion strategies were attempted. Finally, the best-performing solution (i.e., either individual or fused data) was further modelled using an OCC approach with the purpose of reproducing a food authentication scenario and establish whether commercially useful accuracy levels can be reached. All the developed classification models underwent to a "year-by-year" validation strategy that enabled a sound assessment of their long-term robustness and excluded any issue of model overfitting.

2. Materials and Methods

2.1. Sample Collection

Authentic and traceable multifloral honey samples were collected from three main honey-producing provinces of Argentina: Buenos Aires (BA), Catamarca (Cat) and Misiones (Mis) (Figure S1), within the framework of the Argentinean National Projects PICT 3264/2014 and PICT 0774/2017, following the instructions depicted on the Projects' analytical plan, and used for the scope of the present study. The samples (about 1 Kg of raw honey each) were provided directly by beekeepers and/or honey producer cooperatives along with farming information: harvest date and conditions, declared botanical origin, field or hive address and GPS coordinates, agricultural system, treatments, etc. The honeys were harvested between April and August and the sampling was repeated over four harvesting seasons (i.e., 2014, 2015, 2016 and 2017). Collected information on honey samples are to be considered part of the above-mentioned projects, and may be available upon request according to the data protection policy.

From here on, the sample batches (i.e., honeys from each harvest) are referred to as HN2014, HN2015, HN2016, HN2017, respectively. The total number of samples was $n = 502$ and an overview of the sample set is given in Table S1. After collection, the honeys were stored in screw-capped glass containers, in the dark and at 4 °C, until analysis.

2.2. Instrumental Analysis

All the collected samples were fingerprinted by means of FT-MIR, FT-Raman and NIR. After the collection, each sample batch (i.e., harvest) was scanned over a 14-day period. Prior to the analysis, the honeys were incubated at 40 °C and manually stirred in order to dissolve any crystalline residue material. Quality control materials were scanned throughout the whole analysis in order to monitor potential batch-to-batch instrumental drift.

FT-MIR spectra were recorded in attenuated total reflection (ATR) mode, on a Vertex 70 FT-IR spectrometer (Bruker, Billerica, MA, USA), equipped with a Globar source, a DLaTGS detector and a Golden Gate ATR cell (Specac Ltd., Orpington, UK). Analyses were carried out in triplicate, placing the honey samples directly on the ATR crystal. All the spectra were computed at 4 cm^{-1} resolution, across the spectral range 4000–600 cm^{-1} and averaging a total of 64 scans. Data export was performed by Opus 7.2 software (Bruker).

FT-Raman spectra were collected on a Vertex 70 equipped with the RAM II add-on module (Bruker), a laser source emitting at 1064 nm and a Ge$^{(418\text{-}T/R)}$ detector cooled by liquid N_2. The laser power was set to 0.8 W. Honey samples were placed in a glass tube and analyzed in duplicate, across the spectral range 3600–0 cm^{-1}, at a nominal resolution of 4 cm^{-1}. Each spectrum was obtained by averaging 128 scans and exported with Opus 7.2 software (Bruker).

NIR spectroscopic analysis was performed on an XDS Vis/NIR spectrometer (FOSS Analytical, Hilleroed, Denmark) equipped with a tungsten halogen lamp and a dual detector Si (400–1100 nm) and PbS (1100–2500 nm). The spectra were recorded in transflectance mode, directly depositing the honey on the golden reflector. The analysis ran in duplicate and a total of 16 scans were averaged for each spectrum, at a nominal resolution of 2 nm, across the spectral range 400–2500 nm. Signal acquisition and export were performed by ISIscan software (FOSS Analytical).

2.3. Statistical Data Analysis

All the chemometric computations were carried out using Matlab v2019b (The Mathworks, Inc., Natick, MA, USA) and the PLS Toolbox (Eigenvector Research, Inc., Manson, WA, USA).

2.3.1. Data Preprocessing

Prior to any exploratory or classification analysis, spectral preprocessing was applied to reduce the impact of unwanted sources of variability on the overall signal, thus highlighting the chemical information contained in the spectra. Different algorithms for spectral pretreatment, namely 1st and 2nd order derivative according to the Savitzky–Golay method (S-G), multiplicative scatter correction (MSC) and standard normal variate (SNV), were tested both on their own and in combination. The SNV and MSC are both designed to remove from reflectance spectra part of the variability that may be caused by scattering effects. In many cases, these two spectral pretreatment produced very similar results, so that they are widely regarded as exchangeable [25]. S-G derivative filter emphasizes band width, position, and separation while simultaneously reducing baseline and background effects [26].

2.3.2. Unsupervised Pattern Recognition

After the preprocessing, principal component analysis (PCA) was performed as exploratory data analysis for the detection of evident outlying samples and/or potential data structures in a reduced-dimension space. The underlying concept of the PCA is to decrease the dimensionality of a dataset containing a large number of interrelated variables, while retaining as much as possible of the initial data variation. The original descriptors are "compressed", through linear combination, into a new set of uncorrelated variables (i.e., principal components, PCs), which point in the directions of maximal variance. The so-called scores and loadings constitute the main output of the PCA. The scores represent the newly computed latent variables onto which the objects are projected, therefore they can be interpreted in exactly the same way as any other variable. On the other hand, the loadings are the weights given to the original variables during the computation of the PCs; thus, they determine what a PC represent. Both scores and loadings can be graphically plotted as line or scatter plots [27].

2.3.3. Supervised Pattern Recognition and Validation Strategy

The employed spectroscopic techniques were compared on the basis of the classification performance achieved under a supervised chemometric approach, by using partial least squares discriminant analysis (PLS-DA) as classification algorithm. PLS-DA is arguably the most widely used DA technique, particularly suitable for dealing with data matrices characterized by a large number of highly correlated variables, such as spectroscopic data. PLS-DA can be regarded as a linear two-class classifier, although extension to more than two groups is also possible. The method aims to find a linear decision function(s) that divides the multidimensional variable space into as many regions as the number of classes. The objects are then projected onto lines orthogonal to this function and their distance along this discriminator is considered as discriminant score [28].

Binary PLS-DA models were generated on each data block, considering two geographical regions at once (i.e., BA-Mis, BA-Cat, Cat-Mis). At first, the models were built including all the harvesting seasons and optimized through "leave-one-out" cross-validation. Afterwards, the so-called receiver operating characteristic (ROC) curves were derived. ROC curves are widely used in many application fields as they allow a straightforward comparison of binary classifier systems. In the multivariate case, the curves are built varying the criterion threshold at which the classification is performed. Model's sensitivity (i.e., fraction of compliant objects correctly accepted) and specificity (i.e., fraction of alien objects correctly rejected) are computed at each step and graphically represented in a two-axis Cartesian plot, in which 1-specificity is usually reported on the x-axis against the sensitivity on the y-axis. Experimental outcomes are connected by a line that constitutes the ROC curve. The area under the curve (AUC) is often used as summary measure of the general discrimination quality of the model. Intuitively, the larger the AUC, the higher the model classification ability. The ideal situation would be with both sensitivity and specificity equal to 1, which corresponds to a curve passing through the top-left corner of the graph and an AUC = 1; in contrast, a curve lying on the diagonal bisector (corresponding to an AUC = 0.5) suggests no discrimination [23].

Since ROC curves were built upon a cross-validation procedure, which may be prone to overfitting, the results reliability was ensured by the following validation strategy. At first, models were trained on the HN2014 and the provenance of HN2015 was predicted. Afterwards, the training set was augmented with the HN2015 samples and the models, upon re-optimization, were applied for the prediction of HN2016 provenance. As final step, HN2014, HN2015 and HN2016 were included in the training set and the HN2017 samples were classified. In this manner, the whole process involved three external validation steps independent of each other; thus, it can be considered much more reliable than a cross-validation approaches [29]. The validation scheme is summarized in Figure S2.

2.3.4. Data Fusion

Since each honey was fingerprinted by three spectroscopic techniques, three different data matrices for the same sample set were obtained. The process of integrating multiple data blocks into a single global model is called data fusion (DF) and can lead to improvements of the classification accuracy respect to the individual data sources. Essentially, three DF strategies have been proposed in literature according to the degree of information merged: low, mid- and high-level data fusion (LL-, ML- and HL-DF, respectively). In LL-DF, data from all sources are simply concatenated column-wise into a single array. The merged matrix is then processed by the desired chemometric technique. ML-DF operates in a similar way, but relevant features are previously extracted from each data sources, separately. These features can be original descriptors identified as relevant or, more commonly, latent variables (e.g., PCA scores). The so-extracted variables are then concatenated prior to the multivariate data analysis. Lastly, in the HL-DF, separate models are built on the individual data blocks and the fusion occurs at the decision level, i.e., the individual predictions are integrated into a single final response. A more detailed description of DF methodologies employed in food and beverage authentication can be found in [30].

In the present study, LL-, ML- and HL-DF were attempted for the HN2017 prediction (i.e., last step of the year-by-year validation) with the aim of improving the performance of the single techniques. Briefly:

LL-DF: FT-MIR, FT-Raman and NIR data blocks consisted of 1349, 3009 and 751 variables, respectively. Each dataset was preprocessed according to its optimal spectral pretreatment prior to the concatenation. As a result, each sample was described by 5109 predictors. Autoscaling was applied to the fused matrix before further modelling;

ML-DF: PCA was separately performed on the training set of each data block. HN2017 objects were projected onto the PCs space so that both training and test sets were described by the same (latent) variables. Thereafter, PCA scores obtained from the individual blocks were merged and used for subsequent modelling;

HL-DF: The provenance of HN2017 samples was separately predicted carrying out PLS-DA on the individual data blocks as described in Section 2.3.3. Therefore, three column vectors containing the predicted classes were obtained and merged into a single array. The final decision on the class membership was made upon majority vote criterion.

2.3.5. Soft Independent Modelling of Class Analogy

Soft independent modelling of class analogy (SIMCA) was the first class-modelling method introduced in the literature. It is a non-probabilistic distance-based modelling which relies on the assumption that the main systematic variability of the class of interest can be captured by a PCA model of appropriate dimensionality. The results of the PCA decomposition of the target category are used to define the so-called SIMCA inner space. At this point, the membership of the tested objects is decided on the basis of some statistical criterion for outlier detection. A comprehensive tutorial of SIMCA, and OCC methods in general, is provided in [23].

In the present study, being the most represented within the sample set, BA was set as target class whereas Catamarca and Misiones honeys were used as alien objects to challenge the model. The "degree of outlyingness" with respect to the target category was computed as combination of the Mahalanobis distance to the center of the inner space (T^2) and the orthogonal distance (Q). For multivariate models whose assignation rule is based on the combined T^2-Q distances, the classification outcome can be graphically represented in a Cartesian plot reporting the T^2 and Q of the tested objects on the *x*- and *y*-axis, respectively. Roughly, the further from the origin (down-left corner) the sample is, the higher is its degree of outlyingness.

The same validation strategy described in Section 2.3.3 was adopted to ensure the reliability of the obtained classification results.

3. Results

3.1. Data Exploration

Prior to any chemometric manipulation, the recorded raw spectra of all honey samples were plotted and visually inspected (Figure S3). While very consistent FT-MIR and NIR spectra were obtained, FT-Raman spectra exhibited evident baseline drift, likely due to fluorescence phenomena. Therefore, the optimal combination of spectral filters and/or mathematical preprocessing was found to be SNV + S-G derivative (1st order derivative, 2nd order polynomial, 9 points window) + Mean centering for FT-MIR and NIR spectra, whereas a baseline correction step (manually-selected points, 3rd order polynomial, 5 regions) prior to SNV + Mean centering was included in the FT-Raman data preprocessing workflow.

As explained in Section 2.1, each sample batch (i.e., harvest) was scanned within 14 days after the collection. However, the analysis of the whole sample set was performed over a 4-years period. Therefore, the spectra recorded from the quality control materials were both visually examined and inspected through PCA in order to reveal any batch-to-batch instrumental drifts. No substantial spectral differences and/or separation in the scores plot were observed further to the application of SNV as data pretreatment (data not shown).

Once the data consistency had been ensured, PCA was carried out on the preprocessed honey spectra. The first three PCs accounted for more than 87% of the total variance in all the datasets. Regardless of the used platform, the PC1 vs. PC2 scores plot highlighted a noteworthy separation between BA and Mis honeys, whereas Cat samples were more scattered (Figure 1). Visual examination of higher order PCs did not reveal any greater degree of separation. Here too, no apparent clustering related to the harvesting year was noticed.

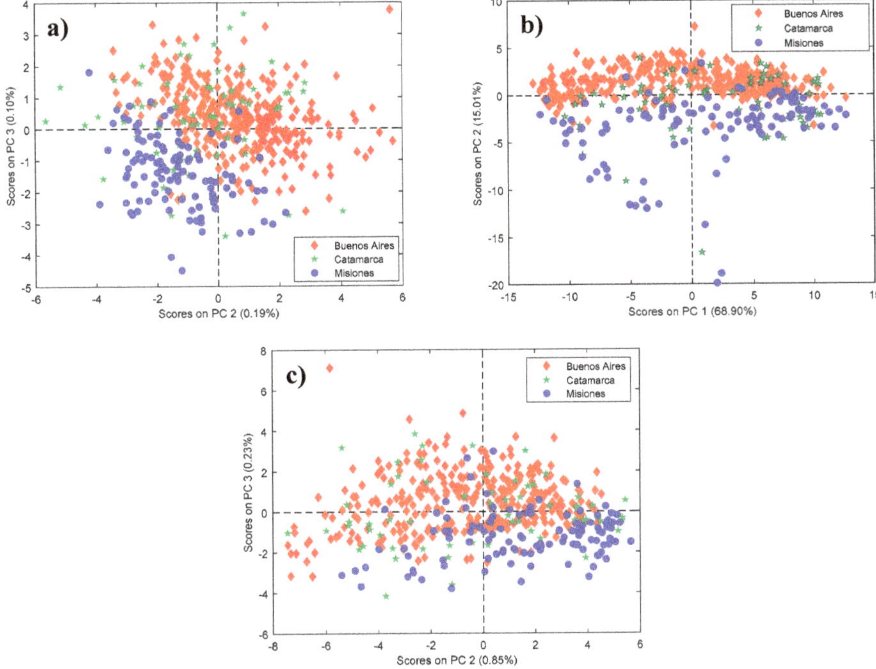

Figure 1. Scatter plot of PC1 vs PC2 scores obtained from FT-MIR (**a**), FT-Raman (**b**) and NIR (**c**) data. Objects are marked according to the provenance region (Red diamond: BA; Green star: Cat; Blue circle: Mis).

As denoted by the PC1 and PC2 loadings (Figure S4), the variables that shown the highest relevance in the PCs definition all corresponded to chemically meaningful spectral intervals. Specifically, most of the dispersion among the samples is explained by the wavelength range 1500–600 cm^{-1} for FT-MIR, 3000–2900 and 1500–0 cm^{-1} for FT-Raman, 480–600 and 1850–2500 nm for NIR.

According to previous reports, carbohydrate moieties are chiefly responsible for absorptions in these ranges of the honey spectra [14,15,31]. Noisy and/or uninformative spectral regions, i.e., CO_2 band and flat regions, were excluded from the subsequent data treatment. As a result, the considered wavelength ranges were, respectively, 3800–2400 cm^{-1} and 1990–600 cm^{-1} for FT-MIR, 3600–2500 cm^{-1} and 1800–0 cm^{-1} for FT-Raman; 400–700 nm and 1300–2500 nm for NIR (Figure S3).

Band assignment was not the main goal of the study as the general tendency in fingerprinting methods is to use the entire spectra in the multivariate data analysis [32]. Nevertheless, description of the main peaks/bands responsible for the sample discrimination might be helpful for future research. Therefore, illustration of the statistically-significant spectral signals and of the three datasets has been reported in Supplementary Materials (Figure S5). Furthermore, assignment of the relevant peaks/bands was carried out based on the literature [6,13,20,31,33–37].

3.2. Techniques Comparison under a Supervised Chemometric Approach

Classification outcomes provided by the individual spectroscopic techniques, as well as the fused datasets, are summarized in this section. ROC curves were constructed as described in Section 2.3.3 and graphically reported in Figure 2.

As expected from the unsupervised pattern recognition, better results were reached in the discrimination of Mis honeys (i.e., BA-Mis and Cat-Mis models). In particular, the BA-Mis model produced nearly perfect classification, with AUC always above 0.99 regardless the spectroscopic technique. In contrast, the BA-Cat model provided slightly lower AUC, ranging from 0.88 (NIR) to 0.93 (FT-MIR), perhaps due to unbalanced number of samples available. Concerning the inter-platforms comparison, FT-MIR provided yielded the largest AUC in all the binary models, while the lowest score was always obtained by FT-Raman spectroscopy.

The results of the validation procedure are summarized as correct classification rates (i.e., ratio between correctly classified and total tested objects, CCRs) in Table 1. For purposes of presentation, only the scores provided by FT-MIR data were reported, while FT-Raman and NIR data are available in Supplementary materials (Tables S2 and S3).

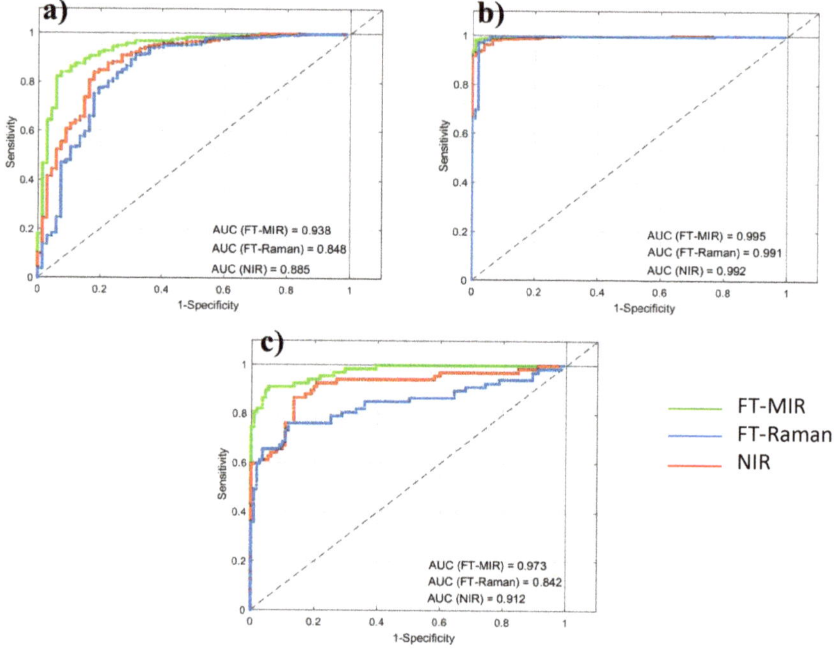

Figure 2. ROC curves related to the binary classification models (**a**) BA vs. Cat; (**b**) BA vs. Mis; (**c**) Cat vs. Mis.

Table 1. PLS-DA prediction results expressed as correct classification rates (FT-MIR data).

Predicted Harvest	Correct Classification Rate (%)		
	BA vs. Cat	BA vs. Mis	Cat vs. Mis
2015	84.6	88.4	92.3
2016	91.8	100.0	92.5
2017	91.9	100.0	95.5

The model validation confirmed what was highlighted by the ROC curves. The best performance was offered by FT-MIR and, here too, the best classification was reached for Mis honeys, whatever the spectroscopic technique. Interestingly, in the case of FT-MIR, an overall improvement of the models' prediction ability was achieved as more harvesting seasons were included in the training set, with all the binary models reaching CCRs > 90% in the prediction of HN2017 (i.e., last step of the validation scheme). It must be pointed out that small differences (e.g., 0.1–0.2%) between the results have to be assessed with caution since these classification outcomes cannot be tested for statistical significance. Nevertheless, the overall trends have been clearly evidenced.

To further enhance the obtained results, the DF strategies described in Section 2.3.4. were attempted and the CCRs achieved in the HN2017 prediction summarized in Table 2.

Table 2. PLS-DA classification results of HN2017, expressed as correct classification rates, according to the different DF strategies.

Predicted Harvest	Correct Classification Rate (%)		
	BA vs. Cat	BA vs. Mis	Cat vs. Mis
LL-DF	85.4	91.7	80.0
ML-DF	87.0	98.6	80.0
HL-DF	93.5	98.6	95.5

All the DF methods provided satisfying classification performance, with HL-DF showing the highest scores, followed by ML-DF and LL-DF. The HL-DF reached comparable results respect to the FT-MIR (Table 1), with slightly better scores in the BA-Cat model and lower CCRs achieved in the BA-Mis honeys discrimination. A further attempt was made by combining the data blocks from two platforms only (i.e., FT-MIR+FT-Raman, FT-MIR+NIR and FT-Raman+NIR). However, no significant classification improvement was achieved (data not shown).

3.3. SIMCA Modelling

FT-MIR dataset underwent to SIMCA modelling as, in the light of the above results, it proved to be the most promising option for a hypothetical fingerprinting method for honey authentication. BA was set as target category to be modelled; thus, Cat and Mis samples represented the alien objects to be rejected by the model. Five PCs were considered sufficient for proper modelling as they accounted for > 95% of the original data variance. The confidence level was set to $\alpha = 0.05$ and the classification rule was based on the so-called T^2-Q augmented distances. The same year-by-year validation was adopted.

SIMCA results are reported as sensitivity, specificity and overall CCRs in Table 3.

Table 3. SIMCA modelling results of class BA (FT-MIR data) according to the different harvesting seasons, expressed as sensitivity, specificity and overall correct classification rates.

Predicted Harvest	Sensitivity (%)	Specificity (%)	CCR (%)
2015	61.0	89.7	69.6
2016	90.6	75.0	85.8
2017	86.7	91.1	88.8

Highly sensitive and specific models were produced, confirming what expected from the excellent classification previously obtained. In accordance with the PLS-DA results (Table 1), the inclusion of 2015 and 2016 harvest in the model training led to an overall enhancement of the model performance. Remarkably, within the prediction of HN2017, 39 out of 45 BA samples were correctly recognized as belonging to the target class (86.7% sensitivity), while 15 out of 17 Cat and 26 on 28 Mis honeys were rightly rejected by the model (91.1% specificity). T^2 and Q distances of the predicted HN2017 samples are graphically represented in Figure 3.

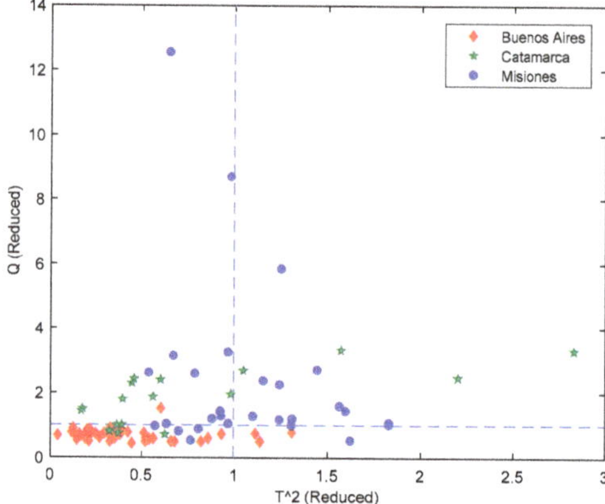

Figure 3. SIMCA modelling on FT-MIR data. Projection of HN2017 objects onto the T^2 (reduced) vs. Q (reduced) model space of class BA.

4. Discussion

Since the analysis of all the collected honeys was carried out over a 4-years period, ensuring the absence of instrumental drift among the analysis batches was the first concern. The routine maintenance of the equipment, typically performed once a year, includes the substitution of overused components (e.g., source) and re-alignment of the interferometer, which may easily result in signal intensity (i.e., absorbance) shifts. Such technical variations, if not properly handled, may give rise to fingerprint deviations that prevent the use of the classification model for its ultimate purpose: the prediction of new harvests [24]. As an example, Woodcock et al. observed a clear separation between honey samples analysed in two consecutive years. However, the authors were not able to definitely attribute such trend to different sample's characteristics, rather than the use of a non-standardized instrument [13].

Both the unsupervised and supervised chemometric approaches evidenced the presence of actual differences between the honeys having diverse provenance. Such differences are unlikely due to random variation or overfitting issues. In fact, it is worth stressing that the employed validation strategy allowed any developed model to be challenged with independent external test sets. With the main factor under investigation being the geographical origin, it is reasonable to ascribe the samples separation to the distinct environmental features of the three Argentinian regions. Variations of soil and weather conditions likely result in different melliferous floras foraged by the bees, which is known to have the greatest influence on the honey's chemical composition [38]. Buenos Aires province is located within the ecoregion *Pampeana*, where the temperate climate and abundant rainfall encourage extensive crop cultivation. Misiones province is characterized by its typical flora, known as "Missionary Forest", favored by the subtropical weather of the ecoregion *Selva Paranaense*. The peculiar characteristics of this ecoregion might underlie differences in the honeys' physiochemical properties, which would explain the better results achieved in the classification of Mis samples. While Buenos Aires and Misiones regions show fairly uniform climate conditions, five different ecoregions are recognized in Catamarca (i.e., *Yungas*, *Chaco Seco*, *Monte de Sierras and Bolsones*, *Puna* and *Altos Andes*) and therefore a number of microclimates can be encountered, from the subtropical rains in the east, to the arid highland in the west [39]. Therefore, the larger overlap of Cat samples over the other classes might be due to this climate, and thus botanical, heterogeneity.

All the employed spectroscopic techniques provided more than satisfying performance, confirming the high potential of vibrational spectroscopy as rapid screening tool for honey authentication. Although lot of research has been done in the application of vibrational spectroscopy for honey testing, cross-platform comparisons have been scarcely documented. Tahir and co-workers observed equivalent performance of FT-MIR and FT-Raman spectroscopy for the prediction of phenolic compounds content and the antioxidant activity in honey [40]. Ballabio et al. recently evaluated five different technologies, including FT-MIR, NIR and FT-Raman spectroscopy, for the botanical origin identification of honeys [41]. The authors reported better classification provided by NIR, respect to FT-MIR and FT-Raman spectroscopy. Nevertheless, the same authors pointed out that such outcomes have to be assessed with caution due to the small size of the sample set. Within the present work, FT-MIR shown to be the best option for honey fingerprinting, providing always the largest AUC within the ROC curves, as well as superior CCRs (>90%) through the validation process. The reason probably lies in the better sensitivity and higher S/N normally provided by FT-MIR instruments respect to NIR and FT-Raman, since fundamental absorptions are being measured in the MIR region [42].

As pointed out in Section 3.2, LL-DF provided the poorest results among the attempted DF strategies. This is consistent with the literature, where LL-DF approach either did not produce substantial classification/prediction improvement over the single techniques or was outperformed by higher-level DF [40,41,43]. The explanation can be found in the high collinearity of vibrational spectroscopy data. In fact, LL-DF introduces, along with useful information, a large number of redundant and irrelevant variables. Such noise is, for example, reduced in the ML-DF by the features extraction prior to the concatenation. Concerning the ML-DF and HL-DF, despite the noteworthy results, no significant classification enhancement was reached respect to the FT-MIR only. On the basis of the present outcomes, the combination of vibrational spectroscopic data cannot be regarded as worthwhile as no evident advantage has been provided over the individual techniques. The authors attributed the ineffectiveness of DF to the lack of information orthogonality between the combined data sources, which is crucial for the successful application of DF [30].

When evaluated under conditions "closer" to a real authentication scenario, FT-MIR still yielded remarkable classification scores. The lower CCRs achieved by SIMCA respect to PLS-DA are not surprising as DA algorithms use information about the modelled classes to maximize the group differences, whereas OOC methods "do not know anything about existence of alternative classes or samples". In fact, despite the widespread opinion that "PLS-DA may go further than SIMCA", performance comparisons of these two algorithms are not even consistent as they employ diverse amounts of modelling information [21]. As mentioned in Section 1, DA algorithms are not suited for one-class problems where only one target category is modeled against a heterogeneous group of off-specification products [23]. For this reason, the authors believe that the SIMCA results (Section 3.3) are more representative of the potential performance of a routine screening method based on FT-MIR fingerprinting. The classification achieved in the HN2017 prediction can be considered excellent for a rapid screening platform and demonstrated that, under a proper characterization of the class of interest, FT-MIR spectroscopy can be a powerful tool for honey authenticity purposes.

In the authors' opinion, the results herein obtained can be sensibly extended to problems of honey's floral origin. In fact, botanical/varietal and geographical origin of food products are often treated as separate issues in food authenticity studies; however, they are highly correlated and hard to be considered individually, especially in the case of natural products such as honey. For example, distinct geographic areas do not only provide different climatic conditions affecting the accumulation of phytochemicals in pollen and nectar, but also normally offer diverse melliferous flora foraged by the bees. All these factors and relationships cannot be ignored in the development of methods for honey's origin confirmation.

Despite the remarkable outcomes, in must be pointed out that the development of a comprehensive model able to identify the geographic origin of an unknown sample is unrealistic; it would require an exhaustive sampling of world honeys over several harvest years. Furthermore, honeys from different localities may not have unique spectral signatures due to similarities in vegetation. Thus, it is unlike to reach similar performance at a world-level. We believe that a fundamental knowledge of the limits and capabilities of the chosen methods is essential for their correct utilization and interpretation. Screening platforms based on spectroscopic fingerprints find the best applicability at a company-level, where the "boundaries" of the application can be clearly defined. Typical examples are internal quality assurance or the management of incoming raw materials from suppliers with established relationships. In these contexts, the target classes can be appropriately outlined and sampled in a representative way.

5. Conclusions

Honey authenticity remains a challenging issue to deal with as reliable and manageable methods for its floral and geographical origin confirmation are still lacking. Several feasibility studies have been reported in literature to demonstrate the capabilities of vibrational spectroscopy for the discrimination of honey's botanical and/or geographical origin.

A key feature of the present work was the realistic and rather large variability included in the sample set. All the collected honeys were multifloral, thus covering differences in nectar sources. Besides, seasonal climate fluctuations were also considered by repeating the sampling over four consecutive harvesting seasons. This extra variation is of great benefit for the robustness of any developed model and crucial to demonstrate its capabilities under real-world conditions.

Excellent classification scores were achieved by all the technologies and the adopted validation strategy allowed to exclude any issue related to model overfitting. The nearly perfect classification results provided by FT-MIR suggested its major potential for honey fingerprinting. DF strategies yielded satisfying outcomes, however, no significant improvement in discrimination power was achieved respect to FT-MIR. Therefore, within an industrial context, a multi-platforms spectroscopic fingerprint is a choice that should be carefully evaluated from a cost/benefit standpoint. In fact, it must be considered that a multiple sample fingerprinting would represent an increased expense in terms of equipment and expertise, making the food control process more time and labour-demanding.

SIMCA modelling was successfully applied on the FT-MIR dataset and demonstrated that the use of large and representative training sets can definitely improve the model robustness over analytical and biological factors. The year-by-year validation not only ensured the results reliability, but also well reproduced a hypothetical quality control context where, reasonably, spectral libraries are gradually enlarged with newly recorded spectra. In the author's opinion, such results can be considered a reliable performance estimation of a potential FT-MIR-based fingerprinting method.

Supplementary Materials: The following are available online at http://www.mdpi.com/2304-8158/9/10/1450/s1, Figure S1: Geographical map of the three Argentinian provinces included in the study, Figure S2: Scheme of the adopted validation strategy, Figure S3: Recorded raw spectra of FT-MIR, FT-Raman and NIR, Figure S4: Line plots of PC1 and PC2 loadings obtained from FT-MIR, FT-Raman and NIR data, Figure S5: Line plots of the VIP scores extracted from the PLS2-DA model generated on FT-MIR, FT-Raman and NIR data, Table S1: Sample set overview, Table S2: PLS-DA prediction results expressed as correct classification rates (FT-Raman data), Table S3: PLS-DA prediction results expressed as correct classification rates (NIR data). Experimental data are openly available as supplementary information.

Author Contributions: T.D.: Conceptualization, Methodology, Validation, Data curation, Investigation, Formal analysis, Visualization, Writing—original draft. R.M.A.-S.: Funding Acquisition, Project Administration, Writing—review & editing. I.A.: Writing—review & editing. V.B.: Funding Acquisition, Project Administration, Resources. Q.A.: Resources, Investigation. C.D.: Supervision, Resources, Writing—review & editing. S.R.F.: Writing—review & editing. J.A.F.P.: Conceptualization, Supervision, Resources, Methodology, Writing—review & editing. All authors have read and agreed to the published version of the manuscript.

Funding: This work was funded by the Argentina-Belgium bilateral Project FNRS–MINCYT (2013), the National Projects PICT 3264/2014 and PICT 0774/2017 within the Program Fondo para la Investigación Científica y Tecnológica (FonCyT) from the Agencia Nacional de Promoción Científica y Tecnológica (ANPCyT)-Ministerio de Ciencia, Tecnología e Innovación (MINCYT) from Argentina, and Comisión de Investigaciones Científicas (CIC) from Provincia de Buenos Aires (Argentina). Inés Aubone thanks Universidad Nacional de Mar del Plata (Mar del Plata, Pcia. Buenos Aires, Argentina) and CIC (Argentina) for her PhD Grant. The authors thank the company NEXCO S.A (Lobos, Pcia. Buenos Aires, Argentina), Fernando Muller (Pcia. Misiones, Argentina), Liliana Gallez and Labea from Universidad Nacional del Sur (Bahía Blanca, Pcia. Buenos Aires, Argentina), Alejandra Villalba (Subsecretaría de la Producción, Pcia. Catamarca, Argentina), Leonardo Dorsch (Mar del Plata, Pcia. Buenos Aires, Argentina), beekeepers and honey producer cooperatives from Argentina for supplying authentic and traceable honey samples.

Acknowledgments: Ines Aubone thanks Universidad Nacional de Mar del Plata (Mar del Plata, Pcia. Buenos Aires, Argentina) and CIC (Argentina) for her PhD Grant. The authors thank the company NEXCO S.A (Lobos, Pcia. Buenos Aires, Argentina), Fernando Muller (Pcia. Misiones, Argentina), Liliana Gallez and Labea from Universidad Nacional del Sur (Bahía Blanca, Pcia. Buenos Aires, Argentina), Alejandra Villalba (Subsecretaría de la Producción, Pcia. Catamarca, Argentina), Leonardo Dorsch (Mar del Plata, Pcia. Buenos Aires, Argentina), beekeepers and honey producer cooperatives from Argentina for supplying authentic and traceable honey samples.

Conflicts of Interest: The authors declare no conflict of interest. The funders had no role in the design of the study; in the collection, analyses, or interpretation of data; in the writing of the manuscript, or in the decision to publish the results.

References

1. The Council of the European Commission. *Council Directive 2001/110/CE Concerning Honey*; 2001. Available online: https://eur-lex.europa.eu/legal-content/EN/TXT/PDF/?uri=CELEX:32001L0110&from=EN (accessed on 12 October 2020).

2. Codex Alimentarius Commission. Codex Alimentarius Commission Standards. *Codex Stan 12-1981* **2001**, 1–8. Available online: http://www.fao.org/input/download/standards/310/cxs_012e.pdf (accessed on 12 October 2020).

3. Soares, S.; Amaral, J.S.; Oliveira, M.B.P.P.; Mafra, I. A Comprehensive Review on the Main Honey Authentication Issues: Production and Origin. *Compr. Rev. Food Sci. Food Saf.* **2017**, *16*, 1072–1100. [CrossRef]

4. Geana, E.I.; Ciucure, C.T. Establishing Authenticity of Honey via Comprehensive Romanian Honey Analysis. *Food Chem.* **2020**, *306*, 125595. [CrossRef] [PubMed]

5. Von Der Ohe, W.; Persano Oddo, L.; Piana, M.L.; Morlot, M.; Martin, P. Harmonized Methods of Melissopalynology. *Apidologie* **2004**, *35* (Suppl. 1), S18–S25. [CrossRef]

6. Latorre, C.H.; Crecente, R.M.P.; Martín, S.G.; García, J.B. A Fast Chemometric Procedure Based on NIR Data for Authentication of Honey with Protected Geographical Indication. *Food Chem.* **2013**, *141*, 3559–3565. [CrossRef] [PubMed]

7. Dong, H.; Xiao, K.; Xian, Y. Isotope Ratio Mass Spectrometry Coupled to Element Analyzer and Liquid Chromatography to Identify Commercial Honeys of Various Botanical Types. *Food Anal. Methods* **2017**, *10*, 2755–2763. [CrossRef]

8. Soares, S.; Grazina, L.; Mafra, I.; Costa, J.; Pinto, M.A.; Duc, H.P.; Oliveira, M.B.P.P.; Amaral, J.S. Novel Diagnostic Tools for Asian (Apis Cerana) and European (Apis Mellifera) Honey Authentication. *Food Res. Int.* **2018**, *105*, 686–693. [CrossRef]

9. Esslinger, S.; Riedl, J.; Fauhl-Hassek, C. Potential and Limitations of Non-Targeted Fingerprinting for Authentication of Food in Official Control. *Food Res. Int.* **2014**, *60*, 189–204. [CrossRef]

10. Pita-Calvo, C.; Vázquez, M. Honeydew Honeys: A Review on the Characterization and Authentication of Botanical and Geographical Origins. *J. Agric. Food Chem.* **2018**, *66*, 2523–2537. [CrossRef]

11. Ellis, D.I.; Brewster, V.L.; Dunn, W.B.; Allwood, J.W.; Golovanov, A.P.; Goodacre, R. Fingerprinting Food: Current Technologies for the Detection of Food Adulteration and Contamination. *Chem. Soc. Rev.* **2012**, *41*, 5706–5727. [CrossRef]

12. McGrath, T.F.; Haughey, S.A.; Patterson, J.; Fauhl-Hassek, C.; Donarski, J.; Alewijn, M.; van Ruth, S.; Elliott, C.T. What Are the Scientific Challenges in Moving from Targeted to Non-Targeted Methods for Food Fraud Testing and How Can They Be Addressed?—Spectroscopy Case Study. *Trends Food Sci. Technol.* **2018**, *76*, 38–55. [CrossRef]

13. Woodcock, T.; Downey, G.; Kelly, J.D.; O'Donnell, C. Geographical Classification of Honey Samples by Near-Infrared Spectroscopy: A Feasibility Study. *J. Agric. Food Chem.* **2007**, *55*, 9128–9134. [CrossRef] [PubMed]

14. Woodcock, T.; Downey, G.; O'Donnell, C.P. Near Infrared Spectral Fingerprinting for Confirmation of Claimed PDO Provenance of Honey. *Food Chem.* **2009**, *114*, 742–746. [CrossRef]

15. Hennessy, S.; Downey, G.; O'Donnell, C. Multivariate Analysis of Attenuated Total Reflection—Fourier Transform Infrared Spectroscopic Data to Confirm the Origin of Honeys. *Appl. Spectrosc.* **2008**, *62*, 1115–1123. [CrossRef] [PubMed]

16. Hennessy, S.; Downey, G.; O'Donnell, C.P. Attempted Confirmation of the Provenance of Corsican PDO Honey Using FT-IR Spectroscopy and Multivariate Data Analysis. *J. Agric. Food Chem.* **2010**, *58*, 9401–9406. [CrossRef] [PubMed]

17. Guelpa, A.; Marini, F.; du Plessis, A.; Slabbert, R.; Manley, M. Verification of Authenticity and Fraud Detection in South African Honey Using NIR Spectroscopy. *Food Control.* **2017**, *73*, 1388–1396. [CrossRef]

18. Ruoff, K.; Luginbühl, W.; Künzli, R.; Iglesias, M.; Bogdanov, S.; Bosset, J.O.; Von Der Ohe, K.; Von Der Ohe, W.; Amadò, R. Authentication of the Botanical and Geographical Origin of Honey by Mid-Infrared Spectroscopy. *J. Agric. Food Chem.* **2006**, *54*, 6873–6880. [CrossRef]

19. Giordano, A.; Retamal, M.; Fuentes, E.; Ascar, L.; Velásquez, P.; Rodríguez, K.; Montenegro, G. Rapid Scanning of the Origin and Antioxidant Potential of Chilean Native Honey Through Infrared Spectroscopy and Chemometrics. *Food Anal. Methods* **2019**, *12*, 1511–1519. [CrossRef]

20. Bisutti, V.; Merlanti, R.; Serva, L.; Lucatello, L.; Mirisola, M.; Balzan, S.; Tenti, S.; Fontana, F.; Trevisan, G.; Montanucci, L.; et al. Multivariate and Machine Learning Approaches for Honey Botanical Origin Authentication Using near Infrared Spectroscopy. *J. Near Infrared Spectrosc.* **2019**, *27*, 65–74. [CrossRef]

21. Rodionova, O.Y.; Titova, A.V.; Pomerantsev, A.L. Discriminant Analysis Is an Inappropriate Method of Authentication. *TrAC—Trends Anal. Chem.* **2016**, *78*, 17–22. [CrossRef]

22. Granato, D.; Putnik, P.; Kovačević, D.B.; Santos, J.S.; Calado, V.; Rocha, R.S.; Cruz, A.G.; Jarvis, B.; Rodionova, O.Y.; Pomerantsev, A. Trends in Chemometrics: Food Authentication, Microbiology, and Effects of Processing. *Compr. Rev. Food Sci. Food Saf.* **2018**, *17*, 663–677. [CrossRef]

23. Oliveri, P. Class-Modelling in Food Analytical Chemistry: Development, Sampling, Optimisation and Validation Issues—A Tutorial. *Anal. Chim. Acta* **2017**, *982*, 9–19. [CrossRef] [PubMed]

24. Riedl, J.; Esslinger, S.; Fauhl-Hassek, C. Review of Validation and Reporting of Non-Targeted Fingerprinting Approaches for Food Authentication. *Anal. Chim. Acta* **2015**, *885*, 17–32. [CrossRef] [PubMed]

25. Fearn, T.; Riccioli, C.; Garrido-Varo, A.; Guerrero-Ginel, J.E. On the Geometry of SNV and MSC. *Chemom. Intell. Lab. Syst.* **2009**, *96*, 22–26. [CrossRef]

26. Zimmermann, B.; Kohler, A. Optimizing Savitzky-Golay Parameters for Improving Spectral Resolution and Quantification in Infrared Spectroscopy. *Appl. Spectrosc.* **2013**, *67*, 892–902. [CrossRef]

27. Bro, R.; Smilde, A.K. Principal Component Analysis. *Anal. Methods* **2014**, *6*, 2812–2831. [CrossRef]

28. Brereton, R.G.; Lloyd, G.R. Partial Least Squares Discriminant Analysis: Taking the Magic Away. *J. Chemom.* **2014**, *28*, 213–225. [CrossRef]

29. Esbensen, K.H.; Geladi, P. Principles of Proper Validation: Use and Abuse of Re-Sampling for Validation. *J. Chemom.* **2010**, *24*, 168–187. [CrossRef]

30. Borràs, E.; Ferré, J.; Boqué, R.; Mestres, M.; Aceña, L.; Busto, O. Data Fusion Methodologies for Food and Beverage Authentication and Quality Assessment—A Review. *Anal. Chim. Acta* **2015**, *891*, 1–14. [CrossRef]

31. Fernández Pierna, J.A.; Abbas, O.; Dardenne, P.; Baeten, V. Discrimination of Corsican Honey by FT-Raman Spectroscopy and Chemometrics. *Biotechnol. Agron. Soc. Environ.* **2011**, *15*, 75–84.

32. Downey, G. Vibrational Spectroscopy in Studies of Food Origin. In *New Analytical Approaches for Verifying the Origin of Food*; Elsevier Ltd.: Amsterdam, The Netherlands, 2013; pp. 94–116. [CrossRef]

33. Subari, N.; Mohamad Saleh, J.; Shakaff, A.Y.M.; Zakaria, A. A Hybrid Sensing Approach for Pure and Adulterated Honey Classification. *Sensors* **2012**, *12*, 14022–14040. [CrossRef] [PubMed]

34. Gallardo-Velázquez, T.; Osorio-Revilla, G.; Zuñiga-de Loa, M.; Rivera-Espinoza, Y. Application of FTIR-HATR Spectroscopy and Multivariate Analysis to the Quantification of Adulterants in Mexican Honeys. *Food Res. Int.* **2009**, *42*, 313–318. [CrossRef]

35. Tewari, J.C.; Irudayaraj, J.M.K. Floral Classification of Honey Using Mid-Infrared Spectroscopy and Surface Acoustic Wave Based z-Nose Sensor. *J. Agric. Food Chem.* **2005**, *53*, 6955–6966. [CrossRef] [PubMed]

36. Gok, S.; Severcan, M.; Goormaghtigh, E.; Kandemir, I.; Severcan, F. Differentiation of Anatolian Honey Samples from Different Botanical Origins by ATR-FTIR Spectroscopy Using Multivariate Analysis. *Food Chem.* **2015**, *170*, 234–240. [CrossRef]

37. Corbella, E.; Cozzolino, D. The Use of Visible and near Infrared Spectroscopy to Classify the Floral Origin of Honey Samples Produced in Uruguay. *J. Near Infrared Spectrosc.* **2005**, *13*, 63–68. [CrossRef]

38. Kaškonienė, V.; Venskutonis, P.R. Floral Markers in Honey of Various Botanical and Geographic Origins: A Review. *Compr. Rev. Food Sci. Food Saf.* **2010**, *9*, 620–634. [CrossRef]

39. Arana, M.D.; Martinez, G.A.; Oggero, A.J.; Natale, E.S.; Morrone, J.J. Map and Shapefile of the Biogeographic Provinces of Argentina. *Zootaxa* **2017**, *4341*, 420. [CrossRef]

40. Tahir, H.E.; Xiaobo, Z.; Zhihua, L.; Jiyong, S.; Zhai, X.; Wang, S.; Mariod, A.A. Rapid Prediction of Phenolic Compounds and Antioxidant Activity of Sudanese Honey Using Raman and Fourier Transform Infrared (FT-IR) Spectroscopy. *Food Chem.* **2017**, *226*, 202–211. [CrossRef]

41. Ballabio, D.; Robotti, E.; Grisoni, F.; Quasso, F.; Bobba, M.; Vercelli, S.; Gosetti, F.; Calabrese, G.; Sangiorgi, E.; Orlandi, M.; et al. Chemical Profiling and Multivariate Data Fusion Methods for the Identification of the Botanical Origin of Honey. *Food Chem.* **2018**, *266*, 79–89. [CrossRef] [PubMed]

42. Abbas, O.; Dardenne, P.; Baeten, V. Near-Infrared, Mid-Infrared, and Raman Spectroscopy. In *Chemical Analysis of Food: Techniques and Applications*; Elsevier Inc.: Amsterdam, The Netherlands, 2012; pp. 59–89. [CrossRef]

43. Huang, F.; Song, H.; Guo, L.; Guang, P.; Yang, X.; Li, L.; Zhao, H.; Yang, M. Detection of Adulteration in Chinese Honey Using NIR and ATR-FTIR Spectral Data Fusion. *Spectrochim. Acta—Part A Mol. Biomol. Spectrosc.* **2020**, *235*, 118297. [CrossRef] [PubMed]

 *foods* 

Review

Dietary Supplement and Food Contaminations and Their Implications for Doping Controls

Katja Walpurgis [1,*], **Andreas Thomas** [1], **Hans Geyer** [1], **Ute Mareck** [1] and **Mario Thevis** [1,2]

[1] Center for Preventive Doping Research/Institute of Biochemistry, German Sport University Cologne, 50933 Cologne, Germany; a.thomas@biochem.dshs-koeln.de (A.T.); h.geyer@biochem.dshs-koeln.de (H.G.); u.mareck@biochem.dshs-koeln.de (U.M.); thevis@dshs-koeln.de (M.T.)

[2] European Monitoring Center for Emerging Doping Agents (EuMoCEDA), 50933 Cologne/Bonn, Germany

* Correspondence: k.walpurgis@biochem.dshs-koeln.de; Tel.: +49-221-4982-7072

Received: 12 June 2020; Accepted: 24 July 2020; Published: 27 July 2020

Abstract: A narrative review with an overall aim of indicating the current state of knowledge and the relevance concerning food and supplement contamination and/or adulteration with doping agents and the respective implications for sports drug testing is presented. The identification of a doping agent (or its metabolite) in sports drug testing samples constitutes a violation of the anti-doping rules defined by the World Anti-Doping Agency. Reasons for such Adverse Analytical Findings (AAFs) include the intentional misuse of performance-enhancing/banned drugs; however, also the scenario of inadvertent administrations of doping agents was proven in the past, caused by, amongst others, the ingestion of contaminated dietary supplements, drugs, or food. Even though controversial positions concerning the effectiveness of dietary supplements in healthy subjects exist, they are frequently used by athletes, anticipating positive effects on health, recovery, and performance. However, most supplement users are unaware of the fact that the administration of such products can be associated with unforeseeable health risks and AAFs in sports. In particular anabolic androgenic steroids (AAS) and stimulants have been frequently found as undeclared ingredients of dietary supplements, either as a result of cross-contaminations due to substandard manufacturing practices and missing quality controls or an intentional admixture to increase the effectiveness of the preparations. Cross-contaminations were also found to affect therapeutic drug preparations. While the sensitivity of assays employed to test pharmaceuticals for impurities is in accordance with good manufacturing practice guidelines allowing to exclude any physiological effects, minute trace amounts of contaminating compounds can still result in positive doping tests. In addition, food was found to be a potential source of unintentional doping, the most prominent example being meat tainted with the anabolic agent clenbuterol. The athletes' compliance with anti-doping rules is frequently tested by routine doping controls. Different measures including offers of topical information and education of the athletes as well as the maintenance of databases summarizing low- or high-risk supplements are important cornerstones in preventing unintentional anti-doping rule violations. Further, the collection of additional analytical data has been shown to allow for supporting result management processes.

Keywords: doping; sport; contamination; SARMs; diuretics

---

1. Introduction

According to the World Anti-Doping Code (WADC), doping is defined as a violation of the Anti-Doping Rules [1], comprising, *inter alia*, the detection of a prohibited substance, its metabolites, or markers in the blood or urine sample of an athlete. However, there are different scenarios where such an Adverse Analytical Finding (AAF) does not necessarily result from a deliberate application of a performance-enhancing/banned drug (*vide infra*). Such cases of inadvertent doping include the ingestion of adulterated or faked dietary supplements, tainted food, and contaminated drugs, as well as passive exposure to doping agents or an insufficient education of the athletes with regards to changes of the Prohibited List annually revised by the World Anti-Doping Agency (WADA) [2–6]. According to WADA's policy of strict liability, an athlete is responsible for the substances found in his/her doping control samples and anti-doping rule violations (ADRVs) occur regardless of his/her intention [1,7]. Possible consequences comprise not only temporary or permanent suspensions, but also loss of medals and/or records, financial sanctions, damage to the athlete's reputation, and failed sponsorships [3,8]. However, the decision-making processes are flexible to consider the circumstances, so that clear evidence about the origin of the detected prohibited substance can potentially lead to reduced sanctions [1,4,7]. On the other hand, it cannot be excluded that athletes occasionally argue with contamination scenarios in an attempt to excuse an AAF in order to avoid impending penalties [2,5]. Consequently, a careful interpretation of the results and, if available, additional data (e.g., from microdose elimination studies) are necessary and desirable.

WADA statistics of the years 2013–2017 demonstrated that between 4 and 19% of the reported AAFs were not sanctioned due to an exoneration of the athlete [9–13]. Reasons included, amongst others, dietary supplement or meat contaminations. In this narrative review, suspected and proven incidences of food and supplement contamination and/or adulteration with doping agents and the respective implications for sports drug testing are presented and discussed. Analytical approaches employed in anti-doping research and routine analysis concerning the presented investigations into presumed contamination scenarios are exclusively based on chromatographic-mass spectrometric methods, offering specificity and sensitivity for conclusive result interpretation. The discussion includes both theoretical and contextual points of view, with an overall aim of indicating the current state of knowledge and the relevance and need for future research into specific areas.

2. Dietary Supplements

2.1. Overview

Since ancient times, athletes try to improve their strength, speed, agility, and bravery by using special diets and products such as lion hearts and deer livers [6,14]. With the growing scientific understanding of exercise physiology in the early 20th century, more specialized dietary supplements and ergogenic aids were employed to increase physical fitness [14].

In general, athletic performance depends on a variety of factors such as talent, motivation, training, and the resistance to injuries, but the individual potential can be optimized by a healthy and appropriate diet [8,15,16]. An additional application of dietary supplements can be reasonable for athletes with nutritional challenges (e.g., vegans) or in certain medical circumstances (e.g., a diagnosed nutrient deficiency); however, for many of them, health and performance enhancing effects are not proven [6,8,15,17,18]. Therefore, they should only be used after consultation of a physician or sports nutritionist [8,15]. Nevertheless, supplement use is nowadays widespread among athletes at all levels of sport, especially as they are readily available without medical prescription [8,18]. According to data obtained from doping controls during the Olympic Games held in Sydney and Athens in 2000 and 2004 [19,20], 78% and 75.7% of the tested athletes used dietary supplements and/or medications during the last three days before testing. The evaluation of 3887 doping control forms collected by the International Association of Athletics Federations (IAAF) both in- and out-of-competition between 2003 and 2008 yielded an average use of 1.7 supplements and 0.8 medications per athlete within the preceding 7 days [21]. Further, during the FIFA World Cups 2002 and 2006, the physicians of the participating teams reported a usage of 1.8 substances per player and match, of which 57.1% were dietary supplements and 42.9% were medications [22]. In 2009, Braun et al. published the results of a questionnaire which was conducted to assess the prevalence of supplement use among 164 young German elite athletes [23]. A total of 80% of the study participants declared the past or present use of at least one supplement, and a significant difference was observed between age groups (older > younger athletes) and performance-levels (in some countries referred to as A/B-level > C/D-level). In addition, in 2019 Baltazar-Martins et al. [24] reported the use of dietary supplements by 64% of 527 surveyed elite athletes.

The reasons for resorting to such aids are manifold: To generally improve health and prevent or cure illnesses/injuries, to promote recovery from training, to directly or indirectly increase athletic performance, to treat a presumed nutrient deficiency due to an unbalanced diet, for weight loss, to enhance mood, or to conveniently provide nutrients and energy when required [6,8,15,17,18,23].

In a recently published consensus statement, dietary supplements are defined as the following: A food, food component, nutrient, or non-food compound that is purposefully ingested in addition to the habitually consumed diet with the aim of achieving a specific health and/or performance benefit. [8]. They comprise sports foods (e.g., sports drinks/bars/gels, protein powders), single nutrients with minerals or vitamins, and ergogenic aids (e.g., caffeine, creatine) as well as superfoods (e.g., chia seeds, goji berry extracts), herbal/botanical products, foods enriched with certain ingredients (e.g., vitamin-/mineral-fortified), and multi-ingredient preparations [8,17].

Even though controversial opinions exist concerning the general effectiveness of dietary supplements in healthy subjects, some products might be beneficial for certain types of athletes when used in appropriate dosing and administration schemes [8,15]. For example, products offering concentrated protein and amino acid supply represent convenient options for strength and power athletes to achieve the necessary level of protein intake without a concurrent fat load [14,15]. Creatine is an organic compound endogenously synthesized from amino acids, which is transported into the muscle and enzymatically converted to creatine phosphate. This, in turn, represents an important source of energy under anaerobic conditions and partially restores muscle ATP content during recovery [8,15,25,26]. Therefore, an additional creatine supplementation is supposed to be favorable especially in strength and team sports involving intermittent high-intensity exercise. Alkalizing agents such as sodium bicarbonate and beta alanine can increase the buffering capacity in muscles when the pH is a limiting factor due to anaerobic glycolysis and a rapid breakdown of glycogen to lactate [8,15,16]. Dietary nitrate improves the bioavailability of nitric oxide (NO), which is an important modulator of skeletal muscle function [8]. The intake of chondroitin and glucosamine, representing main constituents of cartilage, have been mentioned as potentially instrumental in improving joint cartilage conditions of athletes [15], and lastly caffeine, which is a stimulant currently not prohibited in sports, and has been shown to support both physical and mental performance in selected studies [8,15,16].

2.2. Risks Associated with the Use of Dietary Supplements

Athletes using dietary supplements are not only susceptible to acute or long-term damage to their health but also to inadvertent doping [8,15]. While the safety, purity, and efficacy of pharmaceutical products are thoroughly and continuously controlled, no uniform regulations and quality controls exist for the manufacturing of dietary supplements, resulting in a highly variable quality of the available preparations [2,15,27–30].

The main problem for the general population and especially for athletes is an inaccurate labelling of ingredients, which is of concern to all types of dietary supplements including pills, powders, capsules, and liquids [2,8,15,27,28,31–33]. While especially those products featuring comparably expensive components occasionally contain only little (if any) active ingredient [15,27], dietary supplements cross-contaminated or even intentionally fortified with undeclared performance-enhancing substances such as anabolic agents or stimulants in order to increase their efficacy are significantly more worrying [2,8,28,31,34]. Moreover, the use of varying (chemical) synonyms of prohibited substances on product labels adds another level of complexity for athletes to recognize a potential issue [2,8,32].

Cross-contaminations are commonly the result of one of two scenarios: Either inappropriately cleaned containers are used for the transportation or storage of the raw materials or dietary supplements, especially when other preparations such as prohormones are manufactured in the same production line [15,28,31,34–36]. Even though selected reputable manufacturers, working according to Good

Manufacturing Practicing (GMP) regulations, have identified risk factors and installed quality controls accordingly, the situation is further complicated by the fact that the source of some cross-contaminations is not necessarily the facility, where the final products are manufactured [35]. Therefore, product and/or raw material testing needs to be conducted with assays that are applicable to all types of relevant matrices and have limits of detection (LODs) in the low ng/g or parts per billion (ppb) range. Such sensitivities are necessary to account for the excellent detection limits of currently employed analytical methods in sports drug testing and the facts that for many substances any detected amount constitutes an AAF in routine doping controls with some dietary supplements being administered in relatively large amounts [6,31,34,35]. Moreover, batch-to-batch, package-to-package, and even tablet-to-tablet variations can occur.

Even if the resulting concentrations of a prohibited drug are too low to have any physiological effect, they can cause an AAF in sports [8,31,34] Therefore, athletes are advised to use available sources to identify "low-risk" products and prevent unintentional ADRVs due to the administration of contaminated/adulterated dietary supplements [28]. In some countries such as Germany and The Netherlands, athletes can obtain such information from databases cataloguing only tested products from manufacturers performing quality controls on a regular basis, either in-house or by using third-party companies as e.g., analytical laboratories [6,15,28,31,36,37]. Moreover, some anti-doping organizations as for example the US Anti-Doping Agency (USADA) have listed high-risk dietary supplements on a dedicated website [32,38].

2.2.1. Anabolic Agents

Since decades, anabolic agents promising positive effects on muscle mass, strength, and recovery, are the drugs most frequently detected in doping control samples [39]. Their usage is prohibited both in- and out-of-competition and, according to current WADA statistics [40], 44% of the AAFs reported in 2018 were anabolic agents. Besides exogenous anabolic androgenic steroids (AAS) as for example metandienone and stanozolol, this substance class includes also endogenous AAS of exogenous origin such as testosterone and nandrolone, and other anabolic agents as for instance selective androgen-receptor modulators (SARMs) and clenbuterol [36,39,41]. While exogenous AAS are routinely detected in biological samples employing gas chromatography-mass spectrometry (GC-MS) or liquid chromatography-mass spectrometry (LC-MS), abnormal steroid/metabolite concentrations and/or ratios within the steroidal module of the athlete biological passport (ABP) and isotope-ratio mass spectrometry (IRMS) are required to provide evidence for the misuse of endogenous AAS [36,39].

Over the last years, numerous dietary supplements were found to be cross-contaminated with different prohormones or unlabeled AAS such as stanozolol, metandienone, boldenone, and oxandrolone [28].

Prohormones of AAS including dehydroepiandrosterone (DHEA), 4-androstenedione, 4-androstenediol, 5-androstenediol, and different 19-norsteroids are sold as dietary supplements with anabolic properties in the US and several other countries for more than 20 years [36,42,43]. Following ingestion, they are enzymatically converted to testosterone and nandrolone, and therefore also included in the WADA Prohibited List [41]. The misuse of testosterone and its prohormones in sports can be corroborated by an elevated testosterone/epitestosterone ratio (T/E) or abnormal metabolite concentrations/ratios within the steroidal module of the ABP as well as IRMS [36,43–45]. By contrast, the administration of nandrolone and the corresponding prohormones lead to the detection of the urinary metabolite 19-norandrosterone, whose exogenous origin has to be additionally confirmed by means of IRMS if the urinary concentration ranges between 2.5 and 15 ng/mL [43,46]. As many manufacturers of prohormones also produce other non-hormonal dietary supplements, inadequate manufacturing practices and substandard quality controls can result in contaminated products and inadvertent doping in sports [42].

The first cases of dietary supplements contaminated with AAS were reported in 2000 in the context of several AAFs with norandrosterone [43]. The affected athletes used products labeled to contain the flavonoids chrysin/quercetine or plant-derived ingredients attributed to *tribulus terrestris* and *guarana*. However, following extraction, derivatization, and GC-MS analysis, different prohormones of testosterone and nandrolone were identified in these products. As high batch-to-batch and capsule-to-capsule variations were observed and the detected total amounts of 0.3–5100 µg/capsule were significantly lower than in commercially available prohormone preparations (~25 mg per capsule), cross-contaminations appeared more likely than an intentional admixture. Nevertheless, an administration study demonstrated that the ingestion of one capsule only of each of the analyzed products can lead to positive findings with the nandrolone metabolites 19-norandrosterone and 19-noretiocholanolone. Also, the T/E of a female study participant was found elevated.

In the same year, GC-MS analysis of a US supplement labeled to contain different plant extracts, L-carnitine, phenylalanine, vitamin B_6, and other ingredients irrelevant in a doping control context, revealed the presence of the testosterone prohormone 4-androstenedione (0.7 mg/capsule) and the nandrolone precursor 19-norandrostenedione (4.8 mg/capsule) [47]. While the administration of one capsule to five healthy volunteers did not change the T/E or androstenedione/E ratio indicative for an exogenous administration of these agents, the major urinary metabolites of nandrolone (19-norandrosterone and 19-noretiocholanolone) reached levels above the WADA minimum required performance level (MRPL)(WADA TD2019MRPL) of 2 ng/mL for 48–144 h. As the daily dose recommended by the manufacturer is seven capsules, long-term usage of this product could not only be associated with AAFs in sports but also significant health risks.

In 2004, the results of a comprehensive study were published where 634 non-hormonal dietary supplements were purchased from 215 companies located in 15 different countries [42]. A total of 57 of these manufactures were also selling prohormones, and 45.6% of the tested products were obtained from these suppliers. The powders, tablets, fluids, and capsules were homogenized, extracted, derivatized, and finally analyzed by means of GC-MS. Out of the 634 tested products, 14.8% (=94) were found to contain AAS not declared on the label at concentrations between 0.01 and 190 µg/g. While 21.1% of the supplements bought from companies also selling prohormones were tested positive, 9.6% of the products obtained from the remaining suppliers contained AAS. An additional administration study demonstrated that an ingestion of the nandrolone prohormone 19-norandrostenedione at an absolute amount of 1 µg can result in an AAF concerning its metabolite 19-norandrosterone.

The study was repeated several years later and only 4 (=0.7%) of the 597 dietary supplements analyzed by means of GC-MS and LC-MS were found to contain unlabeled AAS, indicating that the prevalence of contaminated products has decreased since 2004 [48]. While the reason(s) for this phenomenon have not been proven, increased awareness and, consequently, improved production processes and/or supplement controls are likely aspects that contributed to the change in identified contaminations.

Shortly thereafter, the analysis of several vitamin and mineral tablets of a manufacturer also selling different prohormone products containing high amounts of unlabeled AAS, revealed the presence of metandienone and stanozolol at concentrations of 0.06–0.2 µg/tablet [49]. Again, it can be assumed that these cross-contaminations originate from using the same production line without proper cleaning. Even though the detected amounts were found to be too low to cause an AAF after the administration of one tablet, other factors such as a long-term application, varying concentrations of the contaminants, and metabolic differences between individuals could potentially lead to inadvertent doping cases.

In the same year, the Swiss anti-doping laboratory reported the findings of different steroids and/or prohormones such as testosterone, androstenedione, norandrostenedione, androstenediol, and DHEA in dietary supplements marketed as creatine and "mental enhancers" [50]. Only trace amounts of 45 ng–300 µg/capsule were detected, but a 3-day administration study with the creatine product containing 1.2 µg of norandrostenedione per capsule showed that the use of this product according to the manufacturer's recommendations can result in the detection of the nandrolone metabolites 19-norandrosterone and 19-noretiocholanolone at concentrations close to the urinary MRPL of 2 ng/mL.

The analysis of 48 dietary supplements marketed as protein concentrates ($n = 29$), creatine preparations ($n = 15$), and "natural fat-burner" extracts from *Citrus aurantium* ($n = 4$) by means of 2D-GC-ToF-MS yielded two positive samples with prohibited AAS: A whey protein gainer was found to contain nandrolone (22 µg/kg), testosterone (70 µg/kg), and DHEA (63 µg/kg), and 5α-androstane-3,17-dione (398 µg/kg) and 19-norandrostenedione (304 µg/kg) were identified in a creatine product [51].

Probable cross-contaminations in the ng/g range with the prohormones 4-androstenedione and 19-norandrostenedione as well as testosterone, testosterone decanoate, and nandrolone decanoate were also detected in dietary supplements labeled to contain L-carnitine, different amino acids, proteins, and carbohydrates [52]

For some products, an intentional manipulation with pharmacologically relevant amounts (>1 mg/g) of unlabeled AAS was assumed [28,53]. Promised an increased strength and muscle growth, attributed to "new" ingredients with imaginary names [39].

For example, a high concentration of unlabeled metandienone was observed in several dietary supplements sold in the UK [53]. In one of these products, the detected amounts were found to vary significantly from capsule to capsule with maximum concentrations of 28.9 mg/g. An administration of these supplements according to the manufacturer's instructions would result in supra-therapeutic doses of a steroid hormone, which has no clinical approval in Germany and several other countries. This would not only result in AAFs in sports, but also be associated with unforeseeable health risks, especially when used by women, children, and adolescents.

In March 2015, the doping control urine samples of 11 Bulgarian weightlifters training for the European Championships were found to contain the stanozolol metabolite 3′-hydroxystanozolol glucuronide [54]. Most of the athletes had declared the use of different supplements and/or non-prescription medications on their doping control form. After the AAFs were reported, all weightlifters as well as their coach stated to have administered a supplement called *Trybest* during training. The analysis of the product revealed the presence of unlabeled stanozolol at amounts of 1.7–4.2 µg per capsule, which can potentially result in the detected urinary concentrations of 3′-hydroxystanozolol. Different scenarios comprising supplement contamination, an intentional adulteration of the product by the manufacturer, and a deliberate sabotage were discussed, and eventually, the athletes were sanctioned as they should have been aware of the risk associated with the administration of dietary supplements.

SARMs are a novel class of anabolic agents, which are not only characterized by a high tissue selectivity and oral bioavailability, but also significantly reduced androgenic side effects [55]. Although no drug candidate has obtained clinical approval yet, different illegal products containing SARMS are available on the black market [56]. Moreover, the U.S. Anti-Doping Agency (USADA) has issued a warning that athletes are at risk of inadvertent doping with different SARMS and especially ostarine, which was found to be an unlabeled or misleadingly labeled ingredient of various dietary supplements and also present as contamination in such products [57]. Since 2017, the AAFs of several U.S. athletes could be linked to the use of such contaminated/adulterated dietary supplements, and reduced sanctions were therefore applied in all these cases [58–64].

2.2.2. Stimulants

The category of stimulants commonly subsumes compounds that increase the activity of the central nervous system (CNS) and thus affect alertness, mood, appetite, and locomotion, as well as the sympathetic nervous system, resulting predominantly in cardiovascular effects [65,66]. They are one of the oldest classes of doping agents and, due to their transient effects, prohibited in-competition only. In the WADA Prohibited List [41], stimulants are divided into two categories: Specified stimulants such as e.g., methylphenidate and pseudoephedrine are widely available (e.g., in pharmaceutical products) and therefore more susceptible to inadvertent doping [65–68]. Consequently, the impending sanctions can potentially be reduced. By contrast, non-specified stimulants comprise strong stimulants as for example amphetamine. Stimulants are routinely identified in doping control samples by using GC-MS or LC-MS [65]. With the exemption of octopamine, the MRPL is set at 100 ng/mL for all stimulants considered as non-threshold substances. AAFs are however communicated only, when the reporting limit defined as 50% of the MRPL (i.e., 50 ng/mL) is exceeded [68]. Although sensitive detection methods are available since several years, stimulants are still popular among athletes [65]: In 2018, 15% of the reported AAFs accounted for these doping agents [40].

Stimulants have also been identified in numerous dietary supplements and, similar to AAS, both cross-contaminations and intentional admixtures have been described, the latter especially in products promoted for weight loss and energy improvement in order to rapidly obtain noticeable effects [6,65]. Additionally, stimulants naturally occurring in plant material can be problematic for athletes, in particular as the content can vary between species and various substance and plant names may exist.

Since 2004, athletes administering caffeine-containing products no longer risk an ADRV as the compound was removed from the WADA Prohibited List [65]. For the natural alkaloid ephedrine, a urinary threshold of 10 µg/mL applies [41], but nevertheless, careful considerations are in order when using *Ephedra sinica* preparations as some products were suspected to contain high amounts of ephedrine, arguably resulting from additions of the drug aiming to achieve significant performance-enhancing or weight-reducing effects [69]. The analysis of nine commercially available *Ephedra* products yielded a highly variable ephedrine content of 1–14 mg per capsule, which can be attributed to the use of different *Ephedra* species. But while natural *Ephedra* preparations usually contain several different alkaloids, two supplements appeared to be artificially fortified with synthetic ephedrine as it was the only detected stimulant (8 and 12 mg/capsule).

This also applies to other weight-loss supplements: In 2007, a Chinese herbal slimming tea and capsules were found to contain the synthetic drug sibutramine at concentrations of 1.8 mg/tea bag and 34 mg/capsule undeclared on the label [70]. Sibutramine is an amphetamine-derivative, which inhibits the re-uptake of the neurotransmitters serotonin and noradrenaline and is known to effectively suppress the appetite [6,70]. The detected amount of 34 mg is significantly higher than the doses administered in clinical studies (10–20 mg) and can therefore not only lead to AAFs in sports but also to unpredictable health risks, especially as the clinical approval of the drug was withdrawn in 2010 due to an increased occurrence of cardiovascular events.

Another stimulant often illegally added to dietary supplements marketed for weight loss and performance-enhancement is 1,3-dimethylamylamine (DMAA), also known as methylhexaneamine [71–76]. The drug is a synthetic aliphatic amine patented by Eli Lilly as nasal decongestant [72–75], but allegedly also a natural ingredient of the plant *Pelargonium graveolens* [77–79]. An extensive debate revolving around the study results published by Ping et al. [77] followed as follow-up studies returned conflicting results [72–76,78,79], and it cannot be excluded that dietary supplements prepared from *Pelargonium graveolens* extract, geranium oil, or geranium stem are artificially fortified with DMAA but labeled as "natural" products [71–75]. But also several entirely unlabeled dietary supplements were found to contain DMAA at concentrations of 136–415 g/kg [71].

The natural monoamine alkaloid phenylethylamine (PEA) and its synthetic derivatives function as neuromodulators in the CNS, resulting in stimulating effects similar to amphetamine [80–82]. Since 2015, these agents are found among the specified stimulants on the WADA Prohibited List [41,81]. PEA and related compounds are widely distributed as dietary supplements promising positive effects on energy and exercise duration [79,80]. Especially "natural" products containing material from the small tree *Acacia rigidula* were found to often contain phenylethylamines [82,83]. As the detected concentrations of PEA in some of these products (0.7–171.6 mg/g) were significantly higher than the natural levels of this compound in *Acacia rigidula* extracts (up to 1.5 µg/g), it can be assumed that also here admixtures of synthetic PEA to these products occurred. But as PEA is also produced by the human body, the differentiation of an illicit administration of the drug from endogenous levels is a complicated analytical task and requires the consideration of PEA metabolite profiles indicative for oral ingestion [81].

Besides PEA, also its derivative β-methylphenethylamine (BMPEA) is claimed to be a natural ingredient of *Acacia rigidula* [82]. However, this postulation was not confirmed in a study analyzing *Acacia rigidula* plant material for the presence of biogenic amines, which was initiated by the U.S. Food and Drug Administration (FDA) [83]. Nevertheless, BMPEA was identified in numerous dietary supplements advertised at metabolic activators and fat-burners at concentrations of 1–61 mg/g [82,83]. With estimated daily doses of up to 146 mg, the administration of such products could not only cause adverse effects but also inadvertent AAFs in sports [83].

In 2013 and 2014, the designer stimulant and PEA analog *N,N*-dimethyl-2-phenylpropan-1-amine (NN-DMPPA) was identified in the doping control urine samples of four athletes as well as a dietary supplement advertised as booster to increase motivation, strength, energy, and endurance, which was labeled to contain adrenergic amines from *Acacia rigidula* and caffeine [84]. The concentration was 122 µg/g, and BMPEA was also detected at an amount of 18 mg/g. The administration of a 3 g single-dose to three healthy volunteers (recommended daily dose by the manufacturer: 1 sachet containing 15 g of powder) resulted in urinary concentrations of more than 50 ng/mL (50% of the MRPL) for 22–23 h (NN-DMPPA) and 3–12 h (BMPEA) [85]. As the MRPL was installed to harmonize the analytical performance of the doping control laboratories and is not a threshold or detection limit, AAFs can also result from lower urinary concentrations [68].

Moreover, several cases of presumably unintentional doping with the PEA derivative *N*-ethyl-α-ethyl-phenylethylamine (ETH)/2-ethylamino-1-phenylbutane (EABP) have been reported, the occurrence of which can at least partially be attributed to inaccurate labeling of dietary supplements, obscuring the presence of this alkaloid [80,86] Different products were found to contain this designer agent at concentrations between 2 and 16 mg/g [80,86,87], and the administration of one of these products to three healthy volunteers resulted in urine levels higher than 50 ng/mL for 46–106 h [87].

In 2018, two AAFs with the specified stimulant heptaminol could be attributed to the use of fat-burners/pre-workout supplements labeled to contain 2-aminoisoheptane, which is an incorrect synonym for octodrine, a psychoactive stimulant of the CNS [88]. Following oral administration, the drug is metabolically converted to heptaminol, but as the misuse of both stimulants is prohibited in competition, these findings are predominantly relevant for an accurate results interpretation.

Furthermore, oxilofrine and the designer stimulant 1,3-dimethylbutylamine have been identified as adulterants in dietary supplements advertised as training boosters and slimming products [89].

2.2.3. Other Substances

Although most of the reported cases on contaminated/faked supplements involve AAS or stimulants, there have been several findings with substances from other classes of doping agents.

In 2018, an athlete was repeatedly tested positive for the diuretic hydrochlorothiazide (HCTZ) [90]. Diuretics are drugs developed for the treatment of hypertension, and their misuse in sports is prohibited both in- and out-of-competition as they can not only interfere with the detection of other doping agents but also be misused to achieve rapid weight losses (relevant in sport disciplines with weight classes). In sports drug testing, they are routinely detected employing LC-MS, which yields urinary detection limits at the picogram level. In the athlete's urine samples, low HCTZ concentrations of 8 and 13 ng/mL were observed, but the administration of any prohibited drug was vehemently denied. However, five different dietary supplements prepared in a compounding pharmacy were used during the period in question, and LC-MS analysis of four of these products revealed the presence of HCTZ at amounts of 2.1–4.6 ng/mL, 0–384 µg/capsule, and 0–147 µg/sachet. A subsequent administration study with three healthy volunteers demonstrated that the ingestion of HCTZ-contaminated powder (6.4 µg/g) can result in urinary HCTZ levels of up to 230 ng/mL, which supported an inadvertent administration of the drug by the athlete. Due to the sub-therapeutic and highly varying amounts of HCTZ detected in the different products, it was assumed that an accidental contamination during product manufacturing or packaging occurred.

Higenamine, or norcoclaurine, is an alkaloid acting as β_2-agonist, whose misuse in sports is prohibited at all times [91–93]. Due to its natural occurrence in numerous plants such as *Annona squamosa*, *Aconitum carmichaelii*, *Plumula nelumbinis*, and *Nelumbo nucifera*, it is often found in pre-workout and fat-burner supplements. However, an unclear or missing labeling of the ingredients of such products has caused several cases of assumed inadvertent doping within the last years [91–93]. LC-MS analysis of different preparations neither listing higenamine or relevant plant extracts on their label yielded the alkaloid at concentrations of 0.02–14 mg/g. As the current reporting limit for urinary higenamine is 10 ng/mL [68], the use of such supplements could definitely cause AAFs in sports.

In 2009, also a peptidic compound called growth hormone releasing peptide 2 (GHRP-2) was detected in two different dietary supplements [94]. GHRP-2 and related peptides are agonists of the ghrelin receptor and thus stimulate the release of growth hormone (GH) from the pituitary. The respective tablets and drinking solution were bought in Cyprus and both correctly labeled to contain GHRP-2, however, the amino acid sequence and chemical structure provided with the tablets were incorrect. Even though the administration of these products cannot result in inadvertent doping in sports, it has to be expected that also unlabeled products contaminated or adulterated with GHRPs are sold on the supplement market. Moreover, the detection of GHRP-2 in such preparations is highly remarkable: Due to their physicochemical properties and enzymatic degradation in the gastrointestinal tract, protein- and peptide-based drugs have usually a poor oral bioavailability and are therefore administered by injection [95]. However, different GHRPs were found to have an unusual high oral activity [96]. Consequently, the administration of dietary supplements containing GHRP-2—Which had no clinical approval at the time of publication—At concentrations of 50 µg/tablet and especially 9 mg/ampoule can potentially result in pharmaceutical effects [94].

3. Contaminations of Drugs and Medical Preparations

Both pharmaceuticals and food are usually tested for the presence of contaminations and impurities at the part per million (ppm) level, which is sufficient to prevent any pharmacological effects, but it cannot rule out entirely implications for sports drug testing [34].

At the end of 2014, the diuretic HCTZ was detected in the in-competition urine sample of a Swiss athlete at an estimated concentration of 5 ng/mL [97]. The athlete had not declared the use of any dietary supplement, but the administration of several tablets containing ibuprofen, a non-steroidal anti-inflammatory drug (NSAID). Surprisingly, the analysis of the ingested analgesic as well as the respective retention sample provided by the manufacturer demonstrated the presence of HCTZ at a concentration of approximately 2 µg per tablet. According to the pharmaceutical company producing the NSAID, the contamination was located in the coating of the tablets and no indications could be found that the 10 ppm cleaning limit defined by current GMP guidelines was exceeded. In order to test the plausibility of the suspected scenario of inadvertent doping, two administration studies with placebo-tablets containing 2.5 µg of HCT were conducted and the collected post-administration samples were found to contain HCTZ at concentrations of up to 16 ng/mL. As these findings supported an accidental ingestion of the doping agent by the athlete, no sanction was imposed.

Another unexpected situation resulting in AAFs triggered by the administration of a permitted medication was published in 2015 [98]. Two athletes tested positive for the diuretic chlorazanil (0.3 and 1.3 ng/mL), an obsolete therapeutic never recorded in anti-doping statistics since the consideration of diuretics as doping agents in 1988. Both athletes denied the administration of the drug but declared the use of Malarone, a malaria chemoprophylaxis drug containing 100 mg of proguanil hydrochloride and 250 mg of atovaquone. While the analysis of the Malarone tablets did not reveal any contaminations with chlorazanil, additional experiments investigating a potential metabolic conversion of proguanil to the structurally related diuretic demonstrated that chlorazanil can be produced from the proguanil metabolite *N*-(4-chlorophenyl)-biguanide if elevated levels of formaldehyde—As it can occur in the course of creatine supplementation—Are present in the urine. Consequently, both AAFs did not proceed to ADRVs.

In contrast to these cross-contamination and unexpected bioconversion scenarios, also cases involving medical preparations intentionally fortified with unlabeled pharmaceuticals were discovered.

For instance, several allegedly herbal preparations were found to contain glucocorticoids such as hydrocortisone, betamethasone, and prednisolone, which were presumed as intentionally added to obtain a higher effectiveness of the therapeutics [99,100]. Glucocorticoids are steroid hormones with anti-inflammatory and immunosuppressive properties used for the treatment of various medical conditions [101]. In sports, their systemic administration is prohibited in-competition and the use of faked supplements could therefore not only cause adverse events but also ADRVs.

Insulin-like growth factor I (IGF-I) is an endogenous cytokine mediating the effects of human growth hormone (hGH), and the misuse of recombinant IGF-I and synthetic analogs in sports is therefore prohibited at all times [102]. In 2013, human IGF-I was detected in four dietary supplements containing deer antler velvet. Such preparations are frequently used in traditional Asian medicine, as the high content of growth factors promises various health benefits. While it remains debatable and certainly depends on the route of administration if any of the IGF-I is eventually bioavailable to the antler velvet consumer, the detection of deer IGF-I in athletes' doping control samples would be reason for reporting an AAF.

Another particularly unusual case resulting in several AAFs with endogenous anabolic-androgenic steroids was reported during the FIFA Women World Cup 2011 [103]. Five members of a soccer team were tested positive after being treated with musk pod formulations. Musk pod extracts are widely used as traditional Asian medicine and known to contain various AAS whose administration in sports is prohibited [103,104]. Therefore, they have been included in "The list of medical products containing prohibited substances employed for doping" published by the State Food and Drug Administration of China. Consequently, sanctions between 14 and 18 months were imposed on the affected soccer players.

4. Food Contaminations

Besides dietary supplements and medical preparations, also food was found to be a potential source of inadvertent doping.

In several countries such as China and Mexico, the sympathomimetic and anabolic agent clenbuterol has been illegally used as growth promoter in animal production [105,106]. As a result, the edible meat is notably lean but was also found to be contaminated with clenbuterol residues, which can pose a health risk for the consumer and lead to AAFs in sports. Due to its anabolic and lipolytic effects, clenbuterol is listed among the anabolic agents in the WADA Prohibited List and is therefore prohibited both in- and out-of-competition [41,106,107]. In routine sports drug testing, clenbuterol can be detected in urine down to concentrations of a few pg/mL by using LC-MS approaches [106,107]. Until the amendment of Article 7.4 of the WADC in 2019, where the option to report atypical findings for clenbuterol if observed below 5 ng/mL of urine was introduced [1,108], no threshold applied for the detection of this drug in doping control samples, and even low concentrations resulted in AAFs and corresponding sanctions [107,109]. In an administration study with meat obtained from calves that were treated with clenbuterol at a dosage of 2 × 5 g/kg over a period of 37/43 days, the consumption by healthy volunteers resulted in urinary drug concentrations of up to 850 pg/mL in some of the participant's urine samples [110].

Although the misuse of clenbuterol in food-producing animals is strictly regulated in most countries, several cases of clenbuterol intoxication following meat consumption have been reported from all over the world [105,107,111]. Symptoms can include tremors, tachycardia, palpitations, hypokalemia, nausea, headache, nervousness, dizziness, fever, chills, peripheral vasodilatation, and—in acute cases—breathing interruptions.

The extent of the clenbuterol problem in some countries was demonstrated by two studies published in 2012 and 2013 [106,109]: In 2011, the analysis of 28 urine samples collected from volunteers returning from or permanently living in China yielded a total of 22 (=79%) positive samples with clenbuterol concentrations between 1 and 51 pg/mL [106]. Moreover, the occurrence of five AAFs with the anabolic agent among athletes of the Mexican national soccer team induced a comprehensive investigation of urine and meat/food samples collected during the FIFA U-17 World Cup held 2011 in Mexico [109]. In 30% (=14/47) of the meat/food sample obtained from the restaurants catering the soccer teams, clenbuterol was detected at amounts of 0.06–11 µg/kg, and 52% (=109/208) of the doping control urine samples were found to contain the drug at concentrations of 1–1556 pg/mL. Due to the obvious problem of contaminated meat, none of the affected athletes were sanctioned.

However, the differentiation between an unintentional clenbuterol ingestion and doping still remains challenging. A promising approach represents the discrimination of clenbuterol enantiomers: While therapeutic clenbuterol is a racemic mixture of (+)- and (-)-enantiomers, animal tissue can be characterized by the enrichment of one of the stereoisomers [112,113]. While (+)-clenbuterol was found to be accumulated in pork and chicken tissue [112–114], the (-)-enantiomer was enriched in cattle and lamb meat [111,113]. Therefore, both the route of administration (pharmaceutical product vs. meat) and the type of ingested meat can potentially influence the ratio of clenbuterol enantiomers in human urine [115,116]. However, the enantiomeric ratio was not only found to vary depending on the analyzed tissue and species of meat-producing animals, but also on the withdrawal period before slaughtering [111–113], and more research on the excretion of clenbuterol enantiomers needs to be conducted before an approach adequate for routine application in sports drug testing is available.

Hair testing is also considered as an alternative strategy to discriminate clenbuterol misuse from contamination [117]. Due to its lipophilic properties, the drug binds permanently to the hair pigment melanin and the segmental analysis of hair can therefore provide valuable additional retrospective information on the time-point of clenbuterol ingestion.

In addition to clenbuterol, also other anabolic agents bear the potential to be misused as growth promoters in livestock production.

In a comprehensive administration study with 50 raw minced beef samples bought in different Belgian butcher shops, two of the participating volunteers were tested positive for the AAS nandrolone and clostebol [118]. As usually lower quality muscle tissue is used for the production of minced meat, it was assumed that the injection sites at the neck or tail base of the animals were processed into the consumed products.

After a Norwegian athlete was tested positive for the major urinary metabolite of the AAS metenolone, a comprehensive administration study was initiated in order to investigate the possibility of inadvertent doping caused by the ingestion of contaminated poultry [119]. For that purpose, chickens were either orally treated (1 mg/day over a period of 21 days) or injected (3 injections with 1 mg of a depot formulation on days 0, 7, and 14) with metenolone and slaughtered on day 22. Subsequently, the resulting meat was administered to eight healthy male volunteers and they were asked to collect urine samples for 24–48 h. GC-MS was employed both for screening and confirmation analysis. While the consumption of the meat obtained from orally treated chickens did not result in any findings with metenolone or its metabolite, half of the volunteers were tested positive for the parent compound 22–24 h following ingestion of the injected chickens. The metabolite could be confirmed in two samples collected 4–6 h post-administration. These findings demonstrate that also contaminated poultry can cause AAFs in sports, however, the respective athlete was still sanctioned as this scenario appeared very unlikely in his case.

Zeranol is a semi-synthetic non-steroidal growth promoter, whose misuse in sports is prohibited at all times [41,120]. Inadvertent doping with this drug can not only occur due to an illegal administration to meat-producing animals, but also due to the natural presence of structurally related mycotoxins in grains: Certain fungi species colonizing in wheat, maize, barley, and oats produce zearalenone, α-, and β-zearalenol, which can be enzymatically converted to zeranol after the consumption of contaminated cereals. As ADRVs with zeranol are very rare, the possibility of an accidental ingestion should be considered in case of AAFs in sports. Metabolic profiling was identified as a potential analytical strategy to distinguish an unintentional ingestion of the mycotoxins from zeranol doping.

A potential source for unintentional doping with the nandrolone metabolites 19-norandrosterone and 19-noretiocholanolone is the consumption of edible tissues (offal and meat) from non-castrated pigs/boars, which are naturally enriched with different steroid hormones [121,122]. After eating 310 g of a meal prepared from boar kidneys, heart, liver, and meat, the urine of three healthy male volunteers was found to contain 19-norandrosterone at maximum concentrations of 3.1–7.5 ng/mL for up to 24 h, which is above the urinary MRPL of 2 ng/mL [121]. The maximal values for 19-noretiocholanolone were 0.5–1.2 ng/mL. In sports drug testing, IRMS is routinely employed to demonstrate the exogenous origin of 19-norandrosterone detected in an athlete's urine sample at low concentrations between 2.5 and 15 ng/mL [46,122]. As such urine levels would also be observed after the consumption of edible tissue from non-castrated pigs, another administration study was conducted in 2018, in order to clarify which impact the ingestion of boar offal has on the $\delta^{13}C$ values of urinary 19-norandrosterone [122]. Two male healthy volunteers consumed a meal prepared from wild boar testicles and subsequently collected urine samples for a period of 24 h. Approximately 4 h following administration, maximum 19-norandrosterone concentrations of 4 and 8 ng/mL were detected employing GC-MS, and IRMS analysis yielded highly enriched $\delta^{13}C$ values, which would constitute an AAF. Consequently, both athletes and anti-doping organizations should be aware of the risk associated with the consumption of boar products [46].

One of the oldest doping agents prohibited in-competition is the narcotic morphine [123]. For the urinary detection of this alkaloid, a threshold of 1 µg/mL applies in order to reduce the risk of inadvertent doping through the administration of pharmaceuticals containing codeine or the ingestion of poppy seeds [123,124]. However, a variety of studies demonstrated that the consumption of products containing poppy seeds can still cause AAFs in sports. In one study, eight poppy seed products commercially available in Germany were analyzed by means of GC-MS and the morphine content was found to vary from below 1 to 152 µg/g [123]. The seeds containing the highest amount of the alkaloid were subsequently used to prepare a poppy seed cake for an administration study including 9 healthy volunteers. Following ingestion, all participants were tested positive for several hours with urinary concentrations of up to 10 µg/mL. Similar results were obtained in a study published in 1990 [125]: While the consumption of 1–3 poppy seed rolls (containing 2 g of Australian seeds with a morphine content of 108 µg/g) did not result in urinary levels higher than 1 µg/mL, the ingestion of poppy seed cake (containing 15 g of Australian seeds with a morphine content of 169 µg/g) yielded concentrations of up to 2 µg/mL.

Due to the undeniable risk of inadvertent doping through the consumption of certain food and meat products, athletes are advised to take precautions and/or avoid certain meals. As there are currently no uniform international regulations or testing programs with regard to the presence of growth promoting agents in meat and the illegal use of such agents strongly varies between countries, this applies in particular to athletes traveling to international sports events [126,127].

5. Practical Aspects—Protection from Inadvertent Doping

The risk of inadvertent doping is predominantly connected to dietary supplements, which are aggressively marketed for muscle gain, fat loss, and boosting effects (mental enhancement). Therefore, athletes are advised to act with caution when intending the use such supplements [128].

If the use of dietary supplements is considered essential, acquiring supplements from low-risk sources is recommended. Information on vendor test results are available at e.g., the Cologne List (www.koelnerliste.com), the Informed Sport list in the UK (www.informed-sport.com), the NZVT list in the Netherlands (www.dopingautoriteit.nl/nzvt), etc.

In addition, dietary supplements produced by pharmaceutical companies are considered to exhibit low contamination risks as such products have not yet been reported as contaminated with doping substances [129].

In general dietary supplements should be considered carefully before use. A guidance for athletes and their advisers to minimize the risk of inadvertent doping is provided in the decision tree of the IOC consensus statement about dietary supplements and the high-performance athlete [8].

6. Conclusions

According to WADA's principle of strict liability, every athlete is responsible for the presence of a prohibited substance or its markers/metabolites in his/her biological samples, irrespective of whether or not the ADRV was committed unintentionally or deliberately. Besides the use of dietary supplements and pharmaceuticals contaminated or artificially fortified with doping agents such as AAS, stimulants, and diuretics, also the consumption of food tainted with anabolic agents or naturally containing high amounts of prohibited substances can cause inadvertent AAFs in sports (summarized in Table 1). Whilst proof for the unequivocal causality between AAF and contaminated food or supplement ingestion is difficult to provide in most instances, plausibility beyond reasonable doubt was demonstrated in selected examples of the listed case studies. The most important strategy to protect athletes from these scenarios is an appropriate education. However, from a laboratory perspective, additional measures include the identification and implementation of novel long-term metabolites for exogenous AAS in order to improve both the retrospectivity and sensitivity of the detection methods, the usage of non-targeted approaches based on high resolution/high mass accuracy mass spectrometry to identify emerging doping agents, the provision of additional analytical data from administration studies, and the development of assays that contribute to a differentiation of an intentional administration from inadvertent doping.

Table 1. Summary of findings. Various prohibited substances were detected as contaminants in dietary supplements, food products, or regular therapeutics that potentially or plausibly resulted in cases of adverse analytical findings.

Confirmed Sources of Prohibited Substances	Risk of Inadvertent Exposure with Prohibited Substance through	Case-Related Explanation Regarding Adverse Analytical Findings	Reference(s)
Dietary supplements contaminated with prohormones of nandrolone (e.g., 19-norandrostenedione)	Supplement consumption	n/a	[42,43,47,50]
Dietary supplement contaminated with prohormones of testosterone (e.g., 4-androstenedione)	Supplement consumption	n/a	[43]
Musk pod formulations naturally containing different anabolic-androgenic steroids	Treatment with traditional Asian medicine	yes	[103]
Meat contaminated with clenbuterol	Food intake	yes	[106,109,110]
		no	[111,116]
Meat contaminated with clostebol	Food intake	yes	[118]
Meat contaminated with nandrolone	Food intake	yes	[118]
Meat contaminated with metenolone	Food intake	yes	[119]
Offal and meat from non-castrated pigs/boars naturally enriched with different steroid hormones	Food intake	yes	[121,122]
Dietary supplement contaminated or adulterated with stanozolol	Supplement consumption	yes	[54]
Dietary supplements contaminated or adulterated with ostarine	Supplement consumption	yes	[58–64]
Dietary supplements contaminated with hydrochlorothiazide	Supplement consumption	yes	[90]
NSAID contaminated with hydrochlorothiazide	Administration of an analgesic	yes	[97]
Malaria chemoprophylaxis drug containing proguanil	*In vesica* conversion of proguanil metabolite	yes	[98]
Dietary supplement containing N,N-dimethyl-2-phenylpropan-1-amine & β-methylphenethylamine	Supplement consumption	no	[85]
Dietary supplement containing N-ethyl-α-ethyl-phenylethylamine	Supplement consumption	yes	[87]
Dietary supplement containing octodrine	Supplement consumption	no	[88]
Poppy seeds naturally containing high amounts of morphine	Food intake	no	[123,125]

Author Contributions: Conceptualization, K.W., H.G., M.T.; investigation, A.T., U.M., H.G., writing–original draft preparation, K.W., H.G., M.T.; All authors have read and agreed to the published version of the manuscript.

Funding: This research received no external funding.

Acknowledgments: The authors thank the Manfred-Donike Institute (Cologne, Germany) and the Federal Ministry of the Interior, Building and Community (Berlin, Germany) for supporting the presented work.

Conflicts of Interest: The authors declare no conflict of interest.

References

1. World Anti-Doping Agency (WADA). World Anti-Doping Code 2015 (with 2019 Amendments). 2019. Available online: www.wada-ama.org (accessed on 20 June 2020).
2. Anderson, J.M. Evaluating the athlete's claim of an unintentional positive urine drug test. *Curr. Sports Med. Rep.* **2011**, *10*, 191–196. [CrossRef]
3. Chan, D.K.; Ntoumanis, N.; Gucciardi, D.F.; Donovan, R.J.; Dimmock, J.A.; Hardcastle, S.J.; Hagger, M.S. What if it really was an accident? The psychology of unintentional doping. *Br. J. Sports Med.* **2016**, *50*, 898–899. [CrossRef]
4. Chan, D.K.C.; Tang, T.C.W.; Yung, P.S.; Gucciardi, D.F.; Hagger, M.S. Is unintentional doping real, or just an excuse? *Br. J. Sports Med.* **2019**, *53*, 978–979. [CrossRef]
5. Yonamine, M.; Garcia, P.R.; De Moraes Moreau, R.L. Non-intentional doping in sports. *Sports Med.* **2004**, *34*, 697–704. [CrossRef]
6. Mathews, N.M. Prohibited Contaminants in Dietary Supplements. *Sports Health* **2018**, *10*, 19–30. [CrossRef]
7. World Anti-Doping Agency (WADA). Q&A: Strict Liability in Anti-Doping. Available online: www.wada-ama.org/en/questions-answers/strict-liability-in-anti-doping (accessed on 20 June 2020).
8. Maughan, R.J.; Burke, L.M.; Dvorak, J.; Larson-Meyer, D.E.; Peeling, P.; Phillips, S.M.; Rawson, E.S.; Walsh, N.P.; Garthe, I.; Geyer, H.; et al. IOC Consensus Statement: Dietary Supplements and the High-Performance Athlete. *Int. J. Sport Nutr. Exerc. Metab.* **2018**, *28*, 104–125. [CrossRef]
9. World Anti-Doping Agency (WADA). Anti-Doping Rule Violations (ADRVs) Report 2013. 2015. Available online: https://www.wada-ama.org/sites/default/files/resources/files/wada-2013-adrv-report-en.pdf (accessed on 27 July 2020).
10. World Anti-Doping Agency (WADA). Anti-Doping Rule Violations (ADRVs) Report 2014. 2016. Available online: https://www.wada-ama.org/sites/default/files/resources/files/wada-2014-adrv-report-en_0.pdf (accessed on 27 July 2020).
11. World Anti-Doping Agency (WADA). Anti-Doping Rule Violations (ADRVs) Report 2015. 2017. Available online: https://www.wada-ama.org/sites/default/files/resources/files/2015_adrvs_report_web_release_0.pdf (accessed on 27 July 2020).
12. World Anti-Doping Agency (WADA). Anti-Doping Rule Violations (ADRVs) Report 2016. 2018. Available online: https://www.wada-ama.org/sites/default/files/resources/files/2016_adrvs_report_web_release_april_2018_0.pdf (accessed on 27 July 2020).
13. World Anti-Doping Agency (WADA). Anti-Doping Rule Violations (ADRVs) Report 2017. 2019. Available online: https://www.wada-ama.org/en/resources/general-anti-doping-information/anti-doping-rule-violations-adrvs-report (accessed on 27 July 2020).
14. Applegate, E.A.; Grivetti, L.E. Search for the competitive edge: A history of dietary fads and supplements. *J. Nutr.* **1997**, *127* (Suppl. 5), 869S–873S. [CrossRef]
15. Maughan, R.J.; Depiesse, F.; Geyer, H. International Association of Athletics Federations. The use of dietary supplements by athletes. *J. Sports Sci.* **2007**, *25* (Suppl. 1), S103–S113. [CrossRef]
16. Pearce, P.Z. Sports supplements: A modern case of caveat emptor. *Curr. Sports Med. Rep.* **2005**, *4*, 171–178. [CrossRef]
17. Garthe, I.; Maughan, R.J. Athletes and Supplements: Prevalence and Perspectives. *Int. J. Sport Nutr. Exerc. Metab.* **2018**, *28*, 126–138. [CrossRef]
18. Savino, G.; Valenti, L.; D'Alisera, R.; Pinelli, M.; Persi, Y.; Trenti, T. Working Group Doping Prevention Project (WDPP). Dietary supplements, drugs and doping in the sport society. *Ann. Ig.* **2019**, *31*, 548–555. [CrossRef]
19. Corrigan, B.; Kazlauskas, R. Medication use in athletes selected for doping control at the Sydney Olympics (2000). *Clin. J. Sport Med.* **2003**, *13*, 33–40. [CrossRef]

20. Tsitsimpikou, C.; Tsiokanos, A.; Tsarouhas, K.; Schamasch, P.; Fitch, K.D.; Valasiadis, D.; Jamurtas, A. Medication use by athletes at the Athens 2004 Summer Olympic Games. *Clin. J. Sport Med.* **2009**, *19*, 33–38. [CrossRef]
21. Tscholl, P.; Alonso, J.M.; Dollé, G.; Junge, A.; Dvorak, J. The use of drugs and nutritional supplements in top-level track and field athletes. *Am. J. Sports Med.* **2010**, *38*, 133–140. [CrossRef]
22. Tscholl, P.; Junge, A.; Dvorak, J. The use of medication and nutritional supplements during FIFA World Cups 2002 and 2006. *Br. J. Sports Med.* **2008**, *42*, 725–730. [CrossRef]
23. Braun, H.; Koehler, K.; Geyer, H.; Thevis, M.; Schänzer, W. Dietary supplement use of elite German athletes and knowledge about the contamination problem. In Proceedings of the 14th Annual Congress of the European College of Sport Sciences, Book of Abstracts, Oslo, Norway, 24–27 June 2009; Loland, S., Bø, K., Fasting, K., Hallén, J., Ommundsen, Y., Roberts, G., Tsolakidis, E., Eds.; Volume 378.
24. Baltazar-Martins, G.; Brito de Souza, D.; Aguilar-Navarro, M.; Munoz-Guerra, J.; Del Mar Plata, M.; Del Coso, J. Prevalence and patterns of dietary supplement use in elite Spanish athletes. *J. Int. Soc. Sports Nutr.* **2019**, *16*, 1–9. [CrossRef]
25. Terjung, R.L.; Clarkson, P.; Eichner, E.R.; Greenhaff, P.L.; Hespel, P.J.; Israel, R.G.; Kraemer, W.J.; Meyer, R.A.; Spriet, L.L.; Tarnopolsky, M.A.; et al. American College of Sports Medicine roundtable. The physiological and health effects of oral creatine supplementation. *Med. Sci. Sports Exerc.* **2000**, *32*, 706–717.
26. Mesa, J.L.; Ruiz, J.R.; González-Gross, M.M.; Gutiérrez Sáinz, A.; Castillo Garzón, M.J. Oral creatine supplementation and skeletal muscle metabolism in physical exercise. *Sports Med.* **2002**, *32*, 903–944. [CrossRef]
27. Maughan, R.J. Contamination of dietary supplements and positive drug tests in sport. *J. Sports Sci.* **2005**, *23*, 883–889. [CrossRef]
28. Geyer, H.; Parr, M.K.; Koehler, K.; Mareck, U.; Schänzer, W.; Thevis, M. Nutritional supplements cross-contaminated and faked with doping substances. *J. Mass Spectrom.* **2008**, *43*, 892–902. [CrossRef]
29. Green, G.A.; Catlin, D.H.; Starcevic, B. Analysis of over-the-counter dietary supplements. *Clin. J. Sport Med.* **2001**, *11*, 254–259. [CrossRef] [PubMed]
30. Ayotte, C.; Lévesque, J.F.; Cléroux, M.; Lajeunesse, A.; Goudreault, D.; Fakirian, A. Sport nutritional supplements: Quality and doping controls. *Can. J. Appl. Physiol.* **2001**, *26* (Suppl. 1), S120–S129. [CrossRef] [PubMed]
31. De Hon, O.; Coumans, B. The continuing story of nutritional supplements and doping infractions. *Br. J. Sports Med.* **2007**, *41*, 800–805. [CrossRef] [PubMed]
32. Helle, C.; Sommer, A.K.; Syversen, P.V.; Lauritzen, F. Doping substances in dietary supplements. *Tidsskr. Nor. Laegeforening* **2019**, *139*. [CrossRef]
33. Van Thuyne, W.; Van Eenoo, P.; Delbeke, F.T. Nutritional supplements: Prevalence of use and contamination with doping agents. *Nutr. Res. Rev.* **2006**, *19*, 147–158. [CrossRef] [PubMed]
34. Judkins, C.; Prock, P. Supplements and inadvertent doping—How big is the risk to athletes. *Med. Sport Sci.* **2012**, *59*, 143–152.
35. Judkins, C.M.; Teale, P.; Hall, D.J. The role of banned substance residue analysis in the control of dietary supplement contamination. *Drug Test. Anal.* **2010**, *2*, 417–420. [CrossRef]
36. Parr, M.K.; Flenker, U.; Schänzer, W. Sports-related issues and biochemistry of natural and synthetic anabolic substances. *Endocrinol. Metab. Clin.* **2010**, *39*, 45–57. [CrossRef]
37. Abbott, A. Dutch set the pace in bid to clean up diet supplements. *Nature* **2004**, *429*, 689. [CrossRef]
38. U.S. Anti-Doping Agency (USADA). Supplement 411. 2020. Available online: www.usada.org/athletes/substances/supplement-411 (accessed on 20 June 2020).
39. Geyer, H.; Schänzer, W.; Thevis, M. Anabolic agents: Recent Strategies for their Detection and Protection from Inadvertent Doping. *Br. J. Sports Med.* **2014**, *48*, 820–826. [CrossRef]
40. World Anti-Doping Agency (WADA). 2018 Anti-Doping Testing Figures. 2019. Available online: www.wada-ama.org/en/resources/laboratories/anti-doping-testing-figures-report (accessed on 20 June 2020).
41. World Anti-Doping Agency (WADA). The World Anti-Doping Code—International Standard: Prohibited List 2020. 2019. Available online: www.wada-ama.org/en/resources/science-medicine/prohibited-list-documents (accessed on 20 June 2020).

42. Geyer, H.; Parr, M.K.; Mareck, U.; Reinhart, U.; Schrader, Y.; Schänzer, W. Analysis of Non-Hormonal Nutritional Supplements for Anabolic-Androgenic Steroids—Results of an International Study. *Int. J. Sports Med.* **2004**, *25*, 124–129. [PubMed]

43. Geyer, H.; Mareck-Engelke, U.; Reinhart, U.; Thevis, M.; Schänzer, W. Positive Doping Cases with Norandrosterone after Application of Contaminated Nutritional Supplements. *Dtsch. Z. Sportmed.* **2000**, *51*, 378–382.

44. World Anti-Doping Agency (WADA). WADA Technical Document TD2018EAAS—Endogenous Anabolic Androgenic Steroids Measurement and Reporting. 2018. Available online: https://www.wada-ama.org/sites/default/files/resources/files/td2018eaas_final_eng.pdf (accessed on 20 June 2020).

45. World Anti-Doping Agency (WADA). WADA Technical Document TD2019IRMS—Detection of Synthetic Forms of Endogenous Anabolic Androgenic Steroids by GC/C/IRMS. 2018. Available online: www.wada-ama.org/sites/default/files/td2019irms_final_eng_clean.pdf (accessed on 20 June 2020).

46. World Anti-Doping Agency (WADA). WADA Technical Document TD2019NA—Harmonization of Analysis and Reporting of 19-Norsteroids Related to Nandrolone. 2018. Available online: https://www.wada-ama.org/sites/default/files/td2019na_final_eng_clean.pdf (accessed on 20 June 2020).

47. De Cock, K.J.; Delbeke, F.T.; Van Eenoo, P.; Desmet, N.; Roels, K.; De Backer, P. Detection and determination of anabolic steroids in nutritional supplements. *J. Pharm. Biomed. Anal.* **2001**, *25*, 843–852. [CrossRef]

48. Koehler, K.; Geyer, H.; Schultze, G.; Parr, M.K.; Guddat, S.; Braun, H.; Mareck, U.; Thevis, M.; Schänzer, W. Nutritional Supplements: Still a Risk of Inadvertent Doping? An Extended International Follow-up. In *Recent Advances in Doping Analysis (25), Proceedings of the Manfred Donike Workshop; 35th Cologne Workshop on Dope Analysis, 35th Cologne Workshop on Dope Analysis, Cologne, Germany, 5–10 March 2017*; Schänzer, W., Thevis, M., Geyer, H., Mareck, U., Eds.; Sport & Buch Strauß: Cologne, Germany, 2017; pp. 63–68.

49. Geyer, H.; Köhler, K.; Mareck, U.; Parr, M.K.; Schänzer, W.; Thevis, M. Cross-contaminations of vitamin- and mineral-tablets with metandienone and stanozolol. In *Recent Advances in Doping Analysis (14), Proceedings of the Manfred Donike Workshop; 24th Cologne Workshop on Dope Analysis, Cologne, Germany, 4–9 June 2006*; Schänzer, W., Geyer, H., Gotzmann, A., Mareck, U., Eds.; Sport & Buch Strauß: Cologne, Germany, 2006; pp. 11–16.

50. Baume, N.; Mahler, N.; Kamber, M.; Mangin, P.; Saugy, M. Research of stimulants and anabolic steroids in dietary supplements. *Scand. J. Med. Sci. Sports* **2006**, *16*, 41–48. [CrossRef] [PubMed]

51. Stepan, R.; Cuhra, P.; Barsova, S. Comprehensive two-dimensional gas chromatography with time-of-flight mass spectrometric detection for the determination of anabolic steroids and related compounds in nutritional supplements. *Food Addit. Contam. Part A* **2008**, *25*, 557–565. [CrossRef]

52. Martello, S.; Felli, M.; Chiarotti, M. Survey of nutritional supplements for selected illegal anabolic steroids and ephedrine using LC-MS/MS and GC-MS methods, respectively. *Food Addit. Contam.* **2007**, *24*, 258–265.

53. Geyer, H.; Bredehöft, M.; Mareck, U.; Parr, M.K.; Schänzer, W. Hohe Dosen des Anabolikums Metandienon in Nahrungsergänzungsmitteln gefunden. *Dtsch. Apoth. Ztg.* **2002**, *29*, 50.

54. Court of Arbitration for Sport. Arbitration CAS 2015/A/4129 Demir Demirev, Stoyan Enev, Ivaylo Filev, Maya Ivanove, Milka Maneva, Ivan Markov, Dian Minchev, Asen Muradiov, Ferdi Nazif, Nadezha-May Nguen & Vladimir Urumov v. International Weightlifting Federation (IWF), Award of 6 October 2015 (Operative Part of 25 August 2015). Available online: https://jurisprudence.tas-cas.org/Shared%20Documents/4129.pdf (accessed on 20 June 2020).

55. Chen, J.; Kim, J.; Dalton, J.T. Discovery and therapeutic promise of selective androgen receptor modulators. *Mol. Interv.* **2005**, *5*, 173–188. [CrossRef]

56. CHAMP Uniformed Services University. Department of Defense Dietary Supplements, Operation Supplement Safety: SARMs in Dietary Supplements. 2020. Available online: http://www.opss.org/infographic/sarms-dietary-supplements (accessed on 20 June 2020).

57. United States Anti-Doping Agency (USADA). Supplement Warning: Athletes at Risk from Ostarine in Supplements. 2017. Available online: http://www.usada.org/athlete-advisory/growing-evidence-ostarine-athlete-risk (accessed on 20 June 2020).

58. United States Anti-Doping Agency (USADA). U.S. Triathlon Athlete Elizabeth Waterstraat Accepts Sanction for Anti-Doping Rule Violation. 2019. Available online: www.usada.org/sanction/u-s-triathlon-athlete-elizabeth-waterstraat-accepts-sanction-for-anti-doping-rule-violation/ (accessed on 20 June 2020).

59. United States Anti-Doping Agency (USADA). U.S. Weightlifting Athlete Trevor Cuicchi Accepts Sanction for Anti-Doping Rule Violation. 2019. Available online: www.usada.org/sanction/trevor-cuicchi-accepts-doping-sanction/ (accessed on 20 June 2020).

60. United States Anti-Doping Agency (USADA). U.S.Wrestling Athlete Victoria Francis Accepts Sanction for Anti-Doping Rule Violation. 2018. Available online: www.usada.org/sanction/victoria-francis-accepts-doping-sanction/ (accessed on 20 June 2020).

61. United States Anti-Doping Agency (USADA). U.S. Weightlifting Athlete Abby Raymond Accepts Sanction for Anti-Doping Rule Violation. 2018. Available online: www.usada.org/sanction/abby-raymond-accepts-doping-sanction/ (accessed on 20 June 2020).

62. United States Anti-Doping Agency (USADA). U.S. Track & Field Athlete, Chris Carter, Accepts Sanction for Anti-Doping Rule Violation. 2017. Available online: www.usada.org/sanction/chris-carter-accepts-doping-sanction/ (accessed on 20 June 2020).

63. United States Anti-Doping Agency (USADA). U.S. Sitting Volleyball Athlete, Roderick Green, Accepts Sanction for Anti-Doping Rule Violation. 2017. Available online: www.usada.org/sanction/roderick-green-accepts-doping-sanction/ (accessed on 20 June 2020).

64. United States Anti-Doping Agency (USADA). US Triathlon Athlete, Paulson, Accepts Sanction for Anti-Doping Rule Violation. 2016. Available online: www.usada.org/sanction/ashley-paulson-accepts-sanction/ (accessed on 20 June 2020).

65. Deventer, K.; Roels, K.; Delbeke, F.T.; Van Eenoo, P. Prevalence of legal and illegal stimulating agents in sports. *Anal. Bioanal. Chem.* **2011**, *401*, 421–432.

66. Docherty, J.R. Pharmacology of stimulants prohibited by the World Anti-Doping Agency (WADA). *Br. J. Pharmacol.* **2008**, *154*, 606–622.

67. World Anti-Doping Agency (WADA). Prohibited List Q&A. Available online: www.wada-ama.org/en/questions-answers/prohibited-list-qa#item-387 (accessed on 20 June 2020).

68. World Anti-Doping Agency (WADA). WADA Technical Document TD2019MRPL—Minimum Required Performance Levels for Detection and Identification of Non-Threshold Substances. 2019. Available online: www.wada-ama.org/sites/default/files/resources/files/td2019mrpl_eng.pdf (accessed on 20 June 2020).

69. Gurley, B.J.; Wang, P.; Gardner, S.F. Ephedrine-type alkaloid content of nutritional supplements containing Ephedra sinica (Ma-huang) as determined by high performance liquid chromatography. *J. Pharm. Sci.* **1998**, *87*, 1547–1553. [CrossRef] [PubMed]

70. Köhler, K.; Geyer, H.; Guddat, S.; Orlovius, A.; Parr, M.K.; Thevis, M.; Mester, J.; Schänzer, W. Sibutramine Found in Chinese Herbal Slimming Tea and Capsules. In *Recent Advances in Doping Analysis (15), Proceedings of the Manfred Donike Workshop; 25th Cologne Workshop on Dope Analysis, Cologne, Germany, 25 February to 2 March 2007*; Schänzer, W., Geyer, H., Gotzmann, A., Mareck, U., Eds.; Sport & Buch Strauß: Cologne, Germany, 2007; pp. 367–370.

71. Monakhova, Y.B.; Ilse, M.; Hengen, J.; El-Atma, O.; Kuballa, T.; Kohl-Himmelseher, M.; Lachenmeier, D.W. Rapid assessment of the illegal presence of 1,3-dimethylamylamine (DMAA) in sports nutrition and dietary supplements using 1H NMR spectroscopy. *Drug Test. Anal.* **2014**, *6*, 944–948. [CrossRef] [PubMed]

72. Di Lorenzo, C.; Moro, E.; Dos Santos, A.; Uberti, F.; Restani, P. Could 1,3 dimethylamylamine (DMAA) in food supplements have a natural origin? *Drug Test. Anal.* **2013**, *5*, 116–121. [CrossRef]

73. Elsohly, M.A.; Gul, W.; Elsohly, K.M.; Murphy, T.P.; Weerasooriya, A.; Chittiboyina, A.G.; Avula, B.; Khan, I.; Eichner, A.; Bowers, L.D. Pelargonium oil and methyl hexaneamine (MHA): Analytical approaches supporting the absence of MHA in authenticated Pelargonium graveolens plant material and oil. *J. Anal. Toxicol.* **2012**, *36*, 457–471. [CrossRef] [PubMed]

74. ElSohly, M.A.; Gul, W.; Tolbert, C.; ElSohly, K.M.; Murphy, T.P.; Avula, B.; Chittiboyina, A.G.; Wang, M.; Khan, I.A.; Yang, M.; et al. Methylhexanamine is not detectable in Pelargonium or Geranium species and their essential oils: A multi-centre investigation. *Drug Test. Anal.* **2015**, *7*, 645–654. [CrossRef]

75. Zhang, Y.; Woods, R.M.; Breitbach, Z.S.; Armstrong, D.W. 1,3-Dimethylamylamine (DMAA) in supplements and geranium products: Natural or synthetic? *Drug Test. Anal.* **2012**, *4*, 986–990. [CrossRef]

76. Lisi, A.; Hasick, N.; Kazlauskas, R.; Goebel, C. Studies of methylhexaneamine in supplements and geranium oil. *Drug Test. Anal.* **2011**, *3*, 873–876. [CrossRef]

77. Ping, Z.; Jun, Q.; Qing, L. A study of the chemical constituents of geranium oil. *J. Guizhou Inst. Technol.* **1996**, *25*, 82–85.

78. Li, J.S.; Chen, M.; Li, Z.C. Identification and quantification of dimethylamylamine in geranium by liquid chromatography tandem mass spectrometry. *Anal. Chem. Insights* **2012**, *7*, 47–58. [CrossRef]

79. Fleming, H.L.; Ranaivo, P.L.; Simone, P.S. Analysis and Confirmation of 1,3-DMAA and 1,4-DMAA in Geranium Plants Using High Performance Liquid Chromatography with Tandem Mass Spectrometry at ng/g Concentrations. *Anal. Chem. Insights* **2012**, *7*, 59–78. [CrossRef]

80. ElSohly, M.A.; Gul, W. LC-MS-MS analysis of dietary supplements for N-ethyl-α-ethyl-phenethylamine (ETH), N, N-diethylphenethylamine and phenethylamine. *J. Anal. Toxicol.* **2014**, *38*, 63–72. [CrossRef]

81. Sigmund, G.; Dib, J.; Tretzel, L.; Piper, T.; Bosse, C.; Schänzer, W.; Thevis, M. Monitoring 2-phenylethanamine and 2-(3-hydroxyphenyl)acetamide sulfate in doping controls. *Drug Test. Anal.* **2015**, *7*, 1057–1062. [CrossRef] [PubMed]

82. Yun, J.; Kwon, K.; Choi, J.; Jo, C.H. Monitoring of the amphetamine-like substances in dietary supplements by LC-PDA and LC-MS/MS. *Food Sci. Biotechnol.* **2017**, *26*, 1185–1190. [CrossRef] [PubMed]

83. Pawar, R.S.; Grundel, E.; Fardin-Kia, A.R.; Rader, J.I. Determination of selected biogenic amines in Acacia rigidula plant materials and dietary supplements using LC-MS/MS methods. *J. Pharm. Biomed. Anal.* **2014**, *88*, 457–466. [CrossRef] [PubMed]

84. Kwiatkowska, D.; Wójtowicz, M.; Jarek, A.; Goebel, C.; Chajewska, K.; Turek-Lepa, E.; Pokrywka, A.; Kazlauskas, R. N,N-dimethyl-2-phenylpropan-1-amine—New Designer Agent Found in Athlete Urine and Nutritional Supplement. *Drug Test. Anal.* **2015**, *4*, 331–335. [CrossRef]

85. Wójtowicz, M.; Jarek, A.; Chajewska, K.; Kwiatkowska, D. N,N-dimethyl-2-phenylpropan-1-amine quantification in urine: Application to excretion study following single oral dietary supplement dose. *Anal. Bioanal. Chem.* **2016**, *408*, 5041–5047. [CrossRef]

86. Cohen, P.A.; Travis, J.C.; Venhuis, B.J. A methamphetamine analog (N,α-diethyl-phenylethylamine) identified in a mainstream dietary supplement. *Drug Test. Anal.* **2014**, *6*, 805–807. [CrossRef]

87. Wójtowicz, M.; Jarek, A.; Chajewska, K.; Turek-Lepa, E.; Kwiatkowska, D. Determination of designer doping agent–2-ethylamino-1-phenylbutane–in dietary supplements and excretion study following single oral supplement dose. *J. Pharm. Biomed. Anal.* **2015**, *115*, 523–533. [CrossRef]

88. Dib, J.; Bosse, C.; Tsivou, M.; Glatt, A.-M.; Geisendorfer, T.; Geyer, H.; Gmeiner, G.; Sigmund, G.; Thevis, M. Is heptaminol a (major) metabolite of octodrine? *Drug Test. Anal.* **2019**, *11*, 1761–1763. [CrossRef]

89. Geyer, H. Adulterated nutritional supplements and unapproved pharmaceuticals as new sources of doping substances for fitness and recreational sports. In *Doping and Public Health*; Ahmadi, N., Ljungqvist, A., Svedsäter, G., Eds.; Routledge: London, UK, 2016; pp. 64–70.

90. Favretto, D.; Visentin, S.; Scrivano, S.; Roselli, E.; Mattiazzi, F.; Pertile, R.; Vogliardi, S.; Tucci, M.; Montisci, M. Multiple incidence of the prescription diuretic hydrochlorothiazide in compounded nutritional supplements. *Drug Test. Anal.* **2019**, *11*, 512–522. [CrossRef]

91. Grucza, K.; Kowalczyk, K.; Wicka, M.; Szutowski, M.; Bulska, E.; Kwiatkowska, D. The use of a valid and straightforward method for the identification of higenamine in dietary supplements in view of anti-doping rule violation cases. *Drug Test. Anal.* **2019**, *11*, 912–917. [CrossRef] [PubMed]

92. Stajić, A.; Anđelković, M.; Dikić, N.; Rašić, J.; Vukašinović-Vesić, M.; Ivanović, D.; Jančić-Stojanović, B. Determination of higenamine in dietary supplements by UHPLC/MS/MS method. *J. Pharm. Biomed. Anal.* **2017**, *146*, 48–52. [CrossRef] [PubMed]

93. Yan, K.; Wang, X.; Wang, Z.; Wang, Y.; Luan, Z.; Gao, X.; Wang, R. The risk of higenamine adverse analytical findings following oral administration of plumula nelumbinis capsules. *Drug Test. Anal.* **2019**, *11*, 1731–1736. [CrossRef] [PubMed]

94. Kohler, M.; Thomas, A.; Geyer, H.; Petrou, M.; Schänzer, W.; Thevis, M. Confiscated black market products and nutritional supplements with non-approved ingredients analyzed in the cologne doping control laboratory 2009. *Drug Test. Anal.* **2010**, *2*, 533–537. [CrossRef]

95. Shaji, J.; Patole, V. Protein and Peptide drug delivery: Oral approaches. *Indian J. Pharm. Sci.* **2008**, *70*, 269–277. [CrossRef] [PubMed]

96. Hartman, M.L.; Farello, G.; Pezzoli, S.S.; Thorner, M.O. Oral administration of growth hormone (GH)-releasing peptide stimulates GH secretion in normal men. *J. Clin. Endocrinol. Metab.* **1992**, *74*, 1378–1384.

97. Helmlin, H.J.; Mürner, A.; Steiner, S.; Kamber, M.; Weber, C.; Geyer, H.; Guddat, S.; Schänzer, W.; Thevis, M. Detection of the diuretic hydrochlorothiazide in a doping control urine sample as the result of a non-steroidal anti-inflammatory drug (NSAID) tablet contamination. *Forensic. Sci. Int.* **2016**, *267*, 166–172. [CrossRef]

98. Thevis, M.; Geyer, H.; Thomas, A.; Tretzel, L.; Bailloux, I.; Buisson, C.; Lasne, F.; Schaefer, M.S.; Kienbaum, P.; Mueller-Stoever, I.; et al. Formation of the diuretic chlorazanil from the antimalarial drug proguanil—implications for sports drug testing. *J. Pharm. Biomed. Anal.* **2015**, *115*, 208–213. [CrossRef]

99. Li, L.; Liang, X.; Xu, T.; Xu, F.; Dong, W. Rapid Detection of Six Glucocorticoids Added Illegally to Dietary Supplements by Combining TLC with Spot-Concentrated Raman Scattering. *Molecules* **2018**, *23*, 1504. [CrossRef]

100. Ahmed, S.; Riaz, M. Quantitation of cortico-steroids as common adulterants in local drugs by HPLC. *Chromatographia* **1991**, *31*, 67–70. [CrossRef]

101. Vernec, A.; Slack, A.; Harcourt, P.R.; Budgett, R.; Duclos, M.; Kinahan, A.; Mjøsund, K.; Strasburger, C.J. Glucocorticoids in elite sport: Current status, controversies and innovative management strategies-a narrative review. *Br. J. Sports Med.* **2020**, *54*, 8–12. [CrossRef] [PubMed]

102. Cox, H.D.; Eichner, D. Detection of human insulin-like growth factor-1 in deer antler velvet supplements. *Rapid Commun. Mass Spectrom.* **2013**, *27*, 2170–2178. [CrossRef]

103. Thevis, M.; Schänzer, W.; Geyer, H.; Thieme, D.; Grosse, J.; Rautenberg, C.; Flenker, U.; Beuck, S.; Thomas, A.; Holland, R.; et al. Traditional Chinese medicine and sports drug testing: Identification of natural steroid administration in doping control urine samples resulting from musk (pod) extracts. *Br. J. Sports Med.* **2013**, *47*, 109–114. [CrossRef]

104. He, Y.; Wang, J.; Liu, X.; Xu, Y.; He, Z. Influences of musk administration on the doping test. *Steroids* **2013**, *78*, 1047–1052. [CrossRef] [PubMed]

105. Prezelj, A.; Obreza, A.; Pecar, S. Abuse of clenbuterol and its detection. *Curr. Med. Chem.* **2003**, *10*, 281–290. [CrossRef]

106. Guddat, S.; Fußhöller, G.; Geyer, H.; Thomas, A.; Braun, H.; Haenelt, N.; Schwenke, A.; Klose, C.; Thevis, M.; Schänzer, W. Clenbuterol—Regional food contamination a possible source for inadvertent doping in sports. *Drug Test. Anal.* **2012**, *4*, 534–538. [CrossRef] [PubMed]

107. Nicoli, R.; Petrou, M.; Badoud, F.; Dvorak, J.; Saugy, M.; Baume, N. Quantification of clenbuterol at trace level in human urine by ultra-high pressure liquid chromatography-tandem mass spectrometry. *J. Chromatogr. A* **2013**, *1292*, 142–150. [CrossRef]

108. World Anti-Doping Ageny (WADA). Stakeholder Notice Regarding Meat Contamination. 2019. Available online: http://www.wada-ama.org/sites/default/files/resources/files/2019-05-30-meat_contamination_notice_final.pdf (accessed on 20 June 2020).

109. Thevis, M.; Geyer, L.; Geyer, H.; Guddat, S.; Dvorak, J.; Butch, A.; Sterk, S.S.; Schänzer, W. Adverse analytical findings with clenbuterol among U-17 soccer players attributed to food contamination issues. *Drug Test. Anal.* **2013**, *5*, 372–376. [CrossRef]

110. Hemmersbach, P.; Tomten, S.; Nilsson, S.; Oftebro, H.; Havrevoll, Ö.; Öen, B.; Birkeland, K. Illegal Use of Anabolic Agents in Animal Fattening—Consequences for Doping Analysis. In *Recent Advances in Doping Analysis (15), Proceedings of the Manfred Donike Workshop; 12th Cologne Workshop on Dope Analysis, Cologne, Germany, 10–15 April 1994*; Schänzer, W., Geyer, H., Gotzmann, A., Mareck, U., Eds.; Sport & Buch Strauß: Cologne, Germany, 1995; pp. 185–191.

111. Parr, M.K.; Blokland, M.H.; Liebetrau, F.; Schmidt, A.H.; Meijer, T.; Stanic, M.; Kwiatkowska, D.; Waraksa, E.; Sterk, S.S. Distinction of clenbuterol intake from drug or contaminated food of animal origin in a controlled administration trial—The potential of enantiomeric separation for doping control analysis. *Food Addit. Contam. Part A* **2017**, *34*, 525–535. [CrossRef]

112. Smith, D.J. Stereochemical composition of clenbuterol residues in edible tissues of swine. *J. Agric. Food Chem.* **2000**, *48*, 6036–6043. [CrossRef]

113. Wang, Z.L.; Zhang, J.L.; Zhang, Y.N.; Zhao, Y. Mass Spectrometric Analysis of Residual Clenbuterol Enantiomers in Swine, Beef and Lamb meat by Liquid Chromatography Tandem Mass Spectrometry. *Anal. Methods* **2016**, *8*, 4127–4133. [CrossRef]

114. Smith, D.J. Total radioactive residues and clenbuterol residues in swine after dietary administration of [14C] clenbuterol for seven days and preslaughter withdrawal periods of zero, three, or seven days. *J. Anim. Sci.* **2000**, *78*, 2903–2912. [CrossRef] [PubMed]

115. Dolores, H.M.; Villaseñor, A.; Piña, O.S.; Mercado Márquez, C.; Bejarano, B.V.; Bonaparte, M.E.G.; López-Arellano, R. Evaluation of R- (-) and S- (+) Clenbuterol enantiomers during a doping cycle or continuous ingestion of contaminated meat using chiral liquid chromatography by LC-TQ-MS. *Drug Test. Anal.* **2019**, *11*, 1238–1247. [CrossRef] [PubMed]

116. Thevis, M.; Thomas, A.; Beuck, S.; Butch, A.; Dvorak, J.; Schänzer, W. Does the analysis of the enantiomeric composition of clenbuterol in human urine enable the differentiation of illicit clenbuterol administration from food contamination in sports drug testing? *Rapid Commun. Mass Spectrom.* **2013**, *27*, 507–512. [CrossRef] [PubMed]

117. Krumbholz, A.; Anielski, P.; Gfrerer, L.; Graw, M.; Geyer, H.; Schänzer, W.; Dvorak, J.; Thieme, D. Statistical significance of hair analysis of clenbuterol to discriminate therapeutic use from contamination. *Drug Test. Anal.* **2014**, *6*, 1108–1116. [CrossRef] [PubMed]

118. Debruyckere, G.; Van Peteghem, C.H.; De Sagher, R. Influence of the consumption of meat contaminated with anabolic steroids on doping tests. *Anal. Chim. Acta* **1993**, *275*, 49–56. [CrossRef]

119. Kicman, A.T.; Cowan, D.A.; Myhre, L.; Nilsson, S.; Tomten, S.; Oftebro, H. Effect on sports drug tests of ingesting meat from steroid (methenolone)-treated livestock. *Clin. Chem.* **1994**, *40*, 2084–2087. [CrossRef]

120. Thevis, M.; Fusshöller, G.; Schänzer, W. Zeranol: Doping offence or mycotoxin? A case-related study. *Drug Test. Anal.* **2011**, *3*, 777–783. [CrossRef]

121. Le Bizec, B.; Gaudin, I.; Monteau, F.; Andre, F.; Impens, S.; De Wasch, K.; De Brabander, H. Consequence of boar edible tissue consumption on urinary profiles of nandrolone metabolites. I. Mass spectrometric detection and quantification of 19-norandrosterone and 19-noretiocholanolone in human urine. *Rapid Commun. Mass Spectrom.* **2000**, *14*, 1058–1065. [CrossRef]

122. Hülsemann, F.; Gougoulidis, V.; Schertel, T.; Fusshöller, G.; Flenker, U.; Piper, T.; Thevis, M. Case Study: Atypical δ13 C values of urinary norandrosterone. *Drug Test. Anal.* **2018**, *10*, 1728–1733. [CrossRef]

123. Thevis, M.; Opfermann, G.; Schänzer, W. Urinary concentrations of morphine and codeine after consumption of poppy seeds. *J. Anal. Toxicol.* **2003**, *27*, 53–56. [CrossRef] [PubMed]

124. World Anti-Doping Agency (WADA). WADA Technical Document TD2019DL v. 2.0—Decision Limits for the Confirmatory Quantification of Threshold Substances. 2019. Available online: https://www.wada-ama.org/en/resources/science-medicine/td2019dl-version-2-0 (accessed on 20 June 2020).

125. ElSohly, H.N.; ElSohly, M.A.; Stanford, D.F. Poppy seed ingestion and opiates urinalysis: A closer look. *J. Anal. Toxicol.* **1990**, *14*, 308–310. [CrossRef] [PubMed]

126. Stephany, R.W. Hormonal growth promoting agents in food producing animals. In *Handbook of Experimental Pharmacology*; Springer: Berlin/Heidelberg, Germany, 2010; pp. 355–367.

127. Braun, H.; Geyer, H.; Koehler, K. Meat products as potential doping traps? *Int. J. Sport Nutr. Exerc. Metab.* **2008**, *18*, 539–542. [CrossRef]

128. Geyer, H.; Braun, H.; Burke, L.M.; Stear, S.J.; Castell, L.M. Inadvertent doping. In: A-Z of nutritional supplements: Dietary supplements, sports nutrition foods and ergogenic aids for health and performance—Part 22. *Br. J. Sports Med.* **2011**, *45*, 752–754. [CrossRef] [PubMed]

129. Geyer, H.; Gülker, A.; Mareck, U.; Parr, M.K.; Schänzer, W. Some good news from the field of nutritional supplements. In *Recent Advances in Doping Analysis*; Schänzer, W., Geyer, H., Gotzmann, A., Mareck, U., Eds.; Sportverlag Strauß: Köln, Germany, 2004; Volume 12, pp. 91–97.

Article

Evaluation of Amyloid β₄₂ Aggregation Inhibitory Activity of Commercial Dressings by A Microliter-Scale High-Throughput Screening System Using Quantum-Dot Nanoprobes

Masahiro Kuragano, **Wataru Yoshinari**, **Xuguang Lin**, **Keiya Shimamori**, **Koji Uwai and Kiyotaka Tokuraku ***

Graduate School of Engineering, Muroran Institute of Technology, Muroran 050-8585, Japan; gano@mmm.muroran-it.ac.jp (M.K.); ysnrwtr0629@icloud.com (W.Y.); linxuguang4000@163.com (X.L.); 19041040@mmm.muroran-it.ac.jp (K.S.); uwai@mmm.muroran-it.ac.jp (K.U.)
* Correspondence: tokuraku@mmm.muroran-it.ac.jp; Tel.: +81-0143-46-5721

Received: 30 April 2020; Accepted: 22 June 2020; Published: 24 June 2020

Abstract: The aggregation and accumulation of amyloid β (Aβ) in the brain is a trigger of pathogenesis for Alzheimer's disease. Previously, we developed a microliter-scale high-throughput screening (MSHTS) system for Aβ₄₂ aggregation inhibitors using quantum-dot nanoprobes. The MSHTS system is seldom influenced by contaminants in samples and is able to directly evaluate Aβ₄₂ aggregation inhibitory activity of samples containing various compounds. In this study, to elucidate whether the MSHTS system could be applied to the evaluation of processed foods, we examined Aβ₄₂ aggregation inhibitory activity of salad dressings, including soy sauces. We estimated the 50% effective concentration (EC₅₀) from serial diluted dressings. Interestingly, all 19 commercial dressings tested showed Aβ₄₂ aggregation inhibitory activity. It was suggested that EC₅₀ differed by as much as 100 times between the dressings with the most (0.065 ± 0.020 *v/v*%) and least (6.737 ± 5.054 *v/v*%) inhibitory activity. The highest activity sample is traditional Japanese dressing, soy sauce. It is known that soy sauce is roughly classified into a heat-treated variety and a non-heat-treated variety. We demonstrated that non-heat-treated raw soy sauce exhibited higher Aβ₄₂ aggregation inhibitory activity than heat-treated soy sauce. Herein, we propose that MSHTS system can be applied to processed foods.

Keywords: Alzheimer's disease; Amyloid β; amyloid β aggregation inhibitor; quantum dot; soy sauce

1. Introduction

One of the problems facing an aging society is the increase of patients with dementia. While various diseases are known to cause dementia, Alzheimer's disease (AD) in particular, accounts for the majority of cases [1–3]. Four AD drugs approved in Japan, donepezil, galantamine, rivastigmine, and memantine, only function by delaying the progression of pathological conditions by temporarily enhancing neurotransmission, and are not fundamental therapeutic agents [4]. The amyloid cascade hypothesis notes that AD is caused by the aggregation and accumulation of 38 to 43 residues of the amyloid β (Aβ) peptide excised from amyloid precursor protein in the brain [5–8]. Recently, Biogen and Eisai reported that a patient's cognitive decline had been blunted in clinical trials using antibodies, aducanumab, that bind specifically to Aβ aggregates [9]. However, in March of 2019, a phase III clinical trial of aducanumab was halted because of insufficient evidence to support its effect in AD [10]. In October of 2019, both companies announced that they would apply for a new drug application of aducanumab to the U.S. Food and Drug Administration in 2020, as the effect was confirmed in some patients who received the drug at a high dose. However, these events remind us of the difficulties in developing AD therapeutics. Therefore, attention is now focused on AD prevention and treatment schemes that target the aggregation and accumulation of Aβ. There is currently a global search for candidate substances that can inhibit Aβ aggregation. Since the aggregation and accumulation of Aβ begins several decades before the expression of AD [11], long-term prevention with functional foods may be more effective than treatment with therapeutic medicine.

Rosmarinic acid (RA) is a polyphenol found in abundance in plants of the *Lamiaceae* such as rosemary, perilla, and lemon balm. RA is a known inhibitor of Aβ aggregation [12,13]. Its Aβ aggregation inhibitory activity was examined using AD model mice and its safety was confirmed in human studies using lemon balm extract [14,15]. Among many other polyphenols, curcumin, which is found in turmeric, is also a famous Aβ aggregation inhibitor [12]. Further, it was reported that importance of functional foods on AD. The extract obtained from miso, a traditional fermented dressing in Japan, suppresses Aβ-induced neuronal damage [16]. Hsu et al., reported that nattokinase degraded amyloid fibrils [17]. Thus, the use of functional foods has attracted attention as a possible AD countermeasure. However, it is technically very difficult to evaluate plant extracts and processed foods as these include various impurities. In general, the Thioflavin T (ThT) method has been used to evaluate Aβ aggregation inhibitory activity of various substances [18]. ThT emits fluorescence when bound to amyloid fibrils. In this method, the level of Aβ aggregation is measured from the fluorescence intensity of ThT. However, the excitation and emission wavelengths of ThT are 455 and 490 nm, respectively, so they compete with the absorption wavelengths of many natural substances. Therefore, the ThT method is unsuitable to evaluate food samples that contain various contaminants. A method of directly observing Aβ aggregates with a transmission electron microscope (TEM) is widely used. Because it is necessary to dry the Aβ aggregates sample when preparing, the observation under physiological conditions is difficult. Further, the amount of aggregates is biased depending on the field of view even in the same sample, suggesting that there is a problem in quantitative. In addition, the ThT and TEM method generally require several steps for sample preparation and observation, and it is difficult to analyze a large amount of the sample at one time. In other words, previous conventional method could not perform accurate and quick high throughput quantitative analysis.

Previously, we succeeded in real-time imaging of the Aβ42 aggregation process with a fluorescence microscope using a quantum dot (QD) nanoprobe and developed a microliter-scale high-throughput screening (MSHTS) system for Aβ42 aggregation inhibitors by applying this imaging method [19,20]. The MSHTS system has some advantages: (1) only a small sample volume of 5 μL is required, (2) high-throughput analysis uses a 1536-well plate, and (3) filter effects due to contaminants in the sample are avoided because the amount of Aβ42 aggregates is quantified from standard deviation (SD) value estimated from the variation in fluorescence intensity of each pixel of obtained images and the emission wavelength of QD605 does not overlap with the absorption of almost natural products [20,21]. Thus, the MSHTS system can evaluate the magnitude of inhibitory activity for Aβ42 aggregation as EC50 values. Before, we evaluated the Aβ42 aggregation inhibitory activity of 52 spices using this method and demonstrated that the herb-based spices of the *Lamiaceae* family exhibited high Aβ42 aggregation inhibitory activity [20]. Then, we found that the activity of boiling water extracts of 11 seaweeds was higher than that of ethanolic extracts and revealed that Aβ42 aggregates morphology was affected with seaweed-derived polysaccharide including in boiling water extracts [22]. Further, we recently developed an automated MSHTS system to evaluate larger numbers of samples at once [21]. Screening 504 plant extracts collected in Hokkaido, Japan, we found that Geraniales and Myrtales within Rosids showed high Aβ42 aggregation inhibitory activity. Thus, MSHTS system is useful for quantitative evaluation of Aβ42 aggregation inhibition ability of various natural products. However, it is unclear whether MSHTS system can evaluate Aβ42 aggregation inhibitory activity in foods including various natural substances with many impurities. In this study, to elucidate whether the MSHTS system is applied to processed foods such as salad dressings, including soy sauces, we evaluated Aβ42 aggregation inhibitory activity of dressings using the MSHTS system. We found that all tested commercial dressings showed $A\beta_{42}$ aggregation inhibitory activity despite there were differences in their activities. Especially raw soy sauce showed the highest inhibitory activity among the tested samples. These results suggest that the MSHTS system is a powerful and useful tool that is expected to be applied and developed in various processed foods.

2. Materials and Methods

2.1. Materials

Human $A\beta_{42}$ (4349-v, Peptide Institute Inc., Osaka, Japan) and Cys-conjugated $A\beta_{40}$ (23519, Anaspec Inc., Fremont, CA, USA) kits were purchased commercially. Twenty different commercially available salad dressings and soy source brands were purchased from Japanese companies (Kewpie, Sameura Foods, Sanyo Coffee Foods, Shiranukacho Shinko Kosha, Shinshu Shizen Okoku, Seijo Ishii, Taiyo Sangyo, Tsukiboshi Foods, Nihon Syoyu Kogyo, Big Chef, Pure Foods Toya, Yamada Bee Farm, Riken Vitamin, H+B Life Science) using catalog shopping in June 2016.

2.2. Preparation of QDAβ Nanoprobe

The QDAβ nanoprobe was prepared using QD-PEG-NH$_2$ (Qdot^{TM} 605 ITK^{TM} Amino (PEG) Quantum dot; Q21501MP, Thermo Fisher Scientific, Waltham, MA USA) according to our previous reports [19–22]. The QDAβ nanoprobe was prepared by first reacting 10 µM QD-PEG-NH$_2$ with 1 mM sulfo-EMCS (22307, Thermo Fisher Scientific, Waltham, MA, USA) in PBS (phosphate-buffered saline) for 1 h at room temperature. QDAβ concentration was determined by comparing absorbance at 350 nm to that of unlabeled QD-PEG-NH$_2$

2.3. Estimation of EC$_{50}$ by the MSHTS System

The EC$_{50}$ values of various dressings were determined by a modified MSHTS system, as was described in our previous reports [20–22]. More specifically, various concentrations of each dressing, 30 nM QDAβ, and 30 µM $A\beta_{42}$ in PBS containing 5% EtOH and 3% DMSO were incubated in a 1536-well plate (782096, Greiner, Kremsmünster, Austria) at 37 °C for 24 h. The QDAβ-$A\beta_{42}$ aggregates that formed in each well were observed by an inverted fluorescence microscope (TE2000, Nikon, Tokyo, Japan). Standard deviation (SD) values of fluorescence intensities of 40,000 pixels (200 × 200 pixels) around the central region of each well were measured by ImageJ software Ver 1.53b (NIH). The SD values, which were approximately proportional to the amount of aggregates [20–22], were plotted against the concentrations of added salad dressings to establish an inhibition curve.

2.4. Fluorescence Microscopy

Aggregates in the 1536-well plate were observed by an inverted fluorescence microscope (TE2000-S, Nikon) using a 4× objective lens (Plan Fluor 4×/0.13 PhL DL, Nikon) equipped with a color CCD camera (DP72, Olympus, Tokyo, Japan).

2.5. Transmission Electron Microscopy

Samples were deposited in 10 µL aliquots onto 200-mesh copper grids and negatively stained with 1% phosphotungstic acid at room temperature. Specimens were examined under an H-7600 TEM (Hitachi, Tokyo, Japan) at 60 kV.

2.6. ThT Assay

The ThT assay was conducted according to the method of Levine modified in our laboratory [18,21]. Statistical analyses between +ThT and −ThT samples were performed with EZR (Saitama Medical Center, Jichi Medical University, Saitama, Japan), which is a graphical user interface for R (The R Foundation for Statistical Computing) [23]. More precisely, it is a modified version of R commander designed to add statistical functions frequently used in biostatistics.

2.7. SDS-Polyacrylamide Gel Electrophoresis

SDS-polyacrylamide gel electrophoresis (SDS-PAGE) was performed using standard techniques. Soy sauce and raw soy sauce were heated by block incubator at 80 °C for 60 min before electrophorese. For dialysis protocol, soy sauce and raw soy sauce were dialyzed against distilled water. Distilled were changed three times for overnight. Then, the gel was silver-stained by staining kit (2D-SLVER STAIN II, COSMO BIO)

3. Results and Discussion

3.1. Evaluation of $A\beta_{42}$ Aggregation Inhibitory Activity of Commercial Dressings by ThT Method

First, in order to assess whether 19 liquid salad dressings could be evaluated for activity by the ThT method, their absorbance spectra were measured using a Nanodrop 2000c (Thermo Fisher Scientific) (Figure 1). The excitation and emission wavelengths of ThT are 450 nm and 490 nm, respectively. As shown in Figure 1, only 3 of the 19 dressings (samples L, N, M) showed no absorbance of 1 or more at each wavelength. Most of the samples contained soy sauce as a raw material, and the color was black or brown depending on the content of soy sauce. In the 15 samples that showed absorption peaks at the ThT excitation and emission wavelengths, the absorption peak shifted to the right in proportion as the color became darker. This indicates that the evaluation of $A\beta_{42}$ aggregation inhibitory activity of the liquid dressings using the ThT method was difficult because the absorption wavelength of almost samples overlapped with the excitation and emission wavelengths of ThT.

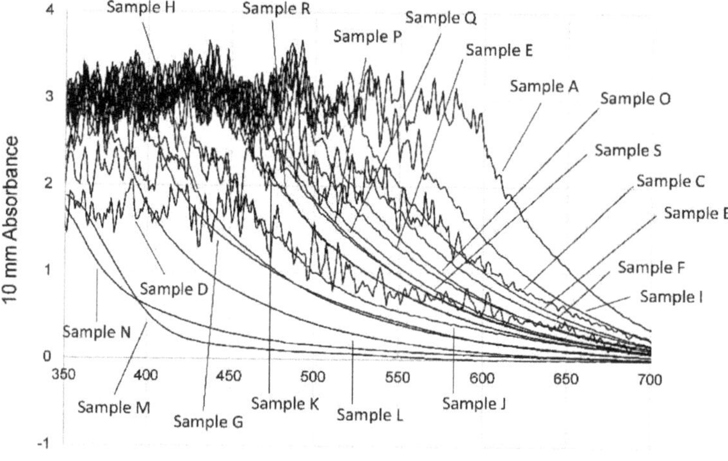

Figure 1. Absorbance of 19 commercial liquid dressings. Absorbance of the 19 commercial liquid dressings shown in Table 1 was measured. As for the absorption wavelength of dressings, three samples (L, M, N) do not show an overlap with the excitation (450 nm) and emission (490 nm) wavelengths of ThT. Two samples (H, G) overlap with the ThT excitation wavelength. The remaining 14 samples had an overlap with both excitation and emission wavelengths.

To confirm whether the evaluation using ThT method was performed correctly, the fluorescence intensity of five samples at a high concentration (40 $v/v\%$) was measured in three conditions; $+A\beta_{42}$ and $+$ThT, $-A\beta_{42}$ and $+$ThT, $-A\beta_{42}$ and $-$ThT (Figure 2). Soy sauce (sample A), soy sauce containing perilla (sample B), Japanese style dressing (sample C), oil dressing (sample D), and Chinese dressing (sample E) were selected and evaluated. Soy sauce (sample A) was used as the control for the other four samples. At 40 $v/v\%$ sample concentration, the fluorescence intensity of the $-A\beta_{42}$ solution, $-A\beta_{42}$ and $-$ThT solution were not significantly different from that of the solution containing $A\beta_{42}$ and ThT. We confirmed that samples A and B showed higher fluorescence intensity than only $A\beta_{42}$

and ThT sample in all conditions (negative control, black line). The fluorescence intensity of samples C, D, and E did not show a higher value than the negative control. The color and some components of the evaluated samples may affect the ThT method by absorbing the excitation or emission of ThT. If the sample solution without $A\beta_{42}$ exhibits a higher fluorescence intensity than the negative control, it is difficult to determine whether the sample solution affects ThT. Therefore, the ThT method might not accurately evaluate the $A\beta_{42}$ aggregation inhibitory activity of a commercial dressing.

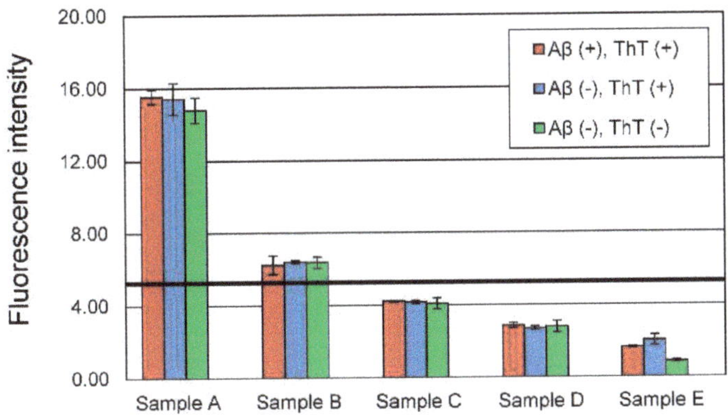

Figure 2. Effect of sample solution on ThT fluorescence intensity. $A\beta_{42}$ (+), ThT (+): $A\beta_{42}$ solution and ThT solution are mixed with sample; $A\beta_{42}$ (−), ThT (+): only ThT solution is mixed with sample solution; $A\beta_{42}$ (−), ThT (−): neither $A\beta_{42}$ solution nor ThT solution are mixed and only the sample solution is used. As a negative control, the sample used was an assay buffer (10% EtOH, 1 × PBS) under the conditions of $A\beta_{42}$ (+) and ThT (+) (black line), and its average value of absorbance was 5.28 (A.U.). There is no significant difference between +ThT (+/− $A\beta_{42}$) and ThT (−) condition in all samples (One way ANOVA, $p > 0.05$).

3.2. Evaluation of $A\beta_{42}$ Aggregation Inhibitory Activity of Commercial Dressings Using MSHTS Sysytem

Previously, we reported the real-time imaging of the $A\beta_{42}$ aggregation process with a fluorescence microscope using QD nanoprobes (Figure 3A) and developed MSHTS system (Figure 3B) for $A\beta_{42}$ aggregation inhibitors by applying this imaging method [19,20]. In the MSHTS system, QD-labeled $A\beta$ co-aggregated with intact $A\beta_{42}$, so that amyloid aggregates were observed by fluorescence microscopy. Since the emission wavelength of QD605 does not overlap with the absorption of almost dressings, it is less susceptible to substances that exhibit an inner filter effect. These aggregates (complex of $A\beta_{42}$ and QD $A\beta$) caused an inhomogeneous distribution of fluorescence intensity in images, resulting in an increased standard deviation (SD) value estimated from the variation in fluorescence intensity of each pixel in the images. Therefore, we could estimate the effects of certain aggregation inhibitors by detecting changes in the SD value. Here, we evaluated the $A\beta_{42}$ aggregation inhibitory activity of 19 commercial dressings using the MSHTS system (Figure 3C), and estimated the EC_{50} values from the SD value of each image (Table 1). In Table 1, EC_{50} values are sorted in ascending order of aggregation inhibitory activity, then each dressing was assigned a letter from A to S. As shown in Figure 3C, all commercial dressings almost completely inhibited $A\beta_{42}$ aggregation at a concentration of 40 $v/v\%$. Samples A to J completely inhibited aggregation even at 4 $v/v\%$, whereas samples K to S formed small aggregates. At a concentration of 0.4 $v/v\%$, a slight change in the shape of aggregates was observed in samples A to E. The EC_{50} value of the sample with the highest activity was 0.065 $v/v\%$, and that with the lowest activity was 6.737 $v/v\%$. There was an about 100-fold difference in activity between these 2 samples. Among the 19 dressings, only three samples (D, K, and N) contained plant oil while the other 16 samples were non-oil type dressings. We compared the mean value of EC_{50} value of non-oil and oil type (non-oil type: 1.330 ± 1.692 $v/v\%$, oil type: 1.019 ± 0.946 $v/v\%$) and performed a statistical analysis. There was no significant difference between non-oil type and oil type (Student's *t*-test, $p > 0.05$). The activity did not depend on the presence of oil, suggesting that oil in the dressing did not affect the $A\beta_{42}$ aggregation inhibitory activity. Next, to confirm whether $A\beta_{42}$ aggregation was inhibited by the effect of the dressing, we observed the $A\beta_{42}$ aggregates at a sample concentration of 0.04 *v/v*% using TEM (Figure 4). Sample A, which had high $A\beta_{42}$ aggregation inhibitory activity, showed a significant decrease in $A\beta_{42}$ aggregates compared to samples J and S. Sample J also significantly decreased aggregates than sample S. These results were consistent with the $A\beta_{42}$ aggregation inhibitory activity calculated by the MSHTS system. Sample A, which showed the highest $A\beta_{42}$ aggregation inhibitory activity, was soy sauce, a traditional Japanese liquid dressing. The remaining 17 samples, except for sample D, contained soy sauce. In other words, soy sauce evidently exhibited high $A\beta_{42}$ aggregation inhibitory activity. Natto, a traditional Japanese food made of fermented soybeans, has antibacterial, as well as a soybean peptide with a neuroprotective effect [24]. It is possible that soybean-derived proteins found in soy sauce, and/or various metabolic products caused by soybean fermentation, may have a positive effect on AD.

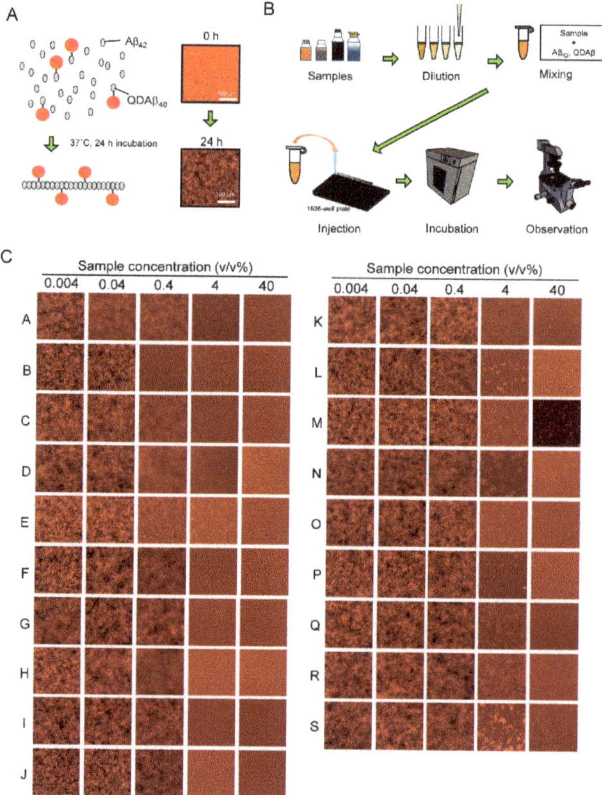

Figure 3. Evaluation of $A\beta_{42}$ aggregation inhibitory activity using the MSHTS system. (**A**) Real-time imaging of $A\beta_{42}$ aggregation using a quantum-dot nanoprobe using fluorescence microscopy. $A\beta_{42}$ and QDAβ were mixed and incubated for 24 h at 37 °C. Co-aggregates of $A\beta_{42}$ and QDAβ formed. (**B**) A scheme of the MSHTS system of $A\beta_{42}$ aggregation inhibitors. (**C**) Fluorescence microscope image of concentration-dependent inhibition of $A\beta_{42}$ aggregation of dressings observed by the MSHTS system. At a sample concentration of 4 *v/v*% or more, the brightness of the image was uniform, indicating that no $A\beta_{42}$ aggregates formed. At a sample concentration of 0.04 *v/v*% or less, $A\beta_{42}$ aggregates were observed in all samples. All images were captured using a conventional fluorescence microscope.

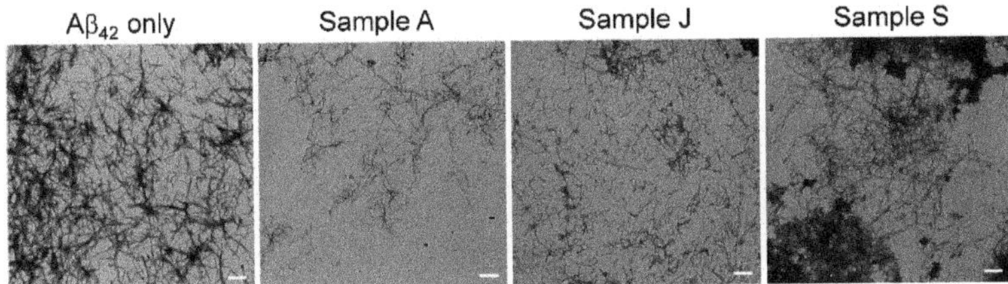

Figure 4. Electron microscopy images of Aβ$_{42}$ aggregates. The Aβ$_{42}$ solution was mixed with each sample (0.04 *v/v*%) and was incubated for 24 h at 37 °C. The images of Aβ$_{42}$ aggregates were captured by TEM at 3000× magnification. Bars: 500 nm.

Table 1. EC$_{50}$ values of 19 commercial dressing samples using MSHTS system.

Sample	EC$_{50}$ (*v/v*%)	Oil Type
RA (positive control)	0.122 ± 0.034 (*w/v*%)	-
A	0.065 ± 0.020	Non-oil
B	0.094 ± 0.017	Non-oil
C	0.133 ± 0.021	Non-oil
D	0.227 ± 0.026	Oil
E	0.230 ± 0.026	Non-oil
F	0.334 ± 0.075	Non-oil
G	0.395 ± 0.130	Non-oil
H	0.413 ± 0.084	Non-oil
I	0.480 ± 0.101	Non-oil
J	0.508 ± 0.025	Non-oil
K	0.763 ± 0.607	Oil
L	1.350 ± 0.247	Non-oil
M	1.360 ± 0.590	Non-oil
N	2.067 ± 0.728	Oil
O	2.132 ± 1.473	Non-oil
P	2.150 ± 0.887	Non-oil
Q	2.313 ± 0.490	Non-oil
R	2.580 ± 0.173	Non-oil
S	6.737 ± 5.054	Non-oil

3.3. Effect of Salt Concentration on Aβ$_{42}$ Aggregation

In general, it is well known that the soy sauce contains a large amount of NaCl and that high salt concentration affects protein aggregation. In order to determine whether the Aβ$_{42}$ aggregation inhibitory activity of soy sauce was due to NaCl, we examined the effect of NaCl concentration on the shape of Aβ$_{42}$ aggregates (Figure 5A). We prepared a 4000 mM NaCl solution. Then, the solution was gradually diluted to six concentrations with five-fold dilutions. These 7 concentrations of NaCl solution was mixed with the 50 μM Aβ$_{42}$ solution and observed using the MSHTS system. The formation of Aβ$_{42}$ aggregates was slightly affected by 200 mM NaCl solution. At 1000 mM of NaCl, Aβ$_{42}$ aggregates were fragmented, and at 2000 mM, abnormal aggregates such as large clumps were observed. The SD values obtained from images were gradually decreased from 2000 mM to 200 mM (Figure 5B). However, 0.32–40 mM NaCl did not affect the SD value, suggesting that Aβ$_{42}$ aggregates were formed. The NaCl concentration in sample A soy sauce is 16.2% (2.77 M). Because the EC$_{50}$ of sample A is 0.065 ± 0.020 *v/v*%, it was indicated that 1.8 mM NaCl is included. The NaCl concentration of the 0.4 *v/v*% and 0.04 *v/v*% solution samples is 11 mM and 1.1 mM, respectively. Therefore, the NaCl in 0.065 *v/v*% (EC$_{50}$ value) soy sauce solution might not affect the formation of Aβ$_{42}$ aggregates. In fact, as shown in Figure 3C, aggregation was inhibited in samples A to J when NaCl was included at 0.4 *v/v*%. Especially, 0.04 *v/v*% of sample A inhibited the aggregation formation. These results suggest that the Aβ$_{42}$ aggregation inhibitory activity by soy sauce was due to not NaCl but other components.

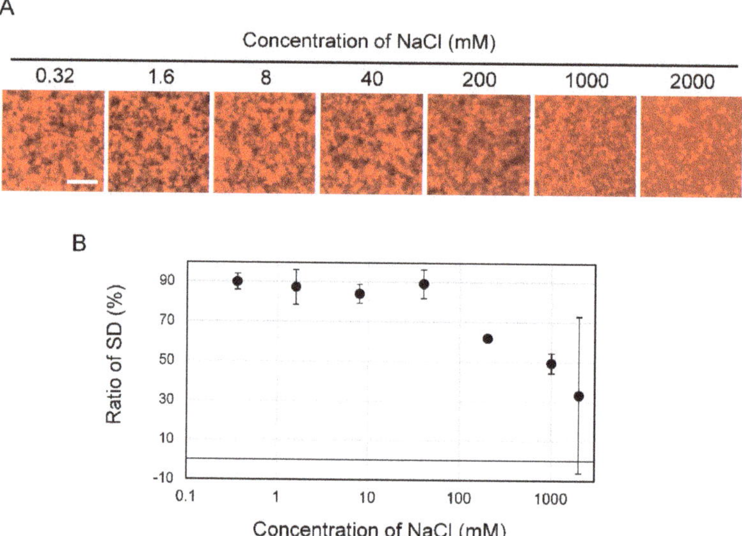

Figure 5. Influence of salt concentration on $A\beta_{42}$ aggregation. (**A**) NaCl solution adjusted to each concentration and the $A\beta_{42}$ solution were mixed then incubated for 24 h at 37 °C. Using the MSHTS system, the influence of NaCl concentration on the formation of $A\beta_{42}$ aggregates was examined. At 2000 mM NaCl, $A\beta_{42}$ and QDAβ were salted out. From their morphology, it is believed that these solids were not $A\beta_{42}$ aggregates. At 1000 mM, aggregates started to form. At 40 mM or less, no significant effect was observed on the aggregates. From 40 to 0.32 mM, normal aggregates were formed. Bars: 100 μm. (**B**) Ratio of SD value at each NaCl concentration. At 40 mM or less, the SD value was not affected by NaCl concentration.

3.4. Influence of Heating and Dialysis Treatment on $A\beta_{42}$ Aggregation Inhibitory Activity

As shown in Figure 3, among the 19 dressings, soy sauce showed the highest activity when the MSHTS system was used. In fact, there are several types of soy sauce, which can be classified according to the sterilization method, the composition of the raw materials, the color, and salt concentration of the product. Among them, we focused on raw soy sauce that was not heat-sterilized. Since raw soy sauce is sterilized by filtration, it has a feature that compounds produced during the fermentation process and enzymes derived from microorganisms are not inactivated. Here, we evaluated $A\beta_{42}$ aggregation inhibitory activity of raw soy sauce purchased from NIHON SYOYU KOGYO using the MSHTS system. The $A\beta_{42}$ aggregation inhibitory activity of raw soy sauce was 0.0045 ± 0.0015 $v/v\%$ (Data not shown). This activity is about 15 times higher than that of soy sauce which showed the highest activity among the 19 dressings. Therefore, it is likely that the difference in the aggregation inhibitory activity of $A\beta_{42}$ between soy sauce and raw soy sauce is caused by the raw material-derived protein and the microorganism-derived enzyme, which are lost by heating.

To examine whether proteins and/or low molecular weight compounds in soy sauce and raw soy sauce affect $A\beta_{42}$ aggregation inhibitory activity, each sample was subjected to SDS-PAGE after heat treatment (80 °C, 60 min) and/or dialysis treatment and their band patterns were compared using silver staining (Figure 6A). As revealed by an SDS-PAGE gel (lanes 1 and 5), the property and amount of proteins in soy sauce and raw soy sauce differed. Bands were detected at 15 and 25 kDa in both samples while 30 and 100–200 kDa bands were detected only in raw soy sauce. Heat treatment did not change the banding pattern of soy sauce, but reduced the intensity of the100–200 kDa band of raw soy sauce (Figure 6A, from lanes 2 and 6). Dialysis treatment reduced the 10–15 and 25 kDa bands in both samples (Figure 6A, lanes 3 and 7). Heat treatment after dialysis reduced the band at about 25 kDa in soy sauce less than the dialyzed sample, and reduced the bands at about 15, 25 and 100–200 kDa in raw soy sauce (Figure 6A, lanes 4 and 8).

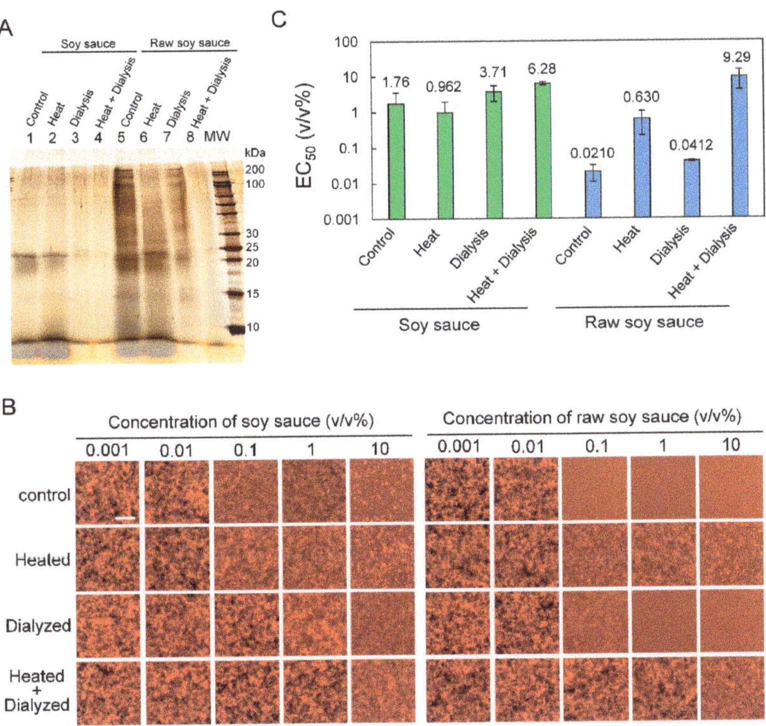

Figure 6. Influence of heat and dialysis treatment on Aβ$_{42}$ aggregation inhibitory activity of soy sauce and raw soy sauce. (**A**) SDS-PAGE analysis demonstrated that heating and dialysis treatments changed the band pattern of soy sauce and raw soy sauce. Lanes 1–4: soy sauce; lanes 5–8: raw soy sauce. Lanes 1 and 5: control; lanes 2 and 6: heat treatment; lanes 3 and 7: dialysis treatment; lanes 4 and 8: dialysis and heat treatment. MW: molecular weight marker. (**B**) Fluorescence images of Aβ$_{42}$ aggregates in each condition. Effect of heating and dialysis treatment on Aβ$_{42}$ aggregation inhibition of soy sauce and raw soy sauce. All images were captured using a conventional fluorescence microscope. Bars: 100 μm. (**C**) Aβ$_{42}$ aggregation inhibitory activity (EC$_{50}$) of soy sauce and raw soy sauce after heating and dialysis calculated by the MSHTS system. Whereas the activity of soy sauce was not changed by heating, the EC$_{50}$ of heated raw soy sauce was about 30 times higher than that of the control sample, suggesting that the activity was greatly reduced. In both soy sauce and raw soy sauce, EC$_{50}$ was approximately double after dialysis.

Next, using the MSHTS system, we assessed whether heating and dialysis treatment would affect the Aβ$_{42}$ aggregation inhibitory activity of soy sauce and raw soy sauce (Figure 6B). Furthermore, the dialyzed soy sauce and raw soy sauce were also heated and analyzed. As shown in Figure 6C, the Aβ$_{42}$ aggregation inhibitory activity of soy sauce was not changed by heating. This result was consistent with the SDS-PAGE banding pattern (Figure 6A, lanes 1 and 2). In addition, the activity of soy sauce decreased even after dialysis treatment, even more so when heat treatment followed dialysis. The Aβ$_{42}$ aggregation inhibitory activity of raw soy sauce was reduced by about 30 times or by about half after heat treatment and dialysis, respectively. Dialysis treatment did not affect the 100–200 kDa band that was reduced by heat treatment, i.e., the 100–200 kDa protein did not contribute to the Aβ$_{42}$ aggregation inhibitory activity of raw soy sauce. The heat treatment after dialysis decreased

Aβ_{42} aggregation inhibitory activity by about 440 times, which was consistent with the SDS-PAGE result in which many bands were reduced (Figure 6A, lane 8). These results suggest that the Aβ_{42} aggregation inhibitor found in soy sauce and raw soy sauce is a low molecular weight compound that is removed by dialysis and a protein that is thermally denatured by heat treatment at 80 °C for 60 min. Since the Aβ_{42} aggregation inhibitory activity of raw soy sauce was greatly reduced after heat treatment, the proteins found in raw soy sauce are considered to be particularly important for the inhibition of Aβ_{42} aggregation.

In general, proteins and peptides are denatured by heating, then lose their physiological activity. It is possible that the presence or absence of heat treatment in the fermentation process may be involved in the physiological activity of dressings such as soy sauce, which contains Aβ_{42} aggregation inhibitory activity. The main raw materials of soy sauce, soy, and wheat, are decomposed into amino acids, peptides and saccharides in the manufacturing process. It is known that soybeans have many physiologically active ingredients such as soy protein and isoflavone [25–27], so these physiologically active ingredients are also present in soy sauce. Soy is widely applied to fermented foods such as miso and natto, Japanese traditional foods. Actually, physiological activity has also been reported for these foods. Miso extract suppresses Aβ-induced neuronal damage [16]. Genistein, one of isoflavone, mitigated Aβ deposition and neuroinflammation in mice [28]. It was reported that natto peptide exhibited antimicrobials effects and that nattokinase has amino residues playing a intramolecular chaperone [24,29] Further, vitamin K2 (menaquinone-7), which is abundant in natto, is an important factor in the synthesis of sphingolipids present in brain cell membranes that support cell signaling function and structure formation [30–32]. We are currently investigating the active compound in raw soy sauce using this analytical method. Soy sauce is used in many traditional dishes in Japan, and its effectiveness against AD would be significant for the prevention of this disease.

4. Conclusions

In this work, we evaluated Aβ_{42} aggregation inhibitory activity of 19 commercial liquid dressings using the MSHTS system. All tested dressings exhibited Aβ_{42} aggregation inhibitory activity, suggesting that the MSHTS system can be applied to processed food containing various impurities. Japanese traditional liquid dressings, soy sauce, exhibited the highest inhibitory activity. However, these findings are limited to in vitro conditions. The physiological activity of the dressings should be clarified through animal experiments, taking into account dynamics such as intestinal absorption and metabolism, particularly the permeability of the blood-brain barrier. Although there is a strong demand for functional food products that help maintain and improve brain function, it is not realistic to subject all food products to animal testing. We are confident that we can progress quickly to a second screening stage such as animal testing by using the MSHTS system as a first screening tool to discover food materials with high Aβ_{42} aggregation inhibitory activity. Recently, we confirmed that aggregation of various amyloid proteins Aβ_{42}, tau, and α-synuclein could be visualized using nonlabelled QD and succeeded in the evaluation of aggregation inhibitory activity of RA [33]. MSHTS system using nonspecific binding of QD to amyloid proteins might bring speed and simplification of the screening of various foods using various amyloids. Furthermore, we expected that the combination of our previously developed automated-MSHTS system [21] and non-specific MSHTS system allows enormous, comprehensive screening of foods, thereby creating the potential for new approaches to overcome AD.

Author Contributions: Conceptualization: K.T.; Methodology: K.T.; Software: M.K., W.Y., and K.S.; Formal analysis: M.K., W.Y., and K.S.; Data curation: M.K., W.Y., and K.S.; Writing—original draft preparation: M.K.; Writing—review and editing: X.L., K.S., K.U., and K.T.; Supervision: K.T. All authors have read and agree to the published version of the manuscript.

Funding: This work was supported by JSPS KAKENHI Grant Number JP16H03288 to K.T. and the Matching Planner Program from Japan Science and Technology Agency (MP28116808300 to K.T.).

Acknowledgments: Part of this work was conducted at the Chitose Institute of Science and Technology, and supported by the Nanotechnology Platform Program (Synthesis of Molecules and Materials) of the Ministry of Education, Culture, Sports, Science and Technology (MEXT), Japan.

Conflicts of Interest: Authors have no conflicts of interest.

Abbreviations

Aβ	amyloid β
MSHTS system	microliter-scale high-throughput screening system
PBS	phosphate buffered saline
QD	quantum dot
RA	rosmarinic acid
SD value	standard deviation value
SDS-PAGE	SDS-polyacrylamide gel electrophoresis
TEM	transmission electron microscope
ThT	thioflavin T

References

1. Ikejima, C.; Hisanaga, A.; Meguro, K.; Yamada, T.; Ouma, S.; Kawamuro, Y.; Hyouki, K.; Nakashima, K.; Wada, K.; Yamada, S.; et al. Multicentre population-based dementia prevalence survey in Japan: A preliminary report. *Psychogeriatrics* **2012**, *12*, 120–123. [CrossRef] [PubMed]
2. Plassman, B.L.; Langa, K.M.; Fisher, G.G.; Heeringa, S.G.; Weir, D.R.; Ofstedal, M.B.; Burke, J.R.; Hurd, M.D.; Potter, G.G.; Rodgers, W.L.; et al. Prevalence of Dementia in the United States: The Aging, Demographics, and Memory Study. *Neuroepidemiology* **2007**, *29*, 125–132. [CrossRef] [PubMed]
3. Lane, C.A.; Hardy, J.; Schott, J.M. Alzheimer's disease. *Eur. J. Neurol.* **2018**, *25*, 59–70. [CrossRef] [PubMed]
4. Kumar, A.; Singh, A. A review on Alzheimer's disease pathophysiology and its management: An update. *Pharmacol. Rep.* **2015**, *67*, 195–203. [CrossRef] [PubMed]
5. Hardy, J. The Amyloid Hypothesis of Alzheimer's Disease: Progress and Problems on the Road to Therapeutics. *Science* **2002**, *297*, 353–356. [CrossRef]
6. Koo, E.H.; Lansbury, P.T.; Kelly, J.W. Amyloid diseases: Abnormal protein aggregation in neurodegeneration. *Proc. Natl. Acad. Sci. USA* **1999**, *96*, 9989–9990. [CrossRef]
7. Pepys, M.B. Amyloidosis. *Annu. Rev. Med.* **2006**, *57*, 223–241. [CrossRef]
8. Selkoe, D.J.; Hardy, J. The amyloid hypothesis of Alzheimer's disease at 25 years. *EMBO Mol. Med.* **2016**, *8*, 595–608. [CrossRef]
9. Sevigny, J.; Chiao, P.; Bussière, T.; Weinreb, P.H.; Williams, L.; Maier, M.; Dunstan, R.; Salloway, S.; Chen, T.; Ling, Y.; et al. The antibody aducanumab reduces Aβ plaques in Alzheimer's disease. *Nature* **2016**, *537*, 50–56. [CrossRef]
10. Selkoe, D.J. Alzheimer disease and aducanumab: Adjusting our approach. *Nat. Rev. Neurol.* **2019**, *15*, 365–366. [CrossRef] [PubMed]
11. Bateman, R.J.; Xiong, C.; Benzinger, T.L.S.; Fagan, A.M.; Goate, A.; Fox, N.C.; Marcus, D.S.; Cairns, N.J.; Xie, X.; Blazey, T.M.; et al. Clinical and Biomarker Changes in Dominantly Inherited Alzheimer's Disease. *N. Engl. J. Med.* **2012**, *367*, 795–804. [CrossRef] [PubMed]
12. Ono, K.; Hasegawa, K.; Naiki, H.; Yamada, M. Curcumin Has Potent Anti-Amyloidogenic Effects for Alzheimer's β-Amyloid Fibrils In Vitro. *J. Neurosci. Res.* **2004**, *75*, 742–750. [CrossRef]
13. Hase, T.; Shishido, S.; Yamamoto, S.; Yamashita, R.; Nukima, H.; Taira, S.; Toyoda, T.; Abe, K.; Hamaguchi, T.; Ono, K.; et al. Rosmarinic acid suppresses Alzheimer's disease development by reducing amyloid β aggregation by increasing monoamine secretion. *Sci. Rep.* **2019**, *9*, 8711. [CrossRef]
14. Hamaguchi, T.; Ono, K.; Murase, A.; Yamada, M. Phenolic Compounds Prevent Alzheimer's Pathology through Different Effects on the Amyloid-β Aggregation Pathway. *Am. J. Pathol.* **2009**, *175*, 2557–2565. [CrossRef] [PubMed]

15. Noguchi-Shinohara, M.; Ono, K.; Hamaguchi, T.; Iwasa, K.; Nagai, T.; Kobayashi, S.; Nakamura, H.; Yamada, M. Pharmacokinetics, Safety and Tolerability of Melissa officinalis Extract which Contained Rosmarinic Acid in Healthy Individuals: A Randomized Controlled Trial. *PLoS ONE* **2015**, *10*, e0126422. [CrossRef] [PubMed]

16. Momoyo, K.; Aiko, S.; Ryuichiro, S. Neuroprotective Action of Miso. *New Food Ind.* **2018**, *60*, 79–83. [CrossRef]

17. Hsu, R.-L.; Lee, K.-T.; Wang, J.-H.; Lee, L.Y.L.; Chen, R.P.Y. Amyloid-Degrading Ability of Nattokinase from Bacillus subtilis Natto. *J. Agric. Food Chem.* **2009**, *57*, 503–508. [CrossRef]

18. LeVine III, H. Thioflavin T interaction with synthetic Alzhei- mers'sdisease β-amyloid peptides: Detection of amyloid aggre- gation in solution. *Prot. Protein Sci.* **1993**, 404–410.

19. Tokuraku, K.; Marquardt, M.; Ikezu, T. Real-Time Imaging and Quantification of Amyloid-β Peptide Aggregates by Novel Quantum-Dot Nanoprobes. *PLoS ONE* **2009**, *4*, e8492. [CrossRef]

20. Ishigaki, Y.; Tanaka, H.; Akama, H.; Ogara, T.; Uwai, K.; Tokuraku, K. A Microliter-Scale High-throughput Screening System with Quantum-Dot Nanoprobes for Amyloid-β Aggregation Inhibitors. *PLoS ONE* **2013**, *8*, e72992. [CrossRef]

21. Sasaki, R.; Tainaka, R.; Ando, Y.; Hashi, Y.; Deepak, H.V.; Suga, Y.; Murai, Y.; Anetai, M.; Monde, K.; Ohta, K.; et al. An automated microliter-scale high-throughput screening system (MSHTS) for real-time monitoring of protein aggregation using quantum-dot nanoprobes. *Sci. Rep.* **2019**, *9*, 2587. [CrossRef]

22. Ogara, T.; Takahashi, T.; Yasui, H.; Uwai, K.; Tokuraku, K. Evaluation of the effects of amyloid β aggregation from seaweed extracts by a microliter-scale high-throughput screening system with a quantum dot nanoprobe. *J. Biosci. Bioeng.* **2015**, *120*, 45–50. [CrossRef]

23. Kanda, Y. Investigation of the freely available easy-to-use software 'EZR' for medical statistics. *Bone Marrow Transplant.* **2013**, *48*, 452–458. [CrossRef]

24. Kitagawa, M.; Shiraishi, T.; Yamamoto, S.; Kutomi, R.; Ohkoshi, Y.; Sato, T.; Wakui, H.; Itoh, H.; Miyamoto, A.; Yokota, S.-I. Novel antimicrobial activities of a peptide derived from a Japanese soybean fermented food, Natto, against Streptococcus pneumoniae and Bacillus subtilis group strains. *AMB Express* **2017**, *7*, 127. [CrossRef]

25. Lee, Y.-B.; Lee, H.J.; Sohn, H.S. Soy isoflavones and cognitive function. *J. Nutr. Biochem.* **2005**, *16*, 641–649. [CrossRef] [PubMed]

26. Katayama, S.; Imai, R.; Sugiyama, H.; Nakamura, S. Oral Administration of Soy Peptides Suppresses Cognitive Decline by Induction of Neurotrophic Factors in SAMP8 Mice. *J. Agric. Food Chem.* **2014**, *62*, 3563–3569. [CrossRef] [PubMed]

27. Sugano, M.; Yamada, Y.; Yoshida, K.; Hashimoto, Y.; Matsuo, T.; Kimoto, M. The hypocholesterolemic action of the undigested fraction of soybean protein in rats. *Atherosclerosis* **1988**, *72*, 115–122. [CrossRef]

28. Park, Y.-J.; Ko, J.; Jeon, S.; Kwon, Y. Protective Effect of Genistein against Neuronal Degeneration in ApoE−/− Mice Fed a High-Fat Diet. *Nutrients* **2016**, *8*, 692. [CrossRef] [PubMed]

29. Weng, Y.; Yao, J.; Sparks, S.; Wang, K. Nattokinase: An Oral Antithrombotic Agent for the Prevention of Cardiovascular Disease. *Int. J. Mol. Sci.* **2017**, *18*, 523. [CrossRef] [PubMed]

30. Ferland, G. Vitamin K and the Nervous System: An Overview of its Actions. *Adv. Nutr.* **2012**, *3*, 204–212. [CrossRef] [PubMed]

31. Tsukamoto, Y.; Kasai, M.; Kakuda, H. Construction of a Bacillus subtilis (natto) with high productivity of vitamin K2 (menaquinone-7) by analog resistance. *Biosci. Biotechnol. Biochem.* **2001**, *65*, 2007–2015. [CrossRef] [PubMed]

32. Schwalfenberg, G.K. Vitamins K1 and K2: The Emerging Group of Vitamins Required for Human Health. *J. Nutr. Metab.* **2017**, *2017*, 1–6. [CrossRef] [PubMed]

33. Lin, X.; Galaqin, N.; Tainaka, R.; Shimamori, K.; Kuragano, M.; Noguchi, T.Q.P.; Tokuraku, K. Real-Time 3D Imaging and Inhibition Analysis of Various Amyloid Aggregations Using Quantum Dots. *Int. J. Mol. Sci.* **2020**, *21*, 1978. [CrossRef] [PubMed]

 *foods*

Article

Food Authentication: Identification and Quantitation of Different Tuber Species via Capillary Gel Electrophoresis and Real-Time PCR

Stefanie Schelm [1], Melanie Siemt [1], Janin Pfeiffer [1], Christina Lang [1], Hans-Volker Tichy [2] and Markus Fischer [1,*]

[1] Hamburg School of Food Science, Institute of Food Chemistry, University of Hamburg, Grindelallee 117, 20146 Hamburg, Germany; stefanie.schelm@chemie.uni-hamburg.de (S.S.); melanie.siemt@studium.uni-hamburg.de (M.S.); janin.pfeiffer@studium.uni-hamburg.de (J.P.); christina.lang@chemie.uni-hamburg.de (C.L.)
[2] LUFA-ITL GmbH, Dr.-Hell-Straße 6, 24107 Kiel, Germany; hans-volker.tichy@agrolab.de
* Correspondence: Markus.Fischer@uni-hamburg.de; Tel.: +49-4042-838-43-57

Received: 23 March 2020; Accepted: 10 April 2020; Published: 16 April 2020

Abstract: Truffles are hypogeous fungi mainly found in Europe and Asia. Due to their special aroma and taste, some truffle species are sold on the international market at an extremely high price. Among the economically relevant species, the white Alba truffle (*Tuber magnatum*) and the black Périgord truffle (*T. melanosporum*) are the most appreciated species. The fruiting bodies of the Asian black truffle are morphologically very similar to *T. melanosporum*, and those of the Bianchetto truffle (*T. albidum* Pico) are similar to *T. magnatum*, but are of little economic value. Highly valued species are adulterated with cheaper ones, especially. Because of this problem, the aim of this study was the development of methods for detecting possible admixtures to protect consumers from fraud. This study is based on seven different truffle species (117 fruiting bodies) from different growing regions. Additionally, selected truffle products were included. Using this material, a real-time PCR (polymerase chain reaction) assay allowing the detection and quantitation of Asian black truffles in *T. melanosporum* up to 0.5% was developed. In addition, a capillary gel electrophoresis assay was designed, which allows the identification and quantitation of different species. The methods can be used to ensure the integrity of truffle products.

Keywords: truffle; *T. melanosporum*; *T. indicum*; real-time PCR; RFLP; quantitative evaluation

1. Introduction

Truffles are underground fungi belonging to the class of the Ascomycetes in the order Pezizales [1,2]. They grow in an ectomycorrhizal symbiosis with roots of different trees and shrubs, e.g., oak, poplar, willow, hazel [3], and Cistus [4]. Tuber spp. are mainly distributed in Europe, Asia, North Africa, and America [5,6]. At least 180 Tuber species exist worldwide [6], 70–75 species have been well described [7], and 32 species are currently listed in Europe [8].

Under specific environmental conditions, such as calcareous soil with a neutral pH [9], truffles produce hypogeous edible ascocarps. The unique aroma and taste emitted from the fruiting bodies are responsible for the gastronomical desirability; therefore, some truffles represent some of the most highly prized edible and valuable mushrooms worldwide [10].

T. magnatum is the most expensive truffle species in general [6]. It is mainly distributed in Italy, but it can also be found in the area around Balkan [11], France, and Switzerland [6]. Another white (or whitish) truffle with lesser economic value, *T. albidum* Pico, is morphologically and biochemically similar to *T. magnatum*, which can be subject to fraud [12]. It is also possible that roots initially colonized by *T. magnatum* have produced other white truffles, such as *T. albidum* Pico [5].

Among the black truffles is the Périgord truffle *T. melanosporum*, the most expensive species which is highly valued for its organoleptic properties [13], and, therefore, there is a risk of fraud. The natural distribution area is mainly France, Spain, and Italy [14]. The Asian black truffles, such as *T. indicum* and *T. himalayense,* are closely related to *T. melanosporum,* and the fruiting bodies are morphologically very similar [15]. Because of the larger production value, *T. indicum* is sold at a lower price and imported from China to Europe, North America, and Australia [16–18]. Cases have been reported where *T. indicum* has been sold as *T. melanosporum,* and incorrect inoculations and incidence of ectomycorrhiza from *T. indicum* in *T. melanosporum* truffle orchards have been found [16,18–20]. Due to the lower price, admixture from the Asian black truffles with *T. melanosporum* is sometimes observed in food products. Since the microscopic identification of truffle fruiting bodies is difficult, molecular methods have been introduced to analyze different truffle species that are morphologically similar.

One region of the DNA suited for the molecular analysis of fungi is the rDNA (ribosomal DNA), which contains two variable non-coding regions, the internal transcribed spacer (ITS) region 1 and 2, between the highly conserved 17S, 5.8S and 25S rRNA (ribosomal RNA) genes [21]. The ITS regions are widely used to analyze ectomycorrhizal communities of mycorrhizal fungi and fungal species in the field, and it is recommended to be used as the primary fungal barcode [22,23]. Another advantage of the ITS region is the repetitive character resulting in a low detection limit [24–26].

Molecular methods based on the ITS region have also been widely used for the identification of truffle species [27–32]. Methods targeting the rDNA region for detecting admixtures from lower prized truffle species in *T. melanosporum* were developed, enabling the qualitative detection of ectomycorrhiza or ascocarps from *T. indicum* in *T. melanosporum* [20,30,33,34]. Different real-time PCR (polymerase chain reaction) methods for truffles were developed, e.g., for the analysis of truffle grounds and the quantitation of mycelium in soil [35–38]. Furthermore, real-time PCR assays for the detection of *T. melanosporum* in processed food products and for the quantitation of *T. aestivum* in mycelium have been developed [35,39,40]. To our knowledge, there is currently no real-time-PCR method available, which can quantify Asian truffles in *T. melanosporum*.

To detect possible admixtures of cheaper truffle species and to protect consumers from fraud, the aim of this study was to develop methods to detect such possible admixtures. The DNA-based methods can be used for quality control in the food industry or in official food control to ensure the integrity of truffle products.

The present paper reports the application of molecular techniques, real-time PCR, capillary gel electrophoresis (CGE), and restriction fragment length polymorphism (RFLP) to identify and quantify admixtures of different truffle species. A specific primer pair (with minor modifications) for *T. indicum* [33] and a new *T. melanosporum* specific primer pair suitable for real-time PCR with hybridization probes were used. The real-time PCR technology with hybridization probes was chosen for the real-time PCR assay. Compared to assays with SYBR Green I, hybridization probes are more specific because the fluorescent signal is derived from a specific probe and thus, is sequence-specific [41]. Moreover, a quantitative CGE based method for species differentiation and a RFLP assay combined with CGE were developed. The RFLP offers an alternative to real-time PCR as an easy to use method. The methods developed were tested on fruit bodies and truffle products from retail outlets.

2. Materials and Methods

2.1. Sample Material

In total, 117 fruiting bodies of different truffle species from distinct origins were analyzed (see Table 1). Upon arrival, all fruiting bodies were frozen in liquid nitrogen and stored at −80 °C. Furthermore, canned truffle fruiting bodies and food products containing truffles purchased at retail locations were used.

Table 1. Sample material used in this study.

Tuber Species	Geographical Origin	Fruiting Bodies Analyzed	
		Numbers with Regard to the Origin	Total Number
T. albidum Pico	Italy	5	5
T. indicum	China	5	5
T. himalayense	Dali, Yunnan, China	20	20
T. brumale	Sarrion, Teruel, Spain	2	2
T. melanosporum	Marche, Italy	2	
	France	1	
	Australia	2	
	Sarrion, Teruel, Spain	8	
	Castello, Valencia, Spain	6	
	unknown	1	20
T. magnatum	Romagna, Italy	2	
	Buzet, Croatia	1	
	Turin, Piemonte, Italy	1	
	Italy	5	
	L'Aquila, Abruzzo, Italy	1	
	Perugia, Umbria, Italy	1	
	Rome, Lazio, Italy	1	
	Naples, Campania, Italy	1	
	Ancona, Marche, Italy	1	
	Campobasso, Molise, Italy	1	15
T. aestivum	unknown	19	
	Romania	15	
	Italy	11	
	Hungary	3	
	Toscana, Florence, Italy	2	50
Processed food containing truffle:			
T. melanosporum fruiting bodies canned in saltwater			6
salt with dried *T. aestivum*			1
T. brumale chopped and cooked in sherry port wine stock			1

2.2. DNA Isolation

For DNA isolation of the matrix mixtures, commercially available kits (QIAGEN DNeasy® Plant Mini Kit (QIAGEN, Hilden, Germany), peqGOLD Fungal DNA Mini Kit (VWR International GmbH, Darmstadt, Germany)) were used. DNA purity was determined photometrically using a DS-11 Spectrophotometer (DeNovix Inc., Wilmington, USA). DNA concentration was determined fluorometrically (Quantus™ Fluorometer, Promega GmbH, Mannheim, Germany).

For a high sample throughput, the simple "alkaline" and the "modified PCI (phenol-chloroform-iso-amyl alcohol)" DNA extraction method, originally developed for tissue samples of chicken embryos [42], were used with slight modifications. In the "alkaline method", approximately 25 mg of sample material was incubated for 20 min at 75 °C in 100 µL 0.2 M NaOH after grinding with a micropistille in a 1.5 mL reaction tube. Afterward, 300 µL 0.04 M Tris/HCl was added. One microliter of the liquid phase was used directly for PCR. Additionally, the "modified PCI method" was used as followed: Approximately 25 mg sample material was ground in 500 µL extraction buffer (0.1 M Tris/HCl, 55 mM CTAB, 1.4 M NaCl, and 20 mM EDTA, pH 8.0) [43] with a micropistille in a 1.5 mL reaction tube and incubated for 30 min at 65 °C. Five hundred microliters chloroform were added and centrifuged at $10,000 \times g$ for 5 min. The supernatant was transferred to a new 2 mL reaction tube and mixed with 500 µL isopropanol and incubated for 30 min at 4 °C. After repeated centrifugation for 15 min, the supernatant was discarded, and the pellet was washed with 500 µL 70% ethanol. The DNA pellet was vacuum-dried and dissolved in 50 µL water. One microliter was used directly for PCR.

2.3. Preparation of Spiked Sample Material

2.3.1. DNA Mixtures

The DNA isolated from different truffle fruiting bodies was adjusted to a concentration of 5 ng/μL and mixed in different ratios (0.1%, 0.5%, 1%, 5%, 10%, 20%, 40%, 70% DNA isolated from *T. indicum* in DNA isolated from *T. melanosporum*; 5%, 20%, 40%, 80% DNA isolated from *T. albidum* Pico in DNA isolated from *T. magnatum*).

Additionally, mixtures of PCR amplicons were prepared by mixing PCR products from different fruiting bodies in different ratios after PCR (20%, 40%, 50% *T. indicum* in *T. melanosporum* PCR amplicons; 5%, 20%, 40%, 80% *T. indicum* in *T. aestivum* and *T. albidum* Pico in *T. magnatum* PCR amplicons).

2.3.2. Matrix Mixtures of Fruiting Bodies

Spiked samples of two distinct truffle species were produced by weighing out ground fruiting bodies of different truffle species in a 2 mL reaction tube (4.3%, 4.6%, 7.4%, 13.5%, 18.28%, 20.4%, 32.2% and 11.2%, 21.7%, 28.3%, 47.5% *T. indicum* in *T. melanosporum*). The ground powder was mixed in 500 μL DNA isolation buffer using a bead ruptor (Bead Ruptor 24; Biolabproducts GmbH, Bebensee, Germany) and the DNA isolation protocol was continued.

2.4. Real-Time PCR

To detect and quantify possible contamination with lower-priced Asian truffles of the *T. indicum* complex (*T. indicum/himalayense*) in higher priced truffles, such as *T. melanosporum*, a specific primer pair (with minor modifications) designed from Paolocci et al. (1997) [32] was used (Indi-fw/ITS4LNG, see Table 2). This primer pair targets the ITS2 region on the rDNA. Additionally, a primer pair specific for *T. melanosporum* (Mela-fw/Mela-rv, see Table 2), also located in the ITS2 region, was designed using the sequences from Paolocci et al. (1997) [32] as templates, which was able to detect the presence of *T. melanosporum*.

Table 2. Primer and probe sequences and size of PCR products.

Primer/Probes	Name	Sequence 5′–3′	Product Size (bp)
specific for *T. melanosporum*	Primer Mela-fw Mela-rv Probe	ACGACGGACTTTATAAACGGTTATAA AGCGGGTATCCCTCCCTGATT Cy5–GACCTGGATCAGTCACAAGTCTTGTCTGGT-BHQ2	141
specific for *T. indicum/ T. himalayense*	Primer Indi-fw ITS4LNG * Probe	AACAACAGACTTTGTAAAGGGTT TGATATGCTTAAGTTCAGCGGG HEX-GGACCTAGATCAGTCACAAGTCATGTCTGG-BHQ2	146

fw = forward; rv = reverse, * Paolocci, et al. (1997) [32]

For real-time PCR assays, hybridization probes labeled with a fluorophore (Hex, Cy5) and a quencher (BHQ2) were designed, taking care that no overlapping of fluorescence maxima occurred. All primer and probe sequences, including the fluorophores and quenchers, used in this work are listed in Table 2.

The real-time PCR assay was performed in a volume of 10 μL including 1× Taq reaction buffer (Biozym Scientific GmbH, Hessisch Oldendorf, Germany), 0.8 mM dNTPs (each 2.5 mM, Bioline GmbH, Luckenwalde, Germany), 0.5 U Taq polymerase (Biozym Taq DNA Polymerase, Biozym Scientific GmbH, Hessisch Oldendorf, Germany), 50 nM of each primer (Life Technologies, Darmstadt, Germany), 40 nM of the fluorescently labeled probe (Eurofins Genomics GmbH, Ebersberg, Germany), and 1 μL of isolated DNA.

For real-time PCR, a CFX96 Touch System thermocycler (Bio-Rad Laboratories GmbH, Munich, Germany) was used. Real-time PCR was performed with the following two-step temperature program:

initial denaturation for 300 s at 95 °C followed by 30 cycles with 20 s denaturation at 95 °C and 60 s annealing and elongation at 60 °C, finished by final elongation for 600 s at 72 °C.

Practical Determination of LoD

To determine the LoD (limit of detection) of the developed real-time PCR DNA, mixtures of *T. melanosporum* and *T. indicum* (0.1%, 0.5%, 1%, 5%, 10%, 20%, 40%, 70% DNA isolated from *T. indicum* in DNA isolated from *T. melanosporum*) were measured in duplicate using the primer pair specific for *T. indicum* with 10 ng DNA in each PCR reaction.

2.5. Isolation of DNA Fragments from Agarose Gels

For the isolation of PCR fragments from agarose gels, the peqGOLD® Gel Extraction Kit (VWR International GmbH, Erlangen, Germany) was used according to the manufacturer's information.

2.6. RFLP and CGE

For the amplification reactions, the primers ITS1 and ITS4 [21] amplifying the ITS region were used. The PCR prior to the RFLP was performed in a volume of 10 μL including 1× Taq reaction buffer (Biozym Scientific GmbH, Hessisch Oldendorf, Germany), 0.8 mM dNTPs (each 2.5 mM, Bioline GmbH, Luckenwalde, Germany), 0.5 U Taq polymerase (Biozym Taq DNA Polymerase, Biozym Scientific GmbH, Hessisch Oldendorf, Germany), 1 μM of each primer (Life Technologies, Darmstadt, Germany), and 1 μL of isolated DNA. The thermal cycle profile was as follows: initial denaturation for 300 s at 95 °C, 35 cycles with denaturation for 20 s at 95 °C, annealing for 20 s at 47.3 °C, elongation for 20 s at 72 °C, and final elongation for 300 s at 72 °C. PCR amplicons were visualized with agarose gel electrophoresis (AGE) on 1.5% agarose gels stained with 0.001% ethidium bromide. The gels were visualized under UV light (254 nm, Biostep, Felix 1040, Biostep GmbH, Jahnsdorf, Germany).

The restriction enzyme *CviQI* (Thermo Fisher Scientific Inc., Waltham, United States; restriction sequence: G/TAC) was used for restriction. To carry out the reaction, 1 U of restriction enzyme, 2 μL of PCR products, 0.8 μL corresponding buffer in a total volume of 8 μL were used. The reaction mixture was incubated at 25 °C for approximately 8 h without heat inactivation.

Detection of PCR products was carried out by AGE (see above). Quantitation by capillary gel electrophoresis was performed according to the manufacturer instructions on a 2100 Bioanalyzer (Agilent Technologies, Santa Clara, United States) using the Agilent DNA 7500 Kit (Agilent Technologies, Santa Clara, CA, United States) and on a Fragment Analyzer™ (Advanced Analytical Technologies, Inc, Ankeny, IA, United States).

3. Results and Discussion

3.1. Real-Time PCR

3.1.1. Primer Specificity

For the detection and quantitation of potential impurities of Asian black truffles in *T. melanosporum* a specific primer pair (with minor modifications) designed from Paolocci et al. (1997) [32] specific for these species was used in combination with a hybridization probe. In addition, a hybridization probe and a primer pair specific for *T. melanosporum* were designed to check the presence of this high prized truffle in the samples. The *T. melanosporum* specific primer pair was designed so that the length for the amplification product was about 140 bp. So the amplification length was similar to the amplification product of the *T. indicum* specific primer pair, and it meets the requirements for the hybridization probes [44] and allows an analysis of fragmented DNA as it could occur in processed food [24].

The specificity of the used primer was tested with DNA isolated from all samples of different *Tuber* spp. listed in Table 1. As can be seen in Table S1, the specific primer pair for *T. indicum* showed only positive PCR results with the Asian black truffles *T. indicum* and *T. himalayense*. Cross contaminations can be ruled out because none of the samples from other *Tuber* species showed positive Cq values. This demonstrates the suitability of the real-time assay for the detection and possible quantitation of *T. indicum/himalayense* and *T. melanosporum* without cross amplifications. This opens up the possibility to use this primer pair to detect possible admixtures of the cheaper Asian black truffles in higher-priced species, such as *T. melanosporum*. A similar performance was observed for the *T. melanosporum* specific primer, which gave only positive signals with the analyzed DNA isolated from *T. melanosporum*. In addition, the *T. melanosporum* fruiting bodies canned in saltwater from food retail showed positive Cq values in real-time PCR, showing that the real-time PCR assay also works with processed food. The ranging Cq values for analyzed fruiting bodies can be explained by the DNA isolation method used ("alkaline method", "modified PCI method" [42]) because the concentration of DNA was not adjusted to a uniform level for specificity test. Since the specificity of the primer pairs should be tested qualitatively, the non-adjusted DNA concentration did not affect the specificity test negatively.

3.1.2. Quantitation

Primer suitability for quantitation was tested by measuring DNA mixtures over the concentration range from 0.1% to 70% *T. indicum* in *T. melanosporum* DNA. The real-time assay of the DNA mixtures showed a reliable amplification over the concentration range from 0.5% to 70% *T. indicum* in *T. melanosporum* DNA (see Figure 1, measuring values are shown in Table S2) with an R^2 of 0.993 (Equitation for R^2 see Equitation S1). Due to the fact that the last measured standard (0.1% *T. indicum* in *T. melanosporum* DNA) gave no measurable signal, the LoD of the real-time PCR assay was set at 0.5% *T. indicum* in *T. melanosporum* for the *T. indicum* specific real-time assay.

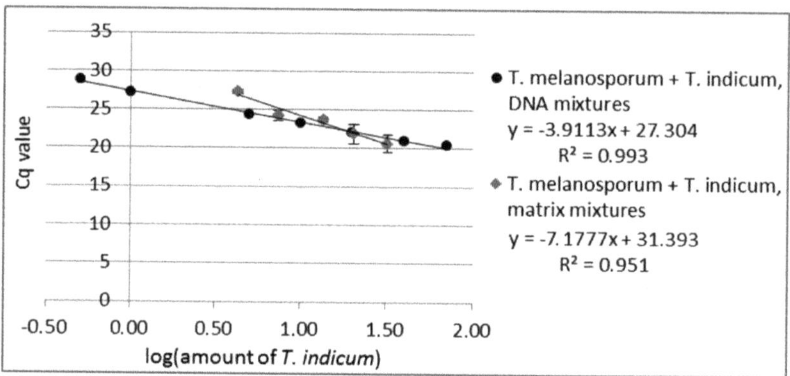

Figure 1. Standard curve of real-time PCR of DNA-mixtures from *T. melanosporum* with *T. indicum* with 0.5%, 1%, 5%, 10%, 20%, 40%, 70% *T. indicum* DNA and standard curve of matrix-mixtures from *T. melanosporum* with *T. indicum* with 4.3%, 7.4%, 13.5%, 20.4, 32.2% *T. indicum*. Each DNA-mixture was analyzed in duplicate and each matrix mixture in triplicate to real-time PCR with the primer pair specific to *T. indicum*. The Cq values are plotted against the logarithm of *T. indicum* amount.

PCR can be influenced by the sample matrix [45], e.g., by coextracted substances. It is also possible that the DNA from some truffle species can be more easily isolated or that the DNA from some species contains more PCR inhibitors. This would lead to an inhomogeneous PCR amplification by samples with more than one truffle species. To assess these effects, five different matrix mixtures of *T. indicum* with *T. melanosporum* were prepared (4.3%, 7.4%, 13.5%, 20.4%, 32.2% *T. indicum*). For the matrix experiments, DNA from each matrix mixture was isolated and analyzed via real-time PCR with both real-time systems, the *T. indicum* and the *T. melanosporum* specific primers, in triplicate with 10 ng DNA pro PCR reaction. As in the case of the analyzed DNA mixtures, the standard curve of the matrix mixtures revealed a linear correlation between the Cq values plotted against the logarithm of *T. indicum* content in *T. melanosporum* with an R^2 of 0.951 (see Figure 1). Eventually, coextracted PCR inhibitors could lead to PCR efficiency under 100%. The results obtained show that the developed real-time PCR opens the possibility to quantify the content of *T. indicum* admixtures in *T. melanosporum* also in matrix mixtures, to determine the rate of fraud by replacing expensive truffles by cheaper ones.

Furthermore, two matrix mixtures of *T. melanosporum* fruiting bodies with different amounts of *T. indicum* (MM1: 18.28%, MM2: 4.60% *T. indicum*) were analyzed in duplicate via real-time PCR with *T. melanosporum* and *T. indicum* specific primer pair. Calculation of the *T. indicum* content was performed using absolute quantitation with an external calibration curve of DNA mixtures (0.1% to 70% *T. indicum* in *T. melanosporum* DNA) and with the external calibration curve of matrix mixtures used above. Using the calibration curve of DNA mixtures for MM1 and MM2, a *T. indicum* amount of 19.44% ± 6.4% or 2.32% ± 0.7%, respectively, was calculated, and using the calibration curve of matrix mixtures a *T. indicum* amount of 17.97% ± 3.16% (MM1) or 5.84% ± 0.92% (MM2) was calculated. The results show that the use of the matrix calibration curve leads to an improvement in quantitative results by compensating matrix effects. These results indicate that it should also be possible to determine the *T. indicum* content in food products or other matrices using a standard curve with a matrix comparable to the sample.

3.2. RFLP and CGE

3.2.1. PCR-Amplification of the ITS Region, Evaluation via CGE

The primers ITS1/ITS4 amplify the ITS1, 5.8S, and ITS2 regions. PCR amplification of DNA with this primer pair from the different *Tuber* spp. resulted in bands on agarose gels with different lengths. *T. himalayense*, *T. indicum*, *T. melanosporum*, and *T. magnatum* generated bands with approximately 630 bp. Amplicons from *T. aestivum* DNA were approximately 50 bp longer. *T. albidum* Pico DNA resulted in bands on agarose gels with about 550 bp, and *T. brumale* DNA resulted in bands with approximately 900 bp (see Figures S1–S6). In contrast to the findings of Paolocci et al. (1995), the analyzed *T. aestivum* samples used in this study showed only one band on agarose gels with approximately 700 bp, which was also shown in, e.g., [46]. The fact that *T. brumale* showed the longest and *T. albidum* Pico the shortest amplicon length (900 bp and 500 bp, respectively) was also detected by other groups [47,48].

Due to the fact that *T. aestivum*, *T. albidum* Pico, and *T. brumale* showed fragments different in size compared with the other truffle species analyzed, an identification and quantitation of these species or another truffle species mixed with *T. aestivum*, *T. albidum* Pico and *T. brumale* should be possible via CGE.

To prove whether the detection and quantitation of *T. indicum* in a mixture with *T. aestivum* or *T. albidum* Pico mixed with *T. magnatum* based on the different amplicon length is possible, different mixtures of PCR amplicons produced with ITS1/ITS4 primers were prepared. Mixtures from *T. indicum* with *T. aestivum* and from *T. magnatum* with *T. albidum* Pico with 5%, 20%, 40%, and 80% *T. indicum/albidum* Pico PCR amplicons were analyzed on CGE and the relative peak area from *T. indicum* and *T. albidum* Pico was integrated. Additionally, isolated DNA of *T. albidum* Pico and *T. magnatum* were mixed prior to PCR to check if the PCR had an influence on the quantitation via the different ITS amplicon length on CGE.

As shown in Figure 2 (measuring values are shown in Table S3), a correlation between the relative peak-area of the measured PCR-amplicons from *T. indicum* or *T. albidum* Pico and the amount of the corresponding truffle species could be detected over the whole range of analyzed samples with an R^2 of 0.999 or 0.985, respectively. The R^2 of 0.872 of the DNA mixture was lower than the R^2 of the amplicon mixtures, but a linear correlation was still visible. The decline in the R^2 can be explained by inhomogeneous samples, or matrix effect occurred during PCR. These results show that a quantitation of a truffle species mixed with another species is possible via just the different lengths of the PCR-amplicons, which makes the analysis fast and simple because no digestion with restriction enzymes is necessary.

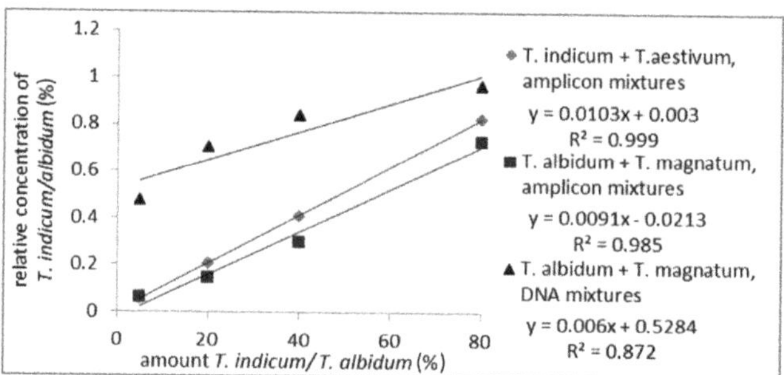

Figure 2. Standard curve of PCR-amplicon mixtures from *T. indicum* with *T. aestivum* and *T. albidum* Pico with *T. magnatum* with 5%, 20%, 40%, 80% *T. indicum/albidum* Pico and standard curve of DNA mixtures from *T. albidum* Pico with *T. magnatum* with 5%, 20%, 40%, 80% *T. albidum* Pico. The detected relative areas of PCR-amplicons are plotted against *T. indicum* or *T. albidum* Pico amount.

3.2.2. RFLP of the ITS Region, Evaluation via CGE

Due to the same length of the region amplified with the ITS1/ITS4 primers, a differentiation of the highly prized black Périgord truffle *T. melanosporum* and the Asian black truffles is not possible by comparing the ITS amplicon length. Thus, a differentiation and quantitation of admixtures from Asian black truffles of the *T. indicum* group in *T. melanosporum* with RFLP and CGE analysis of the ITS1, 5.8S, and ITS2 regions amplified by ITS1/ITS4 primers were performed. As, for example, shown by Roux et al. (1999) [49] and Paolocci et al. (1997) [32], a differentiation between various *Tuber* species via RFLP is possible. But according to our knowledge, this is the first approach to use this technique for a quantitation of possible admixtures from Asian black truffles in *T. melanosporum* via CGE.

It is known from the literature that genetic differences exist in the ITS region of *Tuber* species, especially in *T. aestivum*. Compared to *T. aestivum*, the other analyzed *Tuber* species show a relatively low intraspecific divergence [48,50].

In 2018, Qiao et al. [50] sequenced and analyzed a 500 bp long fragment of the ITS region from 476 truffle ascocarps of the *T. indicum* complex and revealed 54 haplotypes. In the scope of this work, we compared the 476 published ITS sequences with each other, and the restriction site from the endonuclease *CviQI* was examined. The sequences can basically be divided into three groups. (i) The first group with one restriction side after base number 30 includes 258 ascocarps, the majority of analyzed samples, (ii) the second group with two restriction sides after base number 60 and 336, 64 and 340 or 60 and 337 includes 203 ascocarps. (iii) The third group, including 15 ascocarps, just a minority of samples shows no restriction sides. In this work, 25 ascocarps of Asian black truffles were analyzed with RFLP from the ITS region. All fruiting bodies but one showed the same restriction pattern with two bands (150 and 500 bp). To compare the obtained restriction profile with the sequences published by Qiao et al. (2018) [50], some ITS amplicons were sequenced (Sanger sequencing, Eurofins Genomics GmbH, Ebersberg, Germany) showing that they are similar to the first group with one restriction side after base number 30. The fruiting body showing a divergent restriction pattern (see Figure S6) belongs to one ascocarp collected in Yunnan, and the sequence comparisons showed that the obtained sequence is similar to the second group with two restriction sides.

The 20 fruiting bodies from *T. melanosporum* analyzed via RFLP showed a uniform species-specific restriction pattern (see Figure S5), indicating that this should not hinder a differentiation of *T. melanosporum* and *T. indicum*.

To check if quantitative assays for a quantitation of admixtures from the Asian black truffles in *T. melanosporum* were possible mixtures of PCR amplicons from *T. melanosporum* with 20%, 40%, and 50%, Asian black truffles were prepared and incubated with the restriction enzyme *CviQI*. For the mixtures, samples from the Asian black truffles showing one restriction side were used. The restriction fragments were analyzed via AGE (data not shown) and CGE. The CGE-chromatograms from *T. melanosporum* and *T. indicum/himalayense* measured separately are shown in Figure S7. For CGE evaluation, the long restriction fragments (*T. indicum/himalayense*: 500 bp; *T. melanosporum*: 430 bp) were used. Figure 3 presents the results of the quantitative evaluation, which demonstrate a linear relationship between the Asian black truffle content and the relative amount of characteristic restriction fragments (measuring values are shown in Table S4). It is important to note that a linear correlation could only be observed when the concentration of the characteristic restriction fragment was brought into relation with the total concentration of detected restriction fragments. Otherwise, no linear correlation could be observed. So the total concentration of detected restriction fragments and variations in the measured sample volume were considered.

To test the influences of the truffle matrix on the quantitative evaluation mixtures of fruiting bodies from *T. melanosporum*, different amounts of *T. indicum* were prepared. After DNA-isolation, PCR-amplification with the ITS1/ITS4 primer pair, and digestion with *Cvi*QI, PCR fragments were analyzed via AGE (results not shown) and CGE. For each matrix mixture, PCR and enzymatic digestion were performed in a fourfold determination. For a quantitative analysis, the detected concentration of the long restriction fragment relative to the total amount of all detected fragments was plotted against the amount of Asian truffle. As can be seen from Figure 4, the relative peak intensity correlates with the *T. indicum* amount (R^2 of 0.959), demonstrating a possible quantitative determination of possible admixtures with Asian black truffles in *T. melanosporum* samples (measuring values are shown in Table S5). This was comparable with the results of PCR amplicon mixtures. The obtained results show that the CGE assay can be used to determine the amount of admixtures from Asian black truffles in *T. melanosporum*, e.g., for quality control to ensure the integrity of truffle products.

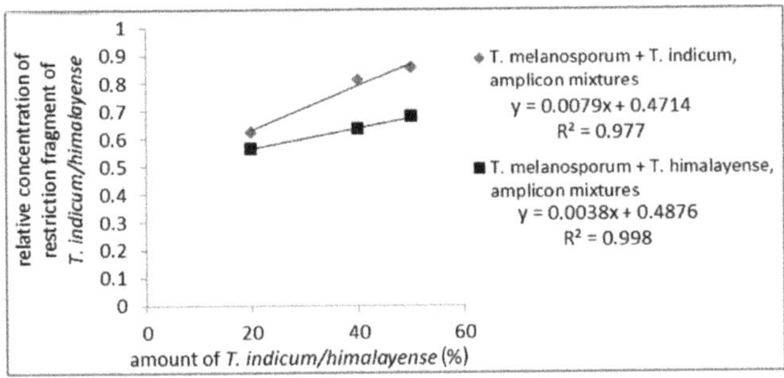

Figure 3. Standard curve of PCR-amplicon mixtures digested with *Cvi*QI from *T. melanosporum* with Asian black truffles added to 50%, 40%, and 20%. The concentration of the long restriction fragment of Asian truffles (500 bp) relative to the total concentration of restriction fragments is plotted against Asian truffle amount.

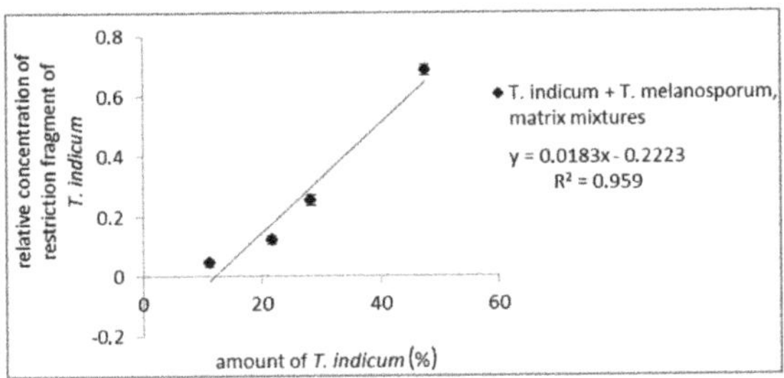

Figure 4. Standard curve of matrix mixtures from fruiting bodies of *T. melanosporum* with Asian black truffles with 11.2%, 21.7%, 28.3%, 47.5% Asian truffle. The concentration of the long restriction fragment (ITS1/ITS4 PCR amplicon digested with *Cvi*QI) of Asian truffles (500 bp) relative to the total concentration of restriction fragments is plotted against Asian truffle amount.

4. Conclusions

The results achieved in the present work show that the developed real-time PCR assay with species-specific primer and the CGE-methods allows the identification of different commercially relevant truffle species. The applied real-time PCR is suited to detect and quantify admixtures from Asian black truffles in *T. melanosporum* up to 0.5%. According to our best knowledge, there is no publication to quantify possible admixtures of *T. indicum* in *T. melanosporum* neither with real-time PCR nor with other molecular biological methods. The developed CGE-method based on the ITS region—with and without restriction digestion—offers a promising alternative to real-time PCR. The molecular biological methods developed can be used for quality control in the food industry or in official food control to ensure the integrity of truffle products.

Supplementary Materials: The following are available online at http://www.mdpi.com/2304-8158/9/4/501/s1, Table S1: Results of the specificity test of the specific real-time primers, Table S2: Measuring values that were used as the basis for preparing the standard curves for Figures 1, Table S3: Measuring values that were used as the basis for preparing the standard curves for Figures 2, Table S4: Measuring values that were used as the basis for preparing the standard curves for Figures 3, Table S5: Measuring values that were used as the basis for preparing the standard curves for Figures 4, Figure S1: Results from the amplification with the ITS1/4 primer pair, DNA isolated of *T. magnatum* fruiting bodies, Figure S2: Results from the amplification with the ITS1/4 primer pair, DNA isolated of *T. albidum* Pico fruiting bodies, Figure S3: Results from the amplification with the ITS1/4 primer pair, DNA isolated of *T. aestivum* fruiting bodies, Figure S4: Results from the amplification with the ITS1/4 primer pair, DNA isolated of *T. brumale* fruiting bodies and food truffle products, Figure S5: Results from the amplification with the ITS1/4 primer pair and the RFLP assay with CviQI, DNA isolated of *T. melanosporum* fruiting bodies, Figure S6: Results from the amplification with the ITS1/4 primer pair and the RFLP assay with CviQI, DNA isolated from *T. indicum/himalayense* fruiting bodies, Figure S7: CGE-chromatogram, PCR-fragments generated with ITS1/ITS4 primer pair after digestion with CviQI, (A) *T. indicum* DNA: 130 bp, 498 bp; (B) *T. melanosporum*: 193 bp, 438 bp.

Author Contributions: Conceptualization, S.S.; methodology, S.S.; validation, S.S. and J.P.; formal analysis, S.S.; investigation, S.S.; resources, M.F.; data curation, S.S., M.S., J.P., and H.-V.T., writing—original draft preparation, S.S.; writing—review and editing, C.L., H.-V.T., and M.F.; visualization, S.S.; supervision, M.F.; project administration, M.F.; funding acquisition, M.F. All authors have read and agreed to the published version of the manuscript.

Funding: This study was performed within the project "Food Profiling-Development of analytical tools for the experimental verification of the origin and identity of food". This Project (Funding reference number: 2816500914) is supported by means of the Federal Ministry of Food and Agriculture (BMEL) by a decision of the German Bundestag (parliament). Project support is provided by the Federal Institute for Agriculture and Food (BLE) within the scope of the program for promoting innovation.

Acknowledgments: The authors gratefully thank our project partner, "LA BILANCIA Trüffelhandels GmbH", for providing sample material.

Conflicts of Interest: The authors declare no conflict of interest. The funders had no role in the design of the study; in the collection, analyses, or interpretation of data; in the writing of the manuscript, or in the decision to publish the results.

References

1.   Trappe, J.M.; Molina, R.; Luoma, D.L.; Cázares, E.; Pilz, D.; Smith, J.E.; Castellano, M.A.; Miller, S.L. Diversity, ecology, and conservation of truffle fungi in forests of the Pacific Northwest. 2009. Available online: https://www.fs.fed.us/pnw/pubs/pnw_gtr772.pdf (accessed on 21 March 2020).

2.   Læssøe, T.; Hansen, K. Truffle trouble: What happened to the Tuberales? *Mycol. Res.* **2007**, *111*, 1075–1099. [CrossRef] [PubMed]

3.   Harley, J.L.; Smith, S.E. *Mycorrhizal symbiosis*; Academic Press Inc.: Lodon, UK, 1983.

4.   Giovannetti, G.; Fontana, A. Mycorrhizal synthesis between Cistaceae and Tuberaceae. *New Phytol.* **1982**, *92*, 533–537. [CrossRef]

5.   Mello, A.; Fontana, A.; Meotto, F.; Comandini, O.; Bonfante, P. Molecular and morphological characterization of Tuber magnatum mycorrhizas in a long-term survey. *Microbiol. Res.* **2001**, *155*, 279–284. [CrossRef]

6.   Zambonelli, A.; Iotti, M.; Murat, C. *True Truffle (Tuber Spp.) in the World: Soil Ecology, Systematics and Biochemistry*; Springer: Cham, Switzerland, 2016.

7.   Bougher, N.L.; Lebel, T. Sequestrate (truffle-like) fungi of Australia and New Zealand. *Aust. Syst. Bot.* **2001**, *14*, 439–484. [CrossRef]

8. Ceruti, A.; Fontana, A.; Nosenzo, C. *Le specie europee del genere Tuber: Una revisione storica*; Museo Regionale di Scienze Naturali di Torino: Torino, Italy, 2003.

9. Chevalier, G.; Frochot, H. Ecology and possibility of culture in Europe of the Burgundy truffle (Tuber uncinatum Chatin). *Agric. Ecosyst. Environ.* **1990**, *28*, 71–73. [CrossRef]

10. Pegler, D. Useful fungi of the world: Morels and truffles. *Mycologist* **2003**, *17*, 174–175. [CrossRef]

11. Marjanovic, Z.; Saljnikov, E.; Milenkovic, M.; Grebenc, T. Ecological features of Tuber magnatum Pico in the conditions of West Balkans–soil characterization. In Proceedings of the 3rd international congress on truffles, Spoleto, Italy, 25–28 November 2008.

12. Lazzari, B.; Gianazza, E.; Viotti, A. Molecular characterization of some truffle species. In *Biotechnology of Ectomycorrhizae*; Plenum Press: New York, NY, USA, 1995; pp. 161–169.

13. Hall, I.R.; Yun, W.; Amicucci, A. Cultivation of edible ectomycorrhizal mushrooms. *Trends Biotechnol.* **2003**, *21*, 433–438. [CrossRef]

14. Mello, A.; Murat, C.; Bonfante, P. Truffles: Much more than a prized and local fungal delicacy. *FEMS Microbiol. Lett.* **2006**, *260*, 1–8. [CrossRef]

15. Favre, J.; Parguey Leduc, A.; Sejalon Delmas, N.; Dargent, R.; Kulifaj, M. The ascocarp of Tuber indicum (Chinese truffle) recently introduced in France: Preliminary study. *C. r. hebd. séances Acad. sci. Serie 3 Sciences de la Vie (France)* **1996**, *319*, 517–521.

16. Bonito, G.; Trappe, J.M.; Donovan, S.; Vilgalys, R. The Asian black truffle Tuber indicum can form ectomycorrhizas with North American host plants and complete its life cycle in non-native soils. *Fungal Ecol.* **2011**, *4*, 83–93. [CrossRef]

17. García-Montero, L.G.; Díaz, P.; Di Massimo, G.; García-Abril, A. A review of research on Chinese Tuber species. *Mycol Prog.* **2010**, *9*, 315–335. [CrossRef]

18. Murat, C.; Zampieri, E.; Vizzini, A.; Bonfante, P. Is the Perigord black truffle threatened by an invasive species? We dreaded it and it has happened! *New Phytol.* **2008**, *178*, 699–702. [CrossRef] [PubMed]

19. Paolocci, F.; Rubini, A.; Riccioni, C.; Granetti, B.; Arcioni, S. Cloning and characterization of two repeated sequences in the symbiotic fungus Tuber melanosporum Vitt. *FEMS Microbiol. Ecol.* **2000**, *34*, 139–146. [CrossRef] [PubMed]

20. Séjalon-Delmas, N.; Roux, C.; Martins, M.; Kulifaj, M.; Bécard, G.; Dargent, R. Molecular tools for the identification of Tuber melanosporum in agroindustry. *J. Agric. Food Chem.* **2000**, *48*, 2608–2613. [CrossRef] [PubMed]

21. White, T.J.; Bruns, T.; Lee, S.; Taylor, J. Amplification and direct sequencing of fungal ribosomal RNA genes for phylogenetics. *PCR Protoc. Guide Methods Appl.* **1990**, *18*, 315–322.

22. Xu, J. Fungal DNA barcoding. *Genome* **2016**, *59*, 913–932. [CrossRef] [PubMed]

23. Schoch, C.L.; Seifert, K.A.; Huhndorf, S.; Robert, V.; Spouge, J.L.; Levesque, C.A.; Chen, W.; Bolchacova, E.; Voigt, K.; Crous, P.W. Nuclear ribosomal internal transcribed spacer (ITS) region as a universal DNA barcode marker for Fungi. *Proc. Natl. Acad. Sci. USA* **2012**, *109*, 6241–6246. [CrossRef]

24. Haase, I.; Brüning, P.; Matissek, R.; Fischer, M. Real-time PCR assays for the quantitation of rDNA from apricot and other plant species in marzipan. *J. Agric. Food Chem.* **2013**, *6*, 3414–3418. [CrossRef]

25. Johnson, S.M.; Carlson, E.L.; Pappagianis, D. Determination of ribosomal DNA copy number and comparison among strains of Coccidioides. *Mycopathologia* **2015**, *179*, 45–51. [CrossRef]

26. López-Calleja, I.M.; de la Cruz, S.; Pegels, N.; González, I.; García, T.; Martín, R. High resolution TaqMan real-time PCR approach to detect hazelnut DNA encoding for ITS rDNA in foods. *Food Chem.* **2013**, *141*, 1872–1880. [CrossRef]

27. Bertini, L.; Potenza, L.; Zambonelli, A.; Amicucci, A.; Stocchi, V. Restriction fragment length polymorphism species-specific patterns in the identification of white truffles. *FEMS Microbiol. Lett.* **1998**, *164*, 397–401. [CrossRef] [PubMed]

28. Gandeboeuf, D.; Dupre, C.; Chevalier, G.; Nicolas, P.; Roeckel-Drevet, P. Typing Tuber ectomycorrhizae by polymerase chain amplification of the internal transcribed spacer of rDNA and the sequence characterized amplified region markers. *Can. J. Microbiol.* **1997**, *43*, 723–728. [CrossRef] [PubMed]

29. Henrion, B.; Chevalier, G.; Martin, F. Typing truffle species by PCR amplification of the ribosomal DNA spacers. *Mycol. Res.* **1994**, *98*, 37–43. [CrossRef]

30. Mabru, D.; Dupré, C.; Douet, J.; Leroy, P.; Ravel, C.; Ricard, J.; Medina, B.; Castroviejo, M.; Chevalier, G. Rapid molecular typing method for the reliable detection of Asiatic black truffle (Tuber indicum) in commercialized products: Fruiting bodies and mycorrhizal seedlings. *Mycorrhiza.* **2001**, *11*, 89–94. [CrossRef]

31. Murat, C.; Vizzini, A.; Bonfante, P.; Mello, A. Morphological and molecular typing of the below-ground fungal community in a natural Tuber magnatum truffle-ground. *FEMS Microbiol. Lett.* **2005**, *245*, 307–313. [CrossRef] [PubMed]

32. Paolocci, F.; Rubini, A.; Granetti, B.; Arcioni, S. Typing Tuber melanosporum and Chinese black truffle species by molecular markers. *FEMS Microbiol. Lett.* **1997**, *153*, 255–260. [CrossRef]

33. Paolocci, F.; Rubini, A.; Granetti, B.; Arcioni, S. Rapid molecular approach for a reliable identification of Tuber spp. ectomycorrhizae. *FEMS Microbiol. Ecol.* **1999**, *28*, 23–30. [CrossRef]

34. Douet, J.; Castroviejo, M.; Mabru, D.; Chevalier, G.; Dupré, C.; Bergougnoux, F.; Ricard, J.; Médina, B. Rapid molecular typing of Tuber melanosporum, T. brumale and T. indicum from tree seedlings and canned truffles. *Anal. Bioanal. Chem.* **2004**, *379*, 668–673. [CrossRef]

35. Gryndler, M.; Trilčová, J.; Hršelová, H.; Streiblová, E.; Gryndlerová, H.; Jansa, J. Tuber aestivum Vittad. mycelium quantified: Advantages and limitations of a qPCR approach. *Mycorrhiza* **2013**, *23*, 341–348. [CrossRef]

36. Zampieri, E.; Rizzello, R.; Bonfante, P.; Mello, A. The detection of mating type genes of Tuber melanosporum in productive and non productive soils. *Appl Soil Ecol.* **2012**, *57*, 9–15. [CrossRef]

37. Iotti, M.; Leonardi, M.; Lancellotti, E.; Salerni, E.; Oddis, M.; Leonardi, P.; Perini, C.; Pacioni, G.; Zambonelli, A. Spatio-temporal dynamic of Tuber magnatum mycelium in natural truffle grounds. *PLoS ONE* **2014**, *9*, e115921. [CrossRef] [PubMed]

38. Salerni, E.; Iotti, M.; Leonardi, P.; Gardin, L.; D'Aguanno, M.; Perini, C.; Pacioni, P.; Zambonelli, A. Effects of soil tillage on Tuber magnatum development in natural truffières. *Mycorrhiza* **2014**, *24*, 79–87. [CrossRef] [PubMed]

39. Parladé, J.; De la Varga, H.; De Miguel, A.M.; Sáez, R.; Pera, J. Quantification of extraradical mycelium of Tuber melanosporum in soils from truffle orchards in northern Spain. *Mycorrhiza* **2013**, *23*, 99–106. [CrossRef] [PubMed]

40. Rizzello, R.; Zampieri, E.; Vizzini, A.; Autino, A.; Cresti, M.; Bonfante, P.; Mello, A. Authentication of prized white and black truffles in processed products using quantitative real-time PCR. *Food Res. Int.* **2012**, *48*, 792–797. [CrossRef]

41. Holland, P.M.; Abramson, R.D.; Watson, R.; Gelfand, D.H. Detection of specific polymerase chain reaction product by utilizing the 5′–3′exonuclease activity of Thermus aquaticus DNA polymerase. *Proc. Natl. Acad. Sci. USA* **1991**, *88*, 7276–7280. [CrossRef]

42. Haunshi, S.; Pattanayak, A.; Bandyopadhaya, S.; Saxena, S.C.; Bujarbaruah, K.M. A simple and quick DNA extraction procedure for rapid diagnosis of sex of chicken and chicken embryos. *J. Poult. Sci.* **2008**, *45*, 75–81. [CrossRef]

43. Bruüning, P.; Haase, I.; Matissek, R.; Fischer, M. Marzipan: Polymerase chain reaction-driven methods for authenticity control. *J. Agric. Food Chem.* **2011**, *59*, 11910–11917. [CrossRef] [PubMed]

44. Debode, F.; Marien, A.; Janssen, É.; Bragard, C.; Berben, G. Influence of the amplicon length on real-time PCR results. *Biotechnol., Agron., Soc. Environ.* **2017**, *21*, 3–11.

45. Cankar, K.; Štebih, D.; Dreo, T.; Žel, J.; Gruden, K. Critical points of DNA quantification by real-time PCR–effects of DNA extraction method and sample matrix on quantification of genetically modified organisms. *BMC Biotechnol.* **2006**, *6*, 37. [CrossRef]

46. Paolocci, F.; Rubini, A.; Riccioni, C.; Topini, F.; Arcioni, S. Tuber aestivum and Tuber uncinatum: Two morphotypes or two species? *FEMS Microbiol. Lett.* **2004**, *235*, 109–115. [CrossRef] [PubMed]

47. Paolocci, F.; Angelini, P.; Cristofari, E.; Granetti, B.; Arcioni, S. Identification of Tuber spp and corresponding ectomycorrhizae through molecular markers. *J. Agric. Food Chem.* **1995**, *69*, 511–517. [CrossRef]

48. Bonito, G.M.; Gryganskyi, A.P.; Trappe, J.M.; Vilgalys, R. A global meta-analysis of Tuber ITS rDNA sequences: Species diversity, host associations and long-distance dispersal. *Mol. Ecol.* **2010**, *19*, 4994–5008. [CrossRef] [PubMed]

49. Roux, C.; Séjalon-Delmas, N.; Martins, M.; Parguey-Leduc, A.; Dargent, R.; Bécard, G. Phylogenetic relationships between European and Chinese truffles based on parsimony and distance analysis of ITS sequences. *FEMS Microbiol. Lett.* **1999**, *180*, 147–155. [CrossRef] [PubMed]
50. Qiao, P.; Tian, W.; Liu, P.; Yu, F.; Chen, J.; Deng, X.; Wan, S.; Wang, R.; Wang, Y.; Guo, H. Phylogeography and population genetic analyses reveal the speciation of the Tuber indicum complex. *Fungal Genet. Biol.* **2018**, *113*, 14–23. [CrossRef] [PubMed]

MDPI
St. Alban-Anlage 66
4052 Basel
Switzerland
Tel. +41 61 683 77 34
Fax +41 61 302 89 18
www.mdpi.com

Foods Editorial Office
E-mail: foods@mdpi.com
www.mdpi.com/journal/foods